Advances in Experimental Medicine and Biology

Volume 855

More information about this series at http://www.springer.com/series/5584

Olga Gursky

Editor

Lipids in Protein Misfolding

Springer

Editor
Olga Gursky
Department of Physiology and Biophysics
Boston University School of Medicine
Boston, MA, USA

ISSN 0065-2598   ISSN 2214-8019   (electronic)
Advances in Experimental Medicine and Biology
ISBN 978-3-319-17343-6   ISBN 978-3-319-17344-3   (eBook)
DOI 10.1007/978-3-319-17344-3

Library of Congress Control Number: 2015941281

Springer Cham Heidelberg New York Dordrecht London

Printed on acid-free paper

Springer International Publishing AG Switzerland is part of Springer Science+Business Media
(www.springer.com)

Preface

This book addresses the role of lipids in the misfolding of a wide range of proteins, from intrinsically disordered peptides as small as 5 kDa, such as Aβ and amylin, to globular, membrane, and lipid surface-binding proteins as large as apolipoprotein B (550 kDa), which is one of the largest proteins found in nature. The major focus of this compendium is on the biophysical and structural aspects of misfolding of clinically important proteins. Admittedly, the book is biased towards apolipoproteins (apos) and is not comprehensive; for example, prions and several other amyloidogenic proteins such as tau, transthyretin, and immunoglobulin have been left out, leaving plenty of room for the authors of future books.

Although volumes have been written on protein misfolding, including two books published by Springer in the Subcellular Biochemistry series, the effects of lipids on this complex process have been relatively less well explored. To our knowledge, there has been just one book on the subject previously published (*Lipids and Cellular Membranes in Amyloid Diseases*, edited by Raz Jelinek, Wiley, 2011), which inspired our effort. The topic seemed broad enough to attract a diverse group of scientists, yet narrow enough to be manageable within the constraints of a single volume. The latter was a naïve notion that dissipated as this volume shaped up and grew large; thankfully, the publisher accommodated.

The main motivation for writing this book was to learn about the molecular mechanisms underlying the mutual effects of the misfolded proteins and lipids. These mechanisms are complex, controversial, and, hence, interesting. The importance of the subject is underscored by the finding that lipids and lipid-associated molecules, such as apolipoproteins and glycosaminoglycans, are ubiquitous components of amyloid deposits in vivo. Moreover, lipids have been reported to promote or prevent the misfolding of many water-soluble proteins. In turn, protein misfolding can perturb lipid assemblies (e.g., poke holes in cell membranes) and thereby contribute to the cytotoxic effects.

What does this book teach us about the influence of lipids on protein misfolding? The short but unsatisfactory answer is "it is complicated." A slightly better answer is that lipids can influence protein misfolding through various specific and nonspecific mechanisms, and this influence goes both ways. Lipid membranes can provide a surface for 2D condensation of proteins, influence the relative orientation of these proteins; induce secondary structure in disordered proteins; destabilize tertiary structure in globular proteins.

Further, direct binding of certain proteins to lipid surfaces may protect the proteins from misfolding by stabilizing their native helical conformation, whereas binding of individual lipid molecules may help nucleate amyloid fibrils. Moreover, cell membranes can interact directly with protein oligomers and their supramolecular assemblies during fibril nucleation and growth, and can also interact directly with mature amyloid fibrils and influence their morphology. For further details, the reader is referred to individual chapters.

In Chap. 1, H. Hong provides a comprehensive overview of the role of lipids in folding, misfolding, function and malfunction of integral membrane proteins. Such misfolding leads to altered protein topology within the lipid bilayer. The misfolding of water-soluble proteins, which leads to protein aggregation and amyloid fibril formation, is addressed in other chapters.

In Chap. 2, V. N. Uversky reviews protein misfolding and aggregation in lipid-mimetic environments, such as alcohols and lipid vesicles. The emphasis is on the effects of the "membrane field" on globular proteins and on intrinsically disordered proteins such as α-synuclein, which are particularly prominent in amyloid diseases.

In Chap. 3, I. Morgado and M. Garvey review the role of lipids in the generation, aggregation, and toxicity of Aβ peptide in Alzheimer's disease. A wide variety of mechanisms is potentially involved in these complex processes. Reviewing them is an onerous task because of the sheer volume of information, at times controversial, reported on the role of lipids in neurodegenerative disorders.

In Chap. 4, A. Jeremic and his team outline the role of lipids in misfolding and aggregation of a small pancreatic hormone, amylin, and the etiology of islet amyloidosis. In their elegant and eye-opening work, the authors use atomic force microscopy to elucidate the molecular mechanism of fiber nucleation and growth.

In Chap. 5, W. Colón and colleagues review biophysical and structural properties of serum amyloid A (SAA), an acute-phase plasma protein that circulates on lipoproteins and is the causative agent of the inflammation-linked amyloidosis. The extensive studies by the authors indicate that the key determinants for in vivo fibril formation by SAA isoforms are factors other than the intrinsic amyloid-forming propensity of the protein.

In Chap. 6, G. Gorbenko, H. Saito, and colleagues address the interactions of protein fibrils with lipid surfaces, including their own detailed fluorescence studies of lysosome and apoA-I fragments. Although amyloid fibrils are less metabolically active than pre-fibrillar aggregates, they are far from being inert and can interact with lipid membranes in vitro and, arguably, in vivo.

In Chap. 7, M. D. W. Griffin, G. J. Howlett, and colleagues review their work on the role of lipids in misfolding and amyloid formation by a small lipid surface-binding protein, apoC-II, which is currently the best available model for understanding apolipoprotein misfolding. Their results reveal distinct and specific effects of lipid-like compounds on fibril nucleation and growth.

In Chap. 8, M. Das and O. Gursky review amyloid formation by human apolipoproteins. Their original studies using current bioinformatics approaches, combined with new insights from the recently solved atomic-resolution structures of several apolipoproteins, help understand why some of these lipid surface-binding proteins readily form amyloid while others do not.

In Chap. 9, N. Dovidchenko and O. Galzitskaya review computational approaches to study amyloid formation, including prediction algorithms to identify amyloidogenic segments in protein sequences, Φ-value analysis to identify the residues critical to the transition states in protein folding and misfolding, and kinetic models to describe fibril nucleation and growth.

In Chap. 10, E. I. Leonova and O. V. Galzitskaya review the roles of syndecans in lipid metabolism and human diseases including Alzheimer's disease. Syndecans and other heparan sulfate proteoglycans, which are ubiquitous components of amyloid plaques, play complex roles in diverse biological processes such as cell signaling, proliferation, and amyloid formation, which are reviewed in this chapter.

As the list of authors suggests, this book is a product of concerted efforts of scientists from the countries around the world, including Australia, India, Korea, Japan, Portugal, Russia, Ukraine, and the USA. The primary authors have diverse backgrounds, from physics to biology, and are at various stages of their career, from PhD candidates to postdocs; to assistant, associate, and full professors; and to a graduate school dean. The unifying attributes of this diverse group of scientists are their demonstrated expertise in the field, commitment to the task, and admirable sportsmanship demonstrated during the review process. I cannot thank the authors enough for their tremendous efforts in putting this book together. Special thanks are due to Dr. Galyna Gorbenko and her team from Ukraine, who pressed on with their work despite the raging war.

Last but not least, this book was made possible by the continuous support of our research by the National Institutes of Health, in particular, by the current grant GM067260. Special thanks are due to Dr. Shobini Jayaraman who is the major driving force behind the research, training, and day-to-day operation of my laboratory. Without Shobini's leadership and tireless efforts, combined with the hard work of Madhurima Das and Dr. Isabel Morgado and the invaluable help with chapter reviews provided by these and other colleagues, including Drs. G. G. Shipley, M. C. Phillips, J. Straub, E. S. Klimtchuk, Nathan Meyers and others, this volume would have never materialized.

Boston, MA, USA   Olga Gursky
February 2015

Contents

Contributors

Emi Adachi Institute of Health Biosciences, Graduate School of Pharmaceutical Sciences, The University of Tokushima, Tokushima, Japan

J. Javier Aguilera Department of Chemistry and Chemical Biology, and Center for Biotechnology and Interdisciplinary Studies, Rensselaer Polytechnic Institute, Troy, NY, USA

Diti Chatterjee Bhowmick Department of Biological Sciences, The George Washington University, Washington, DC, USA

Wilfredo Colón Department of Chemistry and Chemical Biology, and Center for Biotechnology and Interdisciplinary Studies, Rensselaer Polytechnic Institute, Troy, NY, USA

Madhurima Das Department of Physiology and Biophysics, Boston University School of Medicine, Boston, MA, USA

Nikita V. Dovidchenko Institute of Protein Research, Russian Academy of Sciences, Pushchino, Moscow Region, Russia

Oxana V. Galzitskaya Institute of Protein Research, Russian Academy of Sciences, Pushchino, Moscow Region, Russia

Megan Garvey School of Medicine, Deakin University, Waurn Ponds, VIC, Australia

Mykhailo Girych Department of Nuclear and Medical Physics, V.N. Karazin Kharkiv National University, Kharkov, Ukraine

Galyna Gorbenko Department of Nuclear and Medical Physics, V.N. Karazin Kharkiv National University, Kharkov, Ukraine

Michael D.W. Griffin Department of Biochemistry and Molecular Biology, Bio21 Molecular Science and Biotechnology Institute, University of Melbourne, Parkville, VIC, Australia

Olga Gursky Department of Physiology and Biophysics, Boston University School of Medicine, Boston, MA, USA

Heedeok Hong Department of Chemistry and Department of Biochemistry & Molecular Biology, Michigan State University, East Lansing, MI, USA

Geoffrey J. Howlett Department of Biochemistry and Molecular Biology, Bio21 Molecular Science and Biotechnology Institute, University of Melbourne, Parkville, VIC, Australia

Aleksandar M. Jeremic Department of Biological Sciences, The George Washington University, Washington, DC, USA

Elena I. Leonova Institute of Protein Research, Russian Academy of Sciences, Pushchino, Moscow Region, Russia

Institute of Biochemistry and Physiology of Microorganisms, Russian Academy of Sciences, Pushchino, Moscow Region, Russia

Chiharu Mizuguchi Institute of Health Biosciences, Graduate School of Pharmaceutical Sciences, The University of Tokushima, Tokushima, Japan

Yee-Foong Mok Department of Biochemistry and Molecular Biology, Bio21 Molecular Science and Biotechnology Institute, University of Melbourne, Parkville, VIC, Australia

Isabel Morgado Centre of Marine Sciences, University of Algarve, Faro, Portugal

Present Affiliation: Department of Physiology & Biophysics, Boston University School of Medicine, Boston, MA, USA

Timothy M. Ryan The Florey Institute of Neuroscience and Mental Health, The University of Melbourne, Parkville, VIC, Australia

Hiroyuki Saito Institute of Health Biosciences, Graduate School of Pharmaceutical Sciences, The University of Tokushima, Tokushima, Japan

Anjali A. Sarkar Department of Biological Sciences, The George Washington University, Washington, DC, USA

Sanghamitra Singh Department of Biological Sciences, The George Washington University, Washington, DC, USA

Saipraveen Srinivasan Department of Chemistry and Chemical Biology, and Center for Biotechnology and Interdisciplinary Studies, Rensselaer Polytechnic Institute, Troy, NY, USA

Saurabh Trikha Department of Biological Sciences, The George Washington University, Washington, DC, USA

Present Affiliation: Endocrinology Division, Boston Children's Hospital, Harvard Medical School, Boston, MA, USA

Valeriya Trusova Department of Nuclear and Medical Physics, V.N. Karazin Kharkiv National University, Kharkov, Ukraine

Vladimir N. Uversky Department of Molecular Medicine and USF Health Byrd Alzheimer's Research Institute, Morsani College of Medicine, University of South Florida, Tampa, FL, USA

Biology Department, Faculty of Science, King Abdulaziz University, Jeddah, Kingdom of Saudi Arabia

Institute for Biological Instrumentation, Russian Academy of Sciences, Pushchino, Moscow Region, Russia

Laboratory of Structural Dynamics, Stability and Folding of Proteins, Institute of Cytology, Russian Academy of Sciences, St. Petersburg, Russia

Role of Lipids in Folding, Misfolding and Function of Integral Membrane Proteins

Heedeok Hong

Abstract

The lipid bilayer that constitutes cell membranes imposes environmental constraints on the structure, folding and function of integral membrane proteins. The cell membrane is an enormously heterogeneous and dynamic system in its chemical composition and associated physical forces. The lipid compositions of cell membranes not only vary over the tree of life but also differ by subcellular compartments within the same organism. Even in the same subcellular compartment, the membrane composition shows strong temporal and spatial dependence on the environmental or biological cues. Hence, one may expect that the membrane protein conformations and their equilibria strongly depend on the physicochemical variables of the lipid bilayer. Contrary to this expectation, the structures of homologous membrane proteins belonging to the same family but from evolutionary distant organisms exhibit a striking similarity. Furthermore, the atomic structures of the same protein in different lipid environments are also very similar. This suggests that certain stable folds optimized for a specific function have been selected by evolution. On the other hand, there is growing evidence that, despite the overall stability of the protein folds, functions of certain membrane proteins require a particular lipid composition in the bulk bilayer or binding of specific lipid species. Here I discuss the specific and nonspecific modulation of folding, misfolding and function of membrane proteins by lipids and introduce several diseases that are caused by misfolding of membrane proteins.

H. Hong (✉)
Department of Chemistry and Department of
Biochemistry & Molecular Biology, Michigan
State University, Chemistry Building Rm 325,
578 S. Shaw Lane, East Lansing, MI 48824, USA
e-mail: honghd@msu.edu

© Springer International Publishing Switzerland 2015
O. Gursky (ed.), *Lipids in Protein Misfolding*, Advances in Experimental
Medicine and Biology 855, DOI 10.1007/978-3-319-17344-3_1

Keywords

Membrane protein folding and misfolding • Membrane protein misfolding diseases • Membrane protein topology • Lipid-protein interaction • Bilayer curvature stress • Bilayer lateral pressure profile

Abbreviations

APP	Amyloid precursor protein
bR	Bacteriorhodopsin
CCS	Collision-cross section
CFTR	Cystic fibrosis transmembrane conductance regulator
CL	Cardiolipin
ER	Endoplasmic reticulum
GpATM	Glycophorin A transmembrane domain
GPCR	G-protein coupled receptor
H_I	Hexagonal phase
H_{II}	Inverted hexagonal phase
$L\alpha$	Lamellar phase
L_o	Liquid ordered phase
PA	Phosphatidic acid
PC	Phosphatidylcholine
PE	Phosphatidylethanolamine
PG	Phosphatidylglycerol
PI	Phosphatidylinositol
SM	Sphingomyelin
SRP	Signal recognition particle
TM	Transmembrane

1.1   Introduction

At the border between life and environment, membrane proteins carry out numerous critical cellular processes such as energy generation, uptake and secretion of metabolites, maintenance of ion balance, signal transduction, catalysis, cell-cell communication, and more.

Protein folding occurs through a delicate balance of various driving forces, which is largely determined by the folding environment. While water-soluble proteins fold in an isotropic aqueous medium, membrane proteins fold in a heterogeneous anisotropic lipid bilayer. For water-soluble proteins, whose folding is majorly driven by the hydrophobic effect, achieving the

integrity of the hydrophobic core is crucial for the successful folding (Dill 1990). Effective hydration of the protein surface is an important factor that determines the protein's propensity toward misfolding and aggregation (Chong and Ham 2014). The exposed hydrophobic surfaces during the folding and unfolding are major targets for chaperones and degradation machinery (Wickner et al. 1999). On the other hand, membrane proteins are stabilized and function in a highly anisotropic and chemically heterogeneous lipid bilayer. Then, what are the roles of the lipid bilayer and its individual components in the structure, folding and function of membrane proteins? Are lipids just a passive solvent that mediates the assembly of membrane proteins or active modulators of the protein structure, folding and stability?

Membrane proteins can be classified as α-helical or β-barrel types depending on the secondary structural elements within the membrane (Fig. 1.1). α-helical proteins are distributed in the cytoplasmic membranes of prokaryotes and eukaryotes, and in the membranes of subcellular compartments of eukaryotes (Popot and Engelman 2000). β-barrel proteins dominate the outer membranes of Gram-negative bacteria and also exist in the outer membranes of mitochondria and chloroplasts (Tamm et al. 2004). While both types of proteins are crucial in the maintenance of the cellular function, they fold in the membranes by entirely different mechanistic and thermodynamic principles. The main focus of this chapter is on the role of the lipid environment in the folding, stability and function of α-helical membrane proteins, which are more widespread in kingdoms of life. However, valuable general folding principles have been obtained from the folding studies of β-barrel membrane proteins, which are also described.

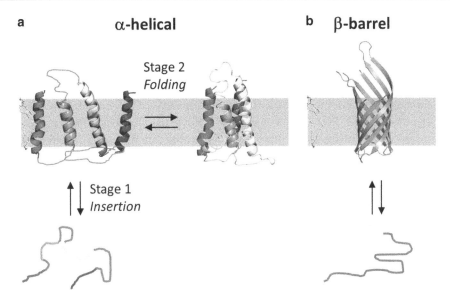

Fig. 1.1 Thermodynamic folding models of integral membrane proteins. (**a**) Two stage model of α-helical membrane protein folding (Engelman et al. 2003; Popot and Engelman 1990). At the first step, individually stable hydrophobic transmembrane segments are integrated into the membrane via the hydrophobic effect. At the second step, individual helices assemble into a functional 3D structure. (**b**) Folding of β-barrel membrane proteins can be described as a cooperative step, where the bilayer insertion and folding are coupled (Hong and Tamm 2004; Moon et al. 2013; Huysmans et al. 2010)

1.2 Folding of α-Helical Membrane Proteins

1.2.1 How Does Lipid Bilayer Constrain the Structure of Membrane Proteins?

Transmembrane (TM) segments of a polypeptide chain are largely composed of hydrophobic amino acids for their favorable partitioning into the nonpolar core of the lipid bilayer (White and Wimley 1999). The hydrophobic nature of α-helical membrane proteins allows the prediction of the TM segments from their amino acid sequences by using the hydropathy plot based on various hydrophobicity scales (Popot and Engelman 2000). In addition, the peptide groups of individual TM segments form ordered α-helical structure to fulfill the hydrogen bond-forming capability, and thereby reduce the desolvation cost in burying the polar peptide group within the nonpolar environment (Bental et al. 1997). Although the hydrophobic thickness of the cell membranes (the distance between phosphate groups in phospholipid bilayers) is maintained within 30–45 Å in various organisms and subcellular compartments (Mitra et al. 2004), the lengths of the TM helices of known structures vary widely from 14 to 36 residues (Bowie 1997). Longer helices are adapted by tilting their axis relative to the bilayer normal or by bending or kinking, while shorter helices can be stabilized by the tertiary contacts with neighboring helices or induce local thinning of the bilayer. The TM helices are packed against each other preferentially at an angle around 20°, while the packing angles for water-soluble proteins widely vary from about −40° to 30° (Bowie 1997). The tertiary contacts are predominantly made by the packing of the TM helices, but the prosthetic groups and folded or unstructured interhelical loops also contribute to the packing.

1.2.2 General Features in the Folding of α-Helical Membrane Proteins

α-helical TM segments are extremely resistant to heat and chemical denaturants (Haltia and Freire 1995). Thus, individual α-helices in the helix bundle membrane proteins can be regarded as independent folding domains packed with one another to form a native 3D structure. Based on the denaturation/refolding studies and structural information of membrane proteins known until 1980s, Popot and Engelman proposed the two-stage model, in which the folding of α-helical membrane proteins was divided into two energetically distinct steps (Fig. 1.1, **left**) (Engelman et al. 2003; Popot and Engelman 1990). This model greatly simplifies the folding problem and has served as a conceptual framework in studying membrane protein folding. The two proposed stages are as follows.

1. *Insertion of independently stable α-helical segments into the membrane*. This step is largely driven by the hydrophobic effect (Hessa et al. 2005, 2007). In cells, the TM segments, which are targeted by the signal recognition particle (SRP) complex, are co-translationally inserted into the membrane through a membrane protein complex called the translocon (SecYEG in the inner membranes of *E. coli*, Sec61αβγ in the endocytoplasmic reticulum (ER) membranes in eukaryotes, and SecYβE in archaea) (Park and Rapoport 2012; du Plessis et al. 2011). The translocon acts as a thermodynamic machine integrating hydrophobic nascent segments into the membrane through the lateral gate or by passing the hydrophilic segments across the membrane through the vertical channel (White and von Heijne 2008).

2. *Association of the TM helices to form a native 3D structure*. In this step, the hydrophobic effect cannot strongly drive the condensation of the TM helices because water molecules are scarce in the core of the lipid bilayer (Joh et al. 2009). Hence, other forces such as van der Waals and polar interactions must drive the folding. Although polar interactions may be strong due to the low dielectric constant in the nonpolar bilayer environment, many experimental results indicate that the strengths of interhelical hydrogen bonds are comparable to those in water-soluble proteins (Bowie 2011). Structural and statistical analyses of

membrane proteins of known structure revealed that membrane proteins tend to bury more fractional area of side chains than water-soluble proteins (Oberai et al. 2009). Extensive packing around small side chains of Gly, Ala, Ser and Thr is an important feature of the interior of membrane proteins, whereas the hydrophobic core packing in water-soluble proteins prefers larger nonpolar side chains and aromatic residues (Eilers et al. 2000; Adamian and Liang 2001; Adamian et al. 2005).

Although our knowledge on the driving forces for membrane protein folding is rapidly growing, it is still at the fledgling stage compared to that of water-soluble proteins. There are relatively unexplored forces such as aromatic-aromatic interactions (Burley and Petsko 1986), π-cation interactions (Gallivan and Dougherty 1999) and the polar nature of the polypeptide backbone that may contribute to the folding of membrane proteins (Wimley and White 1992, 1996).

1.3 Chemical and Physical Properties of Lipid Bilayer

1.3.1 Chemical Properties of Cell Membranes

Since Singer and Nicholson proposed the first unified model of the cell membranes (Singer and Nicolson 1972), the model has been substantially updated (Engelman 2005) (Fig. 1.2). While the membrane is crowded with proteins whose amount is comparable to that of lipids (protein-to-lipid mass ratios vary in the range of 0.2–4) (Lodish et al. 2000), membrane proteins associate transiently or tightly with one another to form structural and functional complexes (Wu et al. 2003). Their diffusion within the membrane plane is significantly affected by cytoskeletons anchored on the membrane and by the "lipid rafts" or membrane domains enriched with cholesterol and sphingomyelin (Kusumi et al. 2005; Winckler et al. 1999; Simons and Sampaio 2011). The hydrophobic thicknesses of integral membrane proteins do not always match those of the

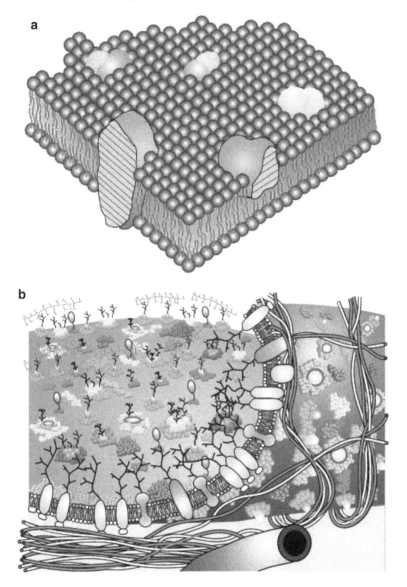

Fig. 1.2 Models of cell membranes. (**a**) Original "fluid-mosaic" model (Singer and Nicolson 1972). The solid bodies represent integral membrane proteins that randomly diffuse in a fluid phase of the lipid bilayer. Some integral membrane proteins form complexes. There are no membrane-associated structures and no phase separation of lipids (domains) with different lipid compositions. (**b**) Updated membrane model showing membrane domains of various sizes with associated cytoskeletons, glycosylated lipids and proteins, and with asymmetric lipid composition (Nicolson 2014). Peripheral and integral membrane proteins often form complexes, and some preferentially partition into membrane domains or domain interfaces (Adopted with permission from Nicolson 2014)

lipid bilayer so that the resultant membrane thickness is reciprocally modulated by both proteins and lipids (Mitra et al. 2004).

Lipid components that play structural roles in cell membranes are classified as phosphoglycerolipids, sphingolipids, cholesterol and cardiolipin, depending on the backbone structure to which fatty acyl chains and polar head groups are attached (Spector and Yorek 1985) (Table 1.1). The lipid composition of cell membranes is highly variable and dynamic depending on the species, subcellular compartment within

Table 1.1 Lipid composition of subcellular organelle membranes of rat liver cell and *E. coli*

	Chol	PC	PE	PS	PI	PG	CL	SM
Rat liver cell[a]								
Plasma membrane	30	18	11	9	4	0	0	14
Golgi complex	8	40	15	4	6	0	0	10
Smooth endoplasmic reticulum	10	50	21	0	7	0	2	12
Rough endoplasmic reticulum	6	55	16	3	8	0	0	3
Nuclear membrane	10	55	20	3	7	0	0	3
Lysosomal membrane	14	25	13	0	7	0	5	24
Mitochondrial membrane								
Inner	3	45	24	1	6	2	18	3
Outer	5	45	23	2	13	3	4	5
E. coli[b]								
Inner membrane	0	0	75	0	0	19	6	0
Inner leaflet of outer membrane	0	0	79	0	0	17	4	0

All values are given as weight %.
Chol cholesterol, *PC* phosphatidylcholine, *PS* phosphatidylserine, *PI* phosphatidylinositol, *PG* phosphatidyl glycerol, *CL* cardiolipin, *SM* sphingomyelin
[a]Taken from (Lehninger et al. 1993)
[b]Taken from (Morein et al. 1996)

a cell, tissue in a body, developmental stage, environmental and physiological conditions, and disease state (Dawidowicz 1987; Chan et al. 2012). Furthermore, even within the same subcellular compartment, the lipid composition is vastly different between the inner and outer leaflets of the membrane (van Meer et al. 2008). This lipid asymmetry results from the balanced action of various flippases, which cause the asymmetry, and scramblases, which equilibrate the lipid distribution between the two leaflets (Clark 2011).

1.3.2 Physical Properties of Lipid Bilayers: Lipid Polymorphism and Lateral Pressure Profile

Isolated lipid components of cell membranes form various forms of mesoscopic assemblies in water (Seddon and Templer 1995). The key concept underlying the "lipid polymorphism" is that the structure of the lipid assembly depends on the shape of the lipid molecules, e. g. the relative volumes of the head group and the hydrocarbon chains (Kumar 1991) (Fig. 1.3). For example, phosphatidylcholines (PC) lipids with relatively large head groups have a cylindrical shape so that

such lipids in water spontaneously form lamellar phase, $L\alpha$, comprised of multiple stacks of relatively flat bilayers. Phosphatidylethanolamine (PE) lipids with a small head group and longer unsaturated acyl chains have the inverted cone shape and prefer the inverted hexagonal phase, H_{II} (Shyamsunder et al. 1988; Gruner et al. 1985). The H_{II} phase can be envisioned as stacks of cylindrical lipid arrays, with hydrocarbon chains pointing outward and polar head groups surrounding the central aqueous pore. Cardiolipin lipids (CLs) form the H_{II} phase in the presence of divalent ions. Single-chain lipids or detergents with a cone shape (i. e. relatively large head group compared to the acyl chain cross-section) form the hexagonal phase, H_I, with the molecular arrangement similar to that in micelles.

The lipid bilayer at its free energy minimum is maintained by several balanced forces (Cantor 1997, 1999) (Fig. 1.3b): (1) Line tension at the interface between nonpolar hydrocarbon chains and polar head groups. This attractive force originates from the hydrophobic effect of burying the nonpolar chains away from water. (2) Repulsions between hydrocarbon chains within the bilayer core and between head groups, which are caused by the dynamic collisions in each region. These

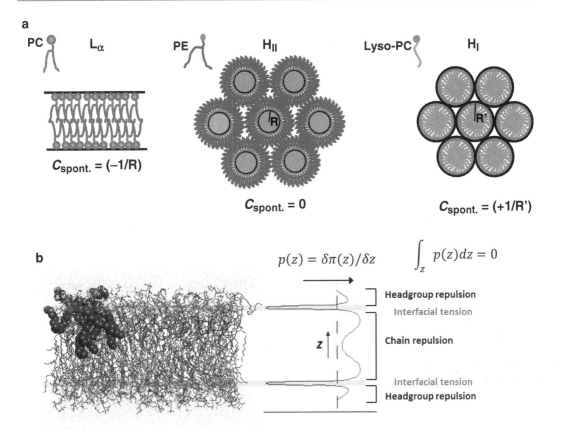

a

PC L_α PE H_{II} Lyso-PC H_I

$C_{spont.} = (-1/R)$

$C_{spont.} = 0$

$C_{spont.} = (+1/R')$

b

$$p(z) = \delta\pi(z)/\delta z \qquad \int_z p(z)dz = 0$$

Headgroup repulsion

Interfacial tension

Chain repulsion

Interfacial tension

Headgroup repulsion

z

Fig. 1.3 Lipid polymorphism and lateral pressure profile. (**a**) Aqueous aggregates of lipids show different mesoscopic phase behavior depending on the relative size of the lipid head group and the acyl chains. PC, which has relatively large head group, forms lamellar phase, $L\alpha$. PE with smaller head group forms inverted hexagonal phase, H_{II}. Lyso-PC, which has only one acyl chain, forms micellar or hexagonal phase, H_I (Escriba et al. 1997). C_{spont} represents spontaneous monolayer curvature. (**b**) Lateral pressure profile (p) along the bilayer normal (z). A tension-free bilayer is maintained by the balanced forces of the head group repulsion, interfacial tension and chain repulsion (The plot for lateral pressure profiles was adopted with permission from Cantor 1997)

balancing forces generate the characteristic lateral pressure profile along the bilayer normal (z-axis). In a tension-free bilayer, the integration over the z-axis yields zero net pressure. The lateral pressure profile has not been determined experimentally but has been calculated using the mean-field theory (Cantor 1997). Surprisingly, the calculated lateral pressure in the bilayer core amounts to several hundred atmospheres.

The lateral pressure profile is the physical origin of an important elastic property of the bilayer called "curvature stress" (Gruner et al. 1985) (Fig. 1.3). For example, in the lamellar phase of a PC bilayer, incorporation of PE reduces the head group repulsion because of the small PE head group and the hydrogen bonding between the ethanolamine group in PE and the phosphate group in an adjacent lipid (Gruner et al. 1988; Tate and Gruner 1987). However, under the constraint of the zero integrated pressure without a change in the attractive line tension, the repulsion between hydrocarbon chains will increase to compensate the reduced head group repulsion. The overall lateral pressure profile will be redistributed to induce the spontaneous negative curvature in each monolayer leaflet (Cantor 1999). Interestingly, the cell membranes, which are regarded as "lamellar phase", are enriched with non-bilayer-forming PE and CL, suggesting an important role of the curvature stress and lateral

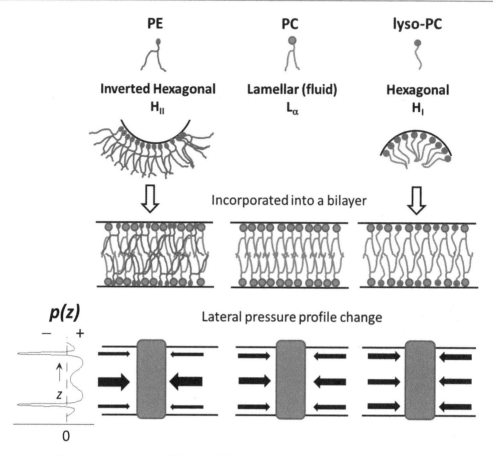

Fig. 1.4 Intrinsic curvature hypothesis (Gruner 1985). When lipids with positive or negative intrinsic spontaneous curvature are incorporated into the bilayer, a curvature stress is induced in each monolayer leaflet. The origin of this phenomenon is the change in the lateral pressure profile depending on the lipid composition (The plot for lateral pressure profiles was adopted with permission from Cantor 1997)

pressure profile in the structure and function of membrane proteins (van den Brink-van der Laan et al. 2004b). Indeed, the lipid composition of *E. coli* cell membranes is maintained near the lamellar-to-inverted hexagonal phase transitions (Morein et al. 1996).

This suggests that integral membrane proteins are constantly subjected to mechanical stress, and the changes in the pressure profile may significantly impact their folding, stability and conformational equilibria. Indeed, modification of the lateral pressure profile was suggested to underlie the mechanisms of general anesthesia by modulating the activity of several ion channels, the sensation of mechanical stimuli by mechanosensitive channels (Milutinovic et al. 2007; Perozo et al. 2002b), the oligomerization and stability of

the pH-gated potassium channel KcsA (van den Brink-van der Laan et al. 2004a) and the stabilization of fusion intermediates in the viral entry and the intracellular vesicular trafficking (Siegel 1993; Yang and Huang 2002). More evidence on this subject will be discussed in Sect. 1.5.

1.4 Lipid-Integral Membrane Protein Interactions for Folding, Structure and Function

In cell membranes, the lipid bilayer serves as a medium that supports the folding, structure and function of membrane proteins. How do constituent lipid molecules in the bilayer interact with proteins? Do they tightly bind to the protein or rapidly exchange with the bulk lipids? Are there binding sites for specific lipid molecules on the surface of membrane proteins? These questions are important for the following reasons. (1) Lipid binding selectivity and specificity could explain the heterogeneous lipid composition of cell membranes that may have evolved for an optimal function of the membrane proteome, or vice versa. (2) Lipid selectivity for a specific membrane protein may be implicated in the protein stability and function. (3) Specific protein-lipid interactions may explain the sorting of membrane proteins between different co-existing membrane domains.

1.4.1   Lipid Contact with Membrane Proteins: Annular Lipids

The first shell of lipids that surround membrane proteins, called "annular lipids" (Contreras et al. 2011), connect the proteins to bulk lipids. Electron paramagnetic resonance (EPR) spectroscopy using spin-labeled lipids revealed that lipid molecules in proteoliposomes exist in restricted (annular) and mobile (bulk) forms (East et al. 1985; Knowles et al. 1979). The two distinct lipid fractions exchange with a time scale of 10–100 ns, while the exchange between unperturbed bulk lipids occurs one to two order of magnitude faster. From the analysis of various-size proteins reconstituted mostly in PCs, approximately two lipid molecules bind to one TM helix on average (Marsh 2008). Thus, the surfaces of integral membrane proteins make favorable but transient contacts with "solvating" lipid molecules.

The selectivity for specific lipid species was estimated by measuring Trp fluorescence quenching (London and Feigenson 1981) or the changes in the immobile lipid fraction from EPR spectra (Powell et al. 1985) by using the competition between a reference lipid (typically PC) and spin-labeled or brominated lipids with various head groups and acyl-chain lengths (Krishnamachary et al. 1994). Lipid selectivity varies for the tested membrane proteins (Marsh

2008). For example, cytochrome c oxidase, which is the last enzyme in the electron transfer pathway in mitochondria and bacteria, strongly prefers CL, while PE, PS and PG bind less selectively (see Sect. 1.4.2 for more details). Mitochondrial ADP/ATP carrier strongly binds CL, PA and stearic acid. While sarcoplasmic reticulum Ca^{2+} ATPase does not show a noticeable preference for any lipid, PE and stearic acid are excluded from the protein surface. On the other hand, rhodopsin does not show any preference for the tested lipids (Marsh 2008).

Further experimental support for the existence of annular lipids was obtained from crystallographic studies of several membrane proteins (Lee 2011). Structural studies of membrane proteins often involve rigorous delipidation steps during purification, such as detergent solubilization and detergent exchange, as well as the addition of external lipids. Reconstitution in lipid environments is necessary to obtain 2D crystals for electron crystallography and 3D crystals for X-ray crystallography in lipid cubic or bicelle phase crystallizations. However, the electron densities of bound lipids obtained in those conditions are often either not observed or weak, hindering a straightforward modeling.

The first detailed pictures of the annular lipids were presented by the structures of bacteriorhodopsin (bR) from Halobacteria. These structures were solved by electron crystallography using 2D crystals in natural purple membranes (Grigorieff et al. 1996) and by X-ray crystallography using 3D crystals prepared by the detergent-induced fusion of 2D crystals and the lipid-cubic phase crystallization (Luecke et al. 1999; Takeda et al. 1998). A total of 18 lipid chains was identified from the structure in the lipid cubic phase at a resolution of 1.55 Å, including four diether lipids and one squalene per monomer, although the unequivocal identification of all lipids was not possible (Fig. 1.5a) (Luecke et al. 1999).

More recent observation of annular lipids bound to aquaporin-0 (AQP0), which is abundant in lens fiber cells, using electron crystallography is intriguing. Walz group solved the structures of AQP0 in 2D crystals reconstituted in two different lipids, DMPC ($diC_{14:0}PC$) and *E. coli*

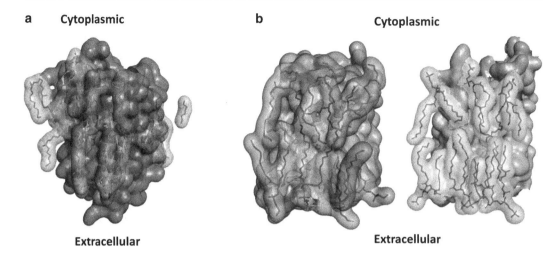

a Cytoplasmic   **b**   Cytoplasmic

Extracellular   Extracellular

Fig. 1.5 Annular lipids observed in the crystal structures of bacteriorhodopsin (bR) and aquaporin-0 (AQP0). (**a**) Atomic structure of bR in the cubic lipid phase determined at 1.55 Å resolution by X-ray crystallography (PDB ID: 1C3W) (Luecke et al. 1999). The annular lipids identified in the structure originate from the native purple membranes. (**b**) *Left*: AQP0 structure in DMPC (*di*C$_{14:0}$PC) solved by 2D electron crystallography at 1.9 Å (PDB ID: 2B6O) (Gonen et al. 2005). *Right*: AQP0 structure in polar lipid extracts from *E. coli* solved by 2D electron crystallography at 2.5 Å (PDB ID: 3M9I) (Figures modified with permission from Luecke et al. 1999 and Hite et al. 2010)

phospholipids whose most abundant lipid species is C$_{16:0}$C$_{16:1c9}$PE (Hite et al. 2010; Gonen et al. 2005) (Fig. 1.4b). Strikingly, the two structures in different lipid environments were essentially the same (0.48 Å root mean square deviation for the backbone atoms) and exhibited the same number of bound lipid molecules (seven lipids per monomer, four in the extracellular leaflet and three in the cytoplasmic leaflet) at similar lipid-protein interfaces.

Although the exact conformations of the corresponding lipids in the two structures were slightly different, the results strongly suggest that there are generic lipid binding motifs on the surface of membrane proteins that can accommodate lipids with varying head groups and acyl chain lengths. Furthermore, the bilayer thicknesses in the two environments deduced from the average distance between the annular lipids in two leaflets are significantly different (27.0 Å *in E. coli* lipids and 31.2 Å in *di*C$_{14:0}$PC) (Hite et al. 2010). This difference may be due to the *cis-unsaturation* of *E. coli* lipids (~52 % of acyl chains) and the resulting thinning of the bilayer. Despite this difference, the protein structure remained unchanged, implying the adaptation of lipid conformation to the protein. Similar lipid

shells around the membrane protein surface have been identified in the cytochrome *bc*$_1$ complex (Lange et al. 2001).

1.4.2 Lipid Contacts with Membrane Proteins: Non-annular Lipids

In contrast to the annular lipids that surround the protein as the first shell, another class of bound lipids, called non-annular, interact with the protein more specifically (Contreras et al. 2011; Lee 2011) (Figs. 1.6 and 1.7). These crystallographically identified lipid molecules often originate from native membranes despite rigorous protein delipidation. Therefore, these lipid molecules bind to the protein with high affinity and specificity and may help stabilize the protein. Although it is difficult to distinguish non-annular from annular lipids just by looking at the structure, several criteria can be applied (Contreras et al. 2011). (1) Non-annular lipids play structural roles within the membrane protein complexes, e. g. in the buried locations of the protein matrix, or at the protein-protein interfaces. (2) Non-annular lipids tightly bind to the protein. (3) Non-annular lipids may play a

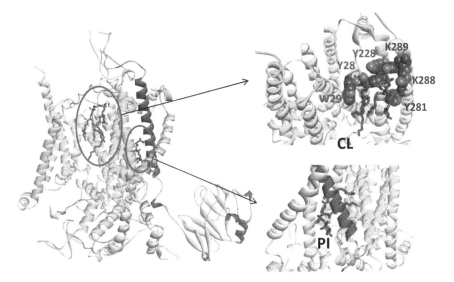

Fig. 1.6 Non-annular lipids in protein structures. *Left*: Non-annular lipids specifically bound to cytochrome bc_1 complex (PDB ID: 1 KB9, 2.5 Å resolution) (Lange et al. 2001). Only the membrane-spanning subunits are shown (solid ribbons). *Upper right*: Cardiolipin binding site (CL in stick, residues in space-filling model) is formed at the subunit interface between cytochrome b (COB, *green*) and cytochrome c_1 (CYT, cyan). Bound CL is stabilized by the electrostatic interactions with Lys288 and Lys289 from CYT and by hydrogen bonding to Tyr28, Trp29 and Lys228 from COB, and to Tyr281 from CYT. *Lower left*: Acyl chain of phosphatidylinositol (PI) is wrapped around the Rieske protein (RIP1) TM domain (*red*). The PI head group makes contacts with the COB, CYT and RIP1 subunits (Figures modified with permission from Lange et al. 2001)

functional role as allosteric effectors of enzymatic activity. In this section, the role of specific binding of a phospholipid in the structural integrity of a membrane protein will be described. Specific binding of cholesterol will be discussed in Sects. 1.6.3, 1.6.4, and 1.6.5.

The crystal structure of yeast cytochrome bc_1 complex, a mitochondrial inner membrane protein in the electron transfer pathway of the respiratory chain, revealed interesting structural and functional role of specific bound lipids (Lange et al. 2001) (Fig. 1.6). Out of five lipid molecules (two PEs, one PC, one PI and one CL), all of which are natural membrane components, the acyl chains of PI are wrapped around the TM helix of the Rieske protein subunit (RIP1 red), and the head group forms extensive hydrogen bonds to the polar side chains in the interfacial TM regions of cytochrome b (COB, green) subunit (Fig. 1.6, **lower right**). The double-negatively charged head group of CL is stabilized by two positively charged "clamp" residues, Lys288 and Lys289 of cytochrome c_1 (CYT1, cyan), packing against Tyr28 of COB subunit (Fig. 1.6, **upper right**); this packing involves a water-mediated hydrogen bond with Lys228, and a hydrogen bond with Trp29 of COB subunit. The four acyl chains are also making extensive van der Waals contacts with the large flat surface comprised of three TM helices from COB and one TM helix from CYT. Double or triple mutations of three Lys residues led to severe growth defects and dramatically reduced the protein expression level, suggesting the role of CL binding in the structural and functional integrity of the protein (Lange et al. 2001).

The 1.8 Å-resolution crystal structure of bovine cytochrome c oxidase is also a wonderful showcase of structurally and functionally important phospholipids (13 identified lipids per monomer in the dimer, including 2 CLs, 1 PC, 3 PEs, 4 PGs and 3 triglycerides) (Shinzawa-Itoh et al. 2007) (Fig. 1.7). Three lipids (two PG in green and one PE in blue) bound to subunit III are the most rigidly bound among all 13 lipid molecules (Fig. 1.7, **bottom right**). On average, the head group and acyl chains of each lipid are stabilized by 12 hydrogen bonds and 98.3 van der Waals

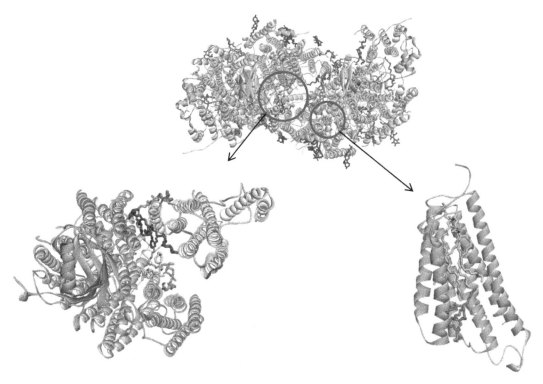

Fig. 1.7 Lipids in the structure of cytochrome *c* oxidase. *Top*: All 26 lipids (shown in sticks) identified in the crystal structure of cytochrome *c* oxidase complex (PDB ID: 1KB9, 1.8 Å resolution) (Shinzawa-Itoh et al. 2007). *Bottom left*: Cardiolipin (*red*), PE (*blue*) and PG (in *green*) stabilize the dimer interface. Subunits from two different monomers are in *grey* and *yellow*. *Bottom right*: Two PGs and one PE are deeply embedded into subunit III, stabilizing the tertiary interactions (Figures modified with permission from Shinzawa-Itoh et al. 2007)

contacts, and the B-factors of these lipids, which reflect molecular ordering, are comparable to those of the protein around the lipid binding site. Structurally, these lipids are buried within the protein matrix, mediating the packing of TM1 and TM2 helices of subunit III with the rest of the subunit. Functionally, the two PG lipids lie in the predicted pathway of molecular oxygen. The conformation of acyl chains was suggested to dynamically modulate the passage of O_2 into the protein. Another class of lipids (one CL in red, two Pes in blue, and one PG in green) participates in the stabilization of the dimer and the subunit interface (Fig. 1.7, **bottom left**). Specially, CL is deeply inserted into the interface and makes favorable contacts with four subunits (subunits III and VIa from one monomer and subunits I and II from the other monomer).

Specific lipid-protein interactions of structural and functional importance have also been inferred from the crystal structures of several other membrane proteins including tetrameric K^+-channel, KcsA (Zhou et al. 2001; Valiyaveetil et al. 2002), photosystem II complex (Guskov et al. 2009), nitrate reductase A (Bertero et al. 2003), photosynthetic reaction center from *Rhodobacter sphaeroides* (Fyfe and Jones 2005), Na^+K^+-ATPase (Shinoda et al. 2009), etc.

1.4.3 Identification of Stabilizing Lipids Using Mass Spectrometry

Recent advances in mass spectrometry made possible the quantitative identification of the high-affinity lipid molecules and their stabilizing role in the gas phase (Laganowsky et al. 2013; Barrera et al. 2008). The key to the success was the combination of the following implements. (1) Optimizing the number of collisions of nano-electrospray droplets (nano-ESI). By this step, detergents are released from protein-detergent-lipid aggregates while maintaining the intact protein complexed with high-affinity lipid molecules. (2) Employment of ion mobility mass spectrometry (IM-MS) for measuring the rotationally-averaged collision cross-section (CSS). CSS is sensitive to the protein shape and, therefore, to the conformational state (e. g. folded or unfolded) (Bush et al. 2010). (3) Use of non-ionic detergents that best support the structure and function of the folded protein. By measuring the changes in CCS values as a function of collision voltages, the "unfolding transition curve" can be constructed for measuring the stability of membrane proteins in the gas phase.

Using IM-MS, Laganowsky et al. (2013) determined the differential effects of various lipid species on the stability of mechanosensitive channels of large conductance (MscL), aquaporin Z (AqpZ) and the ammonia channel (Amt) from *E. coli*. Notably, identification of the high-affinity binding of CL to AqpZ led to the demonstration of the pivotal role of CL in water transport in vitro. Identification of the high-affinity binding of PG to Amt also enabled the elucidation of the lipid role in stabilizing the structure of Amt at the subunit interface. Therefore, this method provides a promising tool for studying the role of specific lipids in the structure, stability and function of membrane proteins. However, it is not clear whether the bound lipids identified in these studies are annular or non-annular.

1.5   Role of Physical Properties of Lipid Bilayers in the Folding and Assembly of Integral Membrane Proteins

How do the chemical and physical properties of cell membranes influence the folding, stability and conformational equilibria of membrane proteins? Researchers have approached this problem from two viewpoints (Lee 2011). First, individual lipid components influence protein conformation by modulating the physical properties of lipid bilayers such as lateral pressure profile, curvature stress, and local lipid deformation by hydrophobic mismatch. Second, lipid components take effects by selective binding to proteins. Differentiating these two distinct phenomena is not an easy task and requires careful control experiments. There is a wealth of experimental data that support either mechanism; the common conclusion is that the lipid bilayer is actively involved in the folding, structure and function of membrane proteins.

1.5.1   Role of Physical Properties in the Function of Membrane Proteins

A classic example connecting physical properties of a bilayer to the conformation and function of membrane proteins is the work by (Keller et al. 1993) who studied the ion channel activities of alamethicin as a function of the curvature stress of the lipid bilayer. Alamethicin is a 20-amino acid antimicrobial peptide from fungus *Trichderma viride*, which acts as a voltage-gated ion channel by forming oligomeric helical bundles (higher than pentamer) (Cafiso 1994; Tieleman et al. 2002). The curvature stress was modulated by increasing the fraction of $diC_{18:1c9}$PE relative to $diC_{18:1c9}$PC in the planar bilayer. $diC_{18:1c9}$PE has a high tendency to form H_{II} phase with a negative spontaneous curvature of the monolayer (lamellar-to-hexagonal transition temperature 15 °C). The spontaneous curvature was strongly correlated with the probability of the higher-conductance open states (Keller et al. 1993). This result suggests that the increased curvature stress induced the formation of higher-order oligomeric states, leading to the enhanced channel activity. The physical origin of the curvature-induced enhancement of channel activity is not well understood. Similar enhancement of membrane transport induced by PE was also observed for sarcoplasmic Ca^{2+}-ATPase reconstituted in $diC_{18:1c9}$PE-containing lipid vesicles (Navarro et al. 1984).

Another example is the dependence of the activity of mechanosensitive channel of large conductance (MscL) on lipid composition. In PC vesicles, the channel exists in a closed state. Addition into the outer leaflet of lysoPC, which by itself forms a micellar phase (H_I), induces the channel opening (Perozo et al. 2002a). When incorporated into the bilayer, lysoPC induces the positive spontaneous monolayer curvature, and the resulting asymmetric distribution of the lateral pressure profile drives the opening of MscL (Perozo et al. 2002a). Further, electron-paramagnetic resonance (EPR) studies suggested that the addition of lysoPC changes the tilt angle of the TM helices. The free energy difference between the closed and open states significantly increases as the bilayer thickness increases, from 4 RT for $diC_{16:1c9}$PC, to 9 RT for $diC_{18:1c9}$PC, and 20 RT for $diC_{20:1c9}$PC, where RT is free energy of thermal motion (Perozo et al. 2002b). This result implies that the hydrophobic thicknesses of the open and closed states of MscL are significantly different, and the lipid deformation by the hydrophobic mismatch between MscL and the bilayer modulates the conformational equilibrium of the channel. Indeed, the estimated lipid deformation energy is similar to the free energy differences between the open and closed states of the channel (Wiggins and Phillips 2004).

1.5.2 Role of Physical Properties in the Folding Kinetics of Membrane Proteins

Physical forces of the lipid bilayer also modulate the kinetics and thermodynamics of membrane proteins folding. The first connection between the two subjects was made by Booth group using bacteriorhodopsin as a model (Booth et al. 1997; Curran et al. 1999). The refolding yield of bacteriorhodopsin (bR) decreased as the fraction of PE in the lipid bilayer increased (Curran et al. 1999). PE induces the negative monolayer curvature and thus, increases the lateral pressure in the bilayer core, which was measured by the increase in the excimer fluorescence of pyrene-labeled lipids at the acyl chain region.

Inhibition of refolding into PE-containing lipid bilayers was also observed for bacterial β-barrel outer membrane proteins. In contrast to the α-helical membrane proteins that are largely composed of stretches of hydrophobic TM segments, β-barrel membrane proteins contain residues that have alternating hydrophobicity in their TM domains. As a consequence, the barrel lumen is composed of hydrophilic residues whereas the lipid-contacting surface is largely hydrophobic (Tamm et al. 2004; Wimley 2003). Because β-barrel membrane proteins are less hydrophobic, they can be solubilized in high concentrations of urea or GdnHCl in the fully unfolded state, and then refolded into lipid vesicles by dilution of denaturants. While a number of OMPs such as OmpA (Kleinschmidt and Tamm 1996, 2002), OmpLA (Moon and Fleming 2011), OmpW (Moon et al. 2013), PagP (Moon et al. 2013; Huysmans et al. 2010), OmpX (Gessmann et al. 2014), etc. can readily refold into thin PC bilayers (Burgess et al. 2008), the incorporation of PE significantly slows down the refolding kinetics (Gessmann et al. 2014; Patel and Kleinschmidt 2013). Probably the increased lateral pressure within the nonpolar core of the bilayer increases its compressibility modulus, thereby decreasing the lipid packing defects necessary for the protein insertion.

Interestingly, incorporation into lipid vesicles of the barrel assembly machinery complex (BAM), which is a conserved assembly factor of OMPs in the bacterial outer membranes, enhanced the refolding rate and yield (Gessmann et al. 2014; Patel and Kleinschmidt 2013). The crystal structures and molecular dynamics simulations of two homologues of BamA, a core component of the BAM complex, from *N. gonorrhoeae* and *H. ducreyi* indicated that the particularly small hydrophobic thickness (9 Å) near the putative lateral gate formed between β-strands 1 and 16 of the barrel domain may induce the local thinning and disorder of the lipid bilayer (Fig. 1.8) (Noinaj et al. 2013). Thus, the lipid perturbation of the outer membrane by BamA was suggested to facilitate the insertion of OMPs by reducing the elastic stiffness of the bilayer (Gessmann et al. 2014). The conformational changes involving the opening of the

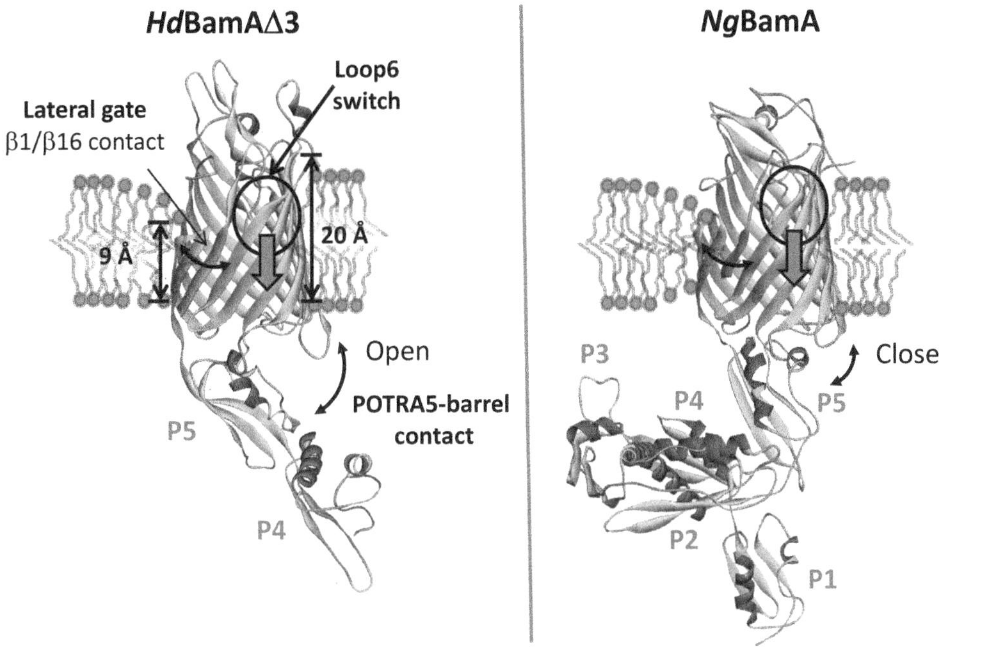

Fig. 1.8 Crystal structures of BamA homologues. Protein structures from and *H. ducreyi* (*Hd*) (PDB code: 4K3C, *left*) and *N. gonorrhoeae* (*Ng*) (PDB code: 4K3B, *right*). The structural features that are predicted to contribute to the insertion of outer-membrane proteins are indicated. The region near the lateral gate formed between strands β1 and β16 in the C-terminal barrel domain has reduced hydrophobic thickness (9 Å), which may perturb the outer membrane. The inter-strand hydrogen bonds are significantly perturbed in the *Ng*BamA (*right*). The local thinning of the bilayer, the opening of the lateral gate, the separation of the POTRA domain 5 (P5) from the barrel, and the deeper insertion of loop6 (*orange arrows*) may collectively induce the direct interaction of nascent polypeptide chains, with the barrel lumen facilitating the insertion of outer membrane proteins (Figure adapted with permission from Noinaj et al. 2013)

contacts between the barrel and the periplasmic POTRA5 domain, and the deeper insertion of the conserved loop 6 into the barrel would also contribute to the opening of the lateral gate and the insertion of nascent polypeptide chains into the outer membrane (Noinaj et al. 2014).

1.5.3 Role of Physical Properties in the Thermodynamic Stability of Membrane Proteins

Thermodynamic study of membrane protein stability, which requires reversibility of the folding-unfolding transition, is more difficult than the kinetic study. Achieving reversibility involves tedious screenings of folding conditions such as temperature, detergents, lipids, denaturants, pH and ionic strength. Despite this challenge, the thermodynamic stability of several β-barrel membrane proteins including OmpA, PagP, OmpLA and OmpW has been studied in lipid bilayers (Hong and Tamm 2004; Huysmans et al. 2010; Moon and Fleming 2011; Moon et al. 2013). The lipid dependence of the OmpA stability in small unilamellar vesicles (SUVs) suggests prominent role of non-specific physical bilayer forces in the membrane protein stability (Hong and Tamm 2004) (Fig. 1.9). While PEs slow down the folding kinetics of OmpA, they increase the stability of the folded protein ($\Delta G^{\circ}_{unfold}$). As the molar fraction of $C_{16:0}C_{18:1c9}$PE increased from 0 to 40 % (the remaining lipid was $C_{16:0}C_{18:1c9}$PC), $\Delta G^{\circ}_{unfold}$ increased from 3.4 to 5.0 kcal/mol. This stability enhancement could be due to the increase in the lateral pressure caused by the negative monolayer curvature, and the favorable contacts of H_{II}-forming PE around the hour-grass shape of the OmpA TM region. Interestingly, the effect of PE can be mimicked by high content of PCs with long unsaturated

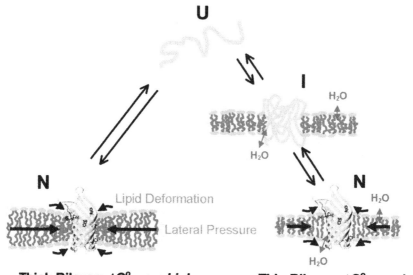

Thick Bilayer: $\Delta G^{o}{}_{u}$, m : high   **Thin Bilayer: $\Delta G^{o}{}_{u}$, m : low**

Fig. 1.9 Elastic coupling of membrane protein stability to lipid bilayer forces. Cartoon depicting structures and bilayer forces acting on OmpA folding/unfolding under equilibrium conditions. Folding into most bilayers is a two-state process (*left path*). *Large black arrows* indicate lateral bilayer pressure imparted on the lipid-protein interface in the hydrophobic core (*red*) of bilayers composed of lipids with negative intrinsic spontaneous curvature. Increasing this pressure increases the thermodynamic stability of the protein. *Small black arrows* indicate lipid deformation forces caused by hydrophobic mismatch between the protein and the unstressed bilayer. These forces decrease the thermodynamic stability of the protein. Folding into thin bilayers is a multistep process (*right path*) with at least one equilibrium intermediate. Water molecules penetrate more easily into the hydrophobic core (*blue arrows*) of more flexible and more dynamic thin bilayers, stabilizing the equilibrium intermediates and decreasing the *m*-value of unfolding. (*m*-value is a linear slope in the dependence of protein thermodynamic stability on denaturant concentration; lower *m*-values indicate smaller change in the solvent-accessible hydrophobic surface area upon unfolding and lower unfolding cooperativity.) Ultimately, in very thin bilayers composed of saturated lipids, complete unfolding (second step) can no longer be observed under any experimental conditions tested (Hong and Tamm 2004) (Figure adapted with permission from Hong and Tamm 2004)

acyl chains ($diC_{16:1c9}$PC to $diC_{20:1c9}$PC), which also induce the negative monolayer curvature (Szule et al. 2002). Thus, the protein stability enhancement by PEs originates from the nonspecific physical properties of the bilayer rather than from the specific head group interactions.

Lipid-dependence of the thermodynamic stability of polytopic TM α-helical membrane proteins has not been studied yet in a lipid bilayer. However, the influence of lipid composition on the stability of TM helix-helix interactions has been studied for the prototype TM helical dimer model system, glycophorin A TM (GpATM) (Hong and Bowie 2011; Hong et al. 2010). The dimer stability of GpATM had been studied mostly in detergent micelles using analytical ultracentrifugation (AUC) (Fleming et al. 1997; Fleming and Engelman 2001) and Förster reso-

nance energy transfer (FRET) (Fisher et al. 1999, 2003; Anbazhagan and Schneider 2010). In these methods, the relative populations of the monomeric and oligomeric states are modulated by diluting the protein with detergents or lipids. However, GpATM dimer is too strong in a bilayer (Adair and Engelman 1994) to be analyzed using conventional dilution methods for measuring the dimer dissociation constant, $K_{d, \, dimer}$. To this end, a new technique called the steric trapping was developed to measure strong protein-protein interactions in lipid bilayers (Hong et al. 2013). The method couples dissociation of biotin-tagged TM dimer to tag-binding monovalent streptavidin. By exploiting the large free energy released from biotin- streptavidin interaction ($\Delta G^{o}{}_{binding}$ ~−19 kcal/mol), the measurable low limit of $K_{d, \, dimer}$ can be extended to 10^{-13}–10^{-12}M, which is three to four orders lower than that measurable by AUC and FRET.

The method revealed striking features of the lipid effects on the TM helix interactions. First, the dimer was much more stable (by ~5 kcal/mol) in a neutral fluid bilayer of $C_{16:0}C_{18:1c9}PC$ than in detergent micelles (Hong et al. 2010). This is largely due to the reduced translational and rotational entropic cost when the dimer equilibrium is confined in liposomes. In addition, the packing contribution of several side chains to the dimer stability increased in the lipid bilayer, i. e. the dimer became more tightly organized. Second, the dimer stability was strongly modulated by the lipid composition and the protein environment (Hong and Bowie 2011). As seen in the case of OmpA, PE lipid, which is a major component of bacterial inner membranes, moderately stabilized the GpATM dimer by increasing the lateral pressure. Strikingly, the dimer was dramatically destabilized by 4–5 kcal/mol in PG, a natural negatively charged lipid in bacterial cell membranes. This destabilization originated from the electrostatic interactions between the cytoplasmic positively charged residues of GpATM and the negatively charged membrane, which distort the optimal dimer structure. Another surprising result was further destabilization of the dimer by the total extracts of *E. coli* inner membrane proteins. It has been predicted that the excluded volume effects in the crowded cellular environment enhance the protein-protein interaction and protein stability (Minton 2000). Contrary to this prediction, anonymous membrane proteins in the bilayer nonspecifically competed with the specific TM helix-helix interactions, destabilizing the GpATM dimer (Hong and Bowie 2011).

1.6 Membrane Protein Misfolding and Diseases

1.6.1 Lipid-Induced Folding, Misfolding and Topogenesis of α-Helical Membrane Proteins

The self-sealed nature of cell membranes and the directionality of polypeptide chain yield a specific pattern of "in" and "out" orientations of TM helices, termed protein "topology" (von Heijne 2006). Topogenesis of polytopic α-helical membrane proteins initially occurs during the co-translational insertion of TM helices through the interaction of the topogenic signals in the polypeptide with the translocon and the membrane, and is completed in the final folding and assembly process (Dowhan and Bogdanov 2009). Thus, the initial topology of individual TM helices is expected to critically affect the mechanism of the subsequent folding and assembly (Fig. 1.1). By this process, ion channels and transporters achieve their correct membrane orientations, and the catalytic domains and the active sites of membrane-bound enzymes are localized in a proper subcellular environment.

The strongest topogenic signal known so far is the "positive-inside" rule, which represents the biased distribution of positively charged amino acids, Arg and Lys, that are roughly five times more abundant in the cytosolic side than in the opposite side of the membrane (Krogh et al. 2001; von Heijne 1992) (see Figs. 1.1 and 1.10a). The N-terminal signal sequence or the first hydrophobic segment of membrane proteins is targeted to the inner membrane of bacteria or to the endoplasmic reticulum (ER) membrane in eukaryotes, and inserted into translocon as a hairpin during translation (Rapoport 2007). The strong electrostatic interaction of the positively charged residues along the hairpin with the negatively charged membrane and with the translocon stabilizes the orientation of the inserted TM helices (Dowhan and Bogdanov 2009). Reversal of the topology may be kinetically inhibited because it involves the translocation of hydrophilic loops and water-soluble domains across the non-polar bilayer core. However, as seen from the homodimeric bacterial multidrug transporter EmrE, the weak bias of the charge distribution can yield a dual topology of subunits and the formation of an antiparallel native protein assembly (Morrison et al. 2012). The topology analysis of *E. coli* inner membranes proteins revealed that, although not frequent, a significant fraction of TM segments does not follow the positive-inside rule (Gray et al. 2011). These findings imply that the topology of membrane proteins may be more

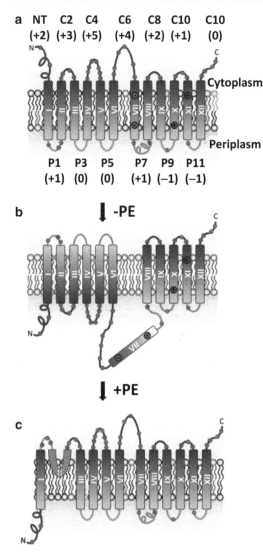

a

NT	C2	C4		C6	C8	C10		C10
(+2)	(+3)	(+5)		(+4)	(+2)	(+1)		(0)

Cytoplasm

Periplasm

P1	P3	P5		P7	P9	P11
(+1)	(0)	(0)		(+1)	(−1)	(−1)

b
↓ -PE

↓ +PE

c

Fig. 1.10 Influence of lipid composition on the topology and folding of polytopic α-helical membrane proteins. (**a**) Topology and distribution of positive charges in the N- (NT) and C- termini (CT), and the cytoplasmic (C2–C10) and periplasmic loops (P1–P11) of LacY expressed in wild type *E. coli* inner membranes. Positive-inside rule is well presented in the native topology of LacY. Locations of positively (*red*) and negatively (*green*) charged residues are indicated. (**b**) Topology of LacY expressed in PE⁻ cells. The topology of TMI-TMVI helices was completely reversed, and TMVII in wild type *E. coli* turned into a cytoplasmic peripheral helix. Negatively and positively charged residues forming salt bridges between TMs are indicated. (**c**) Change in topology of LacY assembled in PE⁻ cells after post-assembly synthesis of PE (Figure modified with permission from Bogdanov et al. 2014)

dynamic than expected, and the reverse charge bias can be counterbalanced by more favorable tertiary or quaternary protein-protein interactions.

Critical roles of lipids in the topogenesis, assembly and function of polytopic membrane proteins have been elucidated through two methodological breakthroughs using the lactose permease (LacY) from *E. coli* as a model system. LacY is a galactoside/H⁺ symporter, which actively translocates lactose into the cytoplasm by exploiting the free energy stored in the electrochemical gradient across the membrane (the active uphill energy-dependent transport). LacY also facilitates the downhill energy-independent transport of lactose (Guan and Kaback 2006). First, Dowhan group developed *E. coli* mutant strains whose lipid compositions can be modified by knocking out and introducing genes involved in the lipid synthesis. Synthesis of PE, a major zwitterionic lipid in the *E. coli* cytoplasmic membrane, can be blocked by disrupting the chromosomal phosphatidylserine synthase gene (*pssA*), which produces PS, a precursor of PE (DeChavigny et al. 1991). Synthesis of PE can be restored by introducing the plasmid-encoded inducible *pssA*. In this system (PE⁻ strain), negatively charged lipids PG and CL substitute PE. PE can also be replaced by another zwitterionic lipid, PC, by introducing the plasmid-encoded inducible phosphatidylcholine synthetase gene (*pcsA*) in the background of the PE⁻ strain (Bogdanov et al. 2010). Second, Kaback group developed a conformation-specific monoclonal antibody (Ab4B1) that recognized the correct folding of a periplasmic P7 loop connecting helices TMVII and TMVIII (Sun et al. 1996). The binding ability of the antibody was well correlated with the correct folding and function of LacY (Fig. 1.10a).

Using those tools, Dowhan team revealed the replacement of PE lipids by PG and CL in the mutant *E. coli* strain led to the loss of binding of Ab4B1, that is, the misfolding of LacY. Strikingly, the topologies of TMI to TMVI were completely reversed relative to those in the wild type strain (Fig. 1.10b). LacY expressed in the membrane without PE still facilitated the downhill energy-independent transport of lactose, but could not support the uphill energy-dependent transport. The expression of PssA, which restored the PE synthesis, reversibly induced the formation of the correctly folded P7 loop and functional LacY (Bogdanov et al. 2008) (Fig. 1.10c). However, the locations of the N-terminus and P1 loop remained unchanged even after the restoration of PE synthesis, while the native-like topologies were regained by TMIII – TMVII, which apparently induced a U-shaped flexible hinge conformation of TMII within the membrane. This result indicates the dynamic nature of membrane topology and the critical role of specific lipids in the folding of polytopic α-helical membrane proteins. Replacement of PE by PC did not cause either topology reversal or loss of the uphill transport function, but the degree of the conformation-specific antibody binding was significantly reduced, further suggesting the fine-tuning role of lipids in the folding and conformational equilibrium of membrane proteins (Bogdanov et al. 2010).

Lipid-dependent control of topogenesis has been also observed for phenylalanine permease (PheP) (Zhang et al. 2003) and γ-aminobutyrate permease (GabP) (Zhang et al. 2005) from *E. coli*. However, the ability of lipids to support the correct topology and folding seems to depend more on the nonspecific role of lipids controlling the overall chemical and physical properties of cell membranes rather than on specific interactions between lipid head groups and certain protein residues. These intriguing findings led to the coining of the term "lipochaperone", a specific lipid facilitating membrane protein folding, by analogy with protein molecular chaperones that assist the folding of water-soluble proteins and prevent their misfolding (Bogdanov and Dowhan 1999).

1.6.2 Membrane Protein Misfolding, Quality Control and Human Diseases

Misfolding and aggregation of membrane proteins in cells are directly linked to a number of human diseases (MacGurn et al. 2012; Houck and Cyr 2012; Reinstein and Ciechanover 2006; Sanders and Myers 2004; Kuznetsov and Nigam 1998). The abnormal folding of membrane proteins seems to be a common event that cells consistently deal with rather than a rare accident. For example, under normal physiological conditions, only <50 % of wild type cystic fibrosis transmembrane regulator (CFTR) and <20 % of wild type peripheral myelin protein 22 (PMP22) that are newly synthesized on the ER membrane reach the cytoplasmic membrane in their correctly folded form (Pareek et al. 1997; Kopito 1999). A wide array of the quality control mechanisms such as chaperones and degradation machinery constantly operates for monitoring the folding, degradation and trafficking of membrane proteomes in all subcellular compartments (MacGurn et al. 2012; Houck and Cyr 2012; Tatsuta and Langer 2008).

There are three major pathways for degrading membrane proteins in eukaryotic cells (Fig. 1.11). In the ER, where most membrane and secreted proteins are synthesized, misfolded membrane proteins are recognized and degraded by the ER-assisted degradation (ERAD) pathways (Hampton 2002). In mammalian cells, three types of the ERAD machines exist in the form of protein complexes (Hrd1/Derlin1,2,3/Sel1L/Ube2G1/Ube2k; DNAJB12/Derlin1/Hsp70/Rma1/Ube2J1; and TEB4/Ube2G/Ube2J) (Kikkert et al. 2004; Grove et al. 2011; Ravid et al. 2006). They are commonly equipped with the membrane-integrated ubiquitin ligase (E3) (Hrd1, Rma1 or TEB4) for the recognition and ubiquitination of misfolded membrane proteins, and the ubiquitin conjugation enzyme (E2) (Ube2G, Ube2K, and Ube2J1). Derlins, DNAJB12 (Hsp40), Sel1L and Hsp70 participate in the recognition process by chaperoning misfolded proteins. Ubiquitinated misfolded proteins are subsequently extracted from the ER membrane by the AAA+ATPase p97 in the cytosol, and are finally degraded in the 26S proteasome. Misfolded proteins in the Golgi and cytoplasmic membranes are recognized by other types of ubiquitination enzymes (Rsp5 in Golgi and Nedd4, Cbl, and CHIP in cytoplasmic membrane), endocytosed into endosomes, and

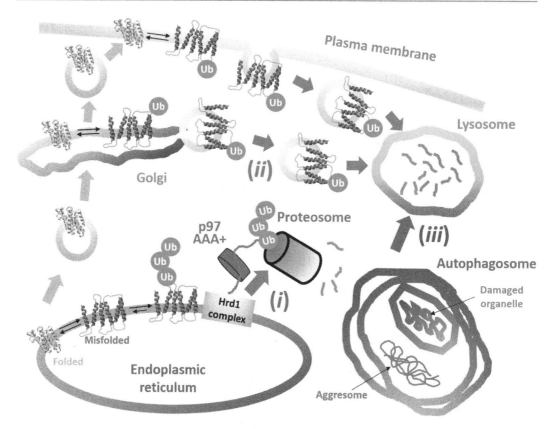

Fig. 1.11 Quality control mechanisms of integral membrane proteins in eukaryotic cells. Correctly folded membrane proteins are ultimately targeted to their final destination (plasma membrane here) through the vesicular transport (*cyan arrows*). (*i*) ERAD pathway (*purple arrow*): the misfolded proteins in the ER membrane are recognized and ubiquitinated by the membrane-bound ubiquitin ligase complex (Hrd1 complex here). The proteins are extracted from the membrane by p97 AAA+ATPase and ultimately degraded in the proteasome (*purple arrow*). (*ii*) Lysosomal degradation pathway (*red arrows*): misfolded proteins in the Golgi and plasma membranes are tagged by ubiquitin and targeted to lysosome. (*iii*) Autophagy degradation pathway (*brown arrow*): damaged subcellular compartments and protein aggregates are engulfed by autophagosome and targeted to lysosome

degraded in lysosomes (MacGurn et al. 2012). Aggregated membrane proteins and damaged subcellular organelles, whose sizes are too large to be degraded by these mechanisms, are engulfed by double-membrane autophagosomes and targeted for lysosomal degradation (He and Klionsky 2009). At most, ~80 % of the membrane protein turnover is carried out by the ERAD pathways (Reinstein and Ciechanover 2006).

The pathogenesis of membrane protein misfolding diseases is partly the consequence of the abnormal interaction between misfolded proteins and the degradation machinery (Sanders and Myers 2004): (1) Proteins with loss-of-function mutations are targeted to their final destinations bypassing the quality control mechanisms. (2) Destabilizing or misfolding mutations enhance proteins' susceptibility to degradation and thus reduce the level of functional proteins. (3) Malfunction of degradation machinery by mutations or stresses causes a failure of clearing toxic misfolded and aggregated proteins. More than 60,000 missense mutations in human genes leading to disease phenotypes have been identified, yet mutations in residues that are directly involved in the protein function are rare (Stenson et al. 2008) (http://www.biobase-international.com/product/hgmd). Rather, a more common disease mechanism seems to involve destabilizing mutations that impair the correct trafficking of the functional proteins by enhancing degradation.

Cystic fibrosis transmembrane conductance regulator (CFTR), which regulates the anion balance across the plasma membranes of epithelial cells (Miller 2010), is an intensely studied model system for membrane protein misfolding diseases. CFTR is composed of 12 TM segments, two nucleotide-binding domains (NBDs) and a regulatory domain. Malfunction of CFTR is a major cause of cystic fibrosis, a severe respiratory and gastrointestinal disorder by abnormal production of viscous mucus (Sheppard and Welsh 1999). Among 1,984 mutations deposited in the cystic fibrosis mutation database (http://www.genet.sickkids.on.ca), 69 % of patients carry the ΔF508 deletion mutation. F508 is located in the NBD on the cytoplasmic side of CFTR, and ΔF508 mutation induces the misfolding of the NBD domain (Lukacs and Verkman 2012). Although ΔF508 mutation partially retains the anion transport function, the mutant is completely degraded by the ERAD pathway and fails to correctly target to the cell surface (Pasyk and Foskett 1995). A correct targeting is also completely aborted in the ERAD pathway (Sun et al. 2006) for CFTR carrying the mutations (G85E, R347P, and R334W) in the transmembrane domain.

The enhanced susceptibility to degradation upon missense mutations is an established mechanism of other diseases. Those include retinitis pigmentosa, a common retinal degeneration disease partly caused by missense mutations in rhodopsin, and Charcot-Marie-Tooth disease, a common inherited neurological disorder, which is related to missense mutations in PMP22, a tetra-membrane spanning integral membrane proteins (Rajan and Kopito 2004; Pareek et al. 1997; Sakakura et al. 2011).

Another critical disease mechanism involves mutations that do not directly affect the function or the conformational stability of membrane proteins, but impair their interactions with the degradation machinery. One example is the Liddle syndrome, a condition of severe hypertension, hypokalemia and metabolic alkalosis. The gene responsible for the Liddle syndrome encodes the β-subunit of renal epithelial sodium channel. A disease mutation occurs at the binding interface with the E3 ubiquitin ligase Nedd4 (Shimkets 1996; Staub et al. 1997). Thus, Nedd4 can neither recognize the sodium channel nor target it for degradation. The accumulation of the channel induces the excessive reabsorption of sodium ions and water, eventually causing hypertension. Impairment of ubiquitin-mediated degradation by mutations on E3 ligases is also implicated in Angelman syndrome (E6-AP E3), Parkinson's disease (Parkin E3), von Hippel-Lindau protein-associated syndromes (von Hippel-Lindau E3), and other diseases (Reinstein and Ciechanover 2006).

Molecular features of misfolded membrane proteins that are selectively recognized by the quality control machines remain unclear. In water-soluble proteins, the integrity of the hydrophobic core is the key determinant for protein interactions with chaperones and degradation machinery. Since the stability of membrane proteins is governed by substantially different thermodynamic principles, the interactions between misfolded membrane proteins and the quality control machines should also be different from those of water-soluble proteins. One possible scenario is that polar interactions of membrane proteins in a low-dielectric lipid bilayer may play a significant role in the folding and stability. Misfolding induces the exposure of polar residues into the bilayer, facilitating their interactions with chaperones and membrane-bound ubiquitin ligases (Houck and Cyr 2012). Indeed, Sato et al. (2009) showed that mutations of the surface polar residues in the TM domain of Hrd1p ubiquitin ligase, a key component of the ERAD machinery, significantly reduced the degradation susceptibility of 3-hydroxy-3-methylglutaryl-coenzyme A reductase.

Still, it is unclear whether the polar interactions are a dominant factor determining the thermodynamic stability of integral membrane proteins. Numerous studies measuring the hydrogen bonding strength in membrane proteins consistently report moderate contributions (0.5–1.5 kcal/mol per bond), not much different from those in water-soluble proteins (Bowie 2011). Furthermore, the analysis of disease mutations mapped on the structures of rhodopsin,

G-protein coupled receptors, and potassium channels did not find a dominant feature of polar-to-nonpolar or nonpolar-to-polar mutations (Oberai et al. 2009). Further biophysical studies on membrane protein folding are necessary to explain disease mechanisms caused by misfolding.

1.6.3 Role of Cholesterol in GPCR Receptor

Lipidomic studies demonstrated that changes in the lipid composition of cell membranes can provide markers of disease; however, a direct link between the lipid-mediated misfolding of membrane proteins and specific diseases is unclear (Hu et al. 2009). Several examples show the pivotal role of lipids in the structure and function of pharmaceutically important membrane proteins. These studies may ultimately help develop the pharmacological chaperones to modulate folding and stability for the optimal function (Mendre and Mouillac 2010).

Although cholesterol has long been known as a critical modulator of membrane fluidity, curvature stress and raft formation, it also acts as a specific "ligand" stabilizing the structure of β2-ardrenergic receptor (β2-AR), a member of the G-protein coupled receptor (GPCR) family (Hanson et al. 2008). β2-AR forms a complex with its effector, L-type Ca^{2+} channel, and senses epinephrine and norepinephrine in muscular tissues, such as bronchial vasculature and blood vessels. The receptor is an important drug target for the treatment of asthma, glaucoma, high blood pressure and cardiac arrhythmias.

The crystal structures of β2-AR were solved for apo- and carazolol (inverse agonist)-bound forms (Cherezov et al. 2007; Rasmussen et al. 2007). Later, the structure was solved in the timolol- (inverse agonist) and cholesterol-bound form (Hanson et al. 2008). This was the first structure in which a cholesterol binding motif was elucidated in atomic detail (Fig. 1.12). Two cholesterol molecules bind to the cleft formed by the helices I, II, III and IV. Two prominent residues that bind cholesterol through CH-π and van der

Waals interactions are $Trp158_{4.50}$ conserved in 94 % of GPCR (Ballesteros and Weinstein 1995) and $Ile154_{4.46}$ conserved in 60 % of GPCR by homology and 35 % by identity (the subscripts specify an GPCR amino acid according to Ballesteros-Weinstein numbering system allowing comparison of equivalent positions in the GPCR family). Two other bulky residues, $Tyr70_{2.41}$ and $Arg151_{4.43}$, which are less conserved, cap cholesterol in the cytoplasmic interfacial region by van der Waals and hydrogen bonding interactions. The four-residue cholesterol binding consensus motif, $[4.39–4.43(R,K)]—[4.50(W,Y)]—[4.46(I,V,L)]—[2.41(F,Y)]$, is widely distributed in the GPCR family, suggesting that cholesterol binding may be a crucial feature of many GPCRs. Indeed, the subsequently solved GPCR structures of $A2_A$

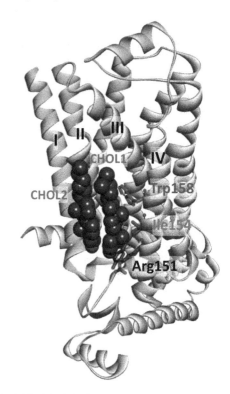

Fig. 1.12 Binding of cholesterol modulates stability and function of GPCR receptors. Binding of two cholesterol molecules (CHOL1 and CHOL2) to β2-aderenergic receptor (β2-AR) (Hanson et al. 2008). Binding site of CHOL1 represents a cholesterol consensus motif. Trp158 and Ile154, which are 95 and 35 % conserved by identity, are central in cholesterol binding (Figure modified with permission from Hanson et al. 2008)

adenosine receptor (Jaakola et al. 2008), μ-type opioid receptor (Manglik et al. 2012) and hydroxytryptoamine (serotonin) receptor 2B (Wang et al. 2013) revealed cholesterol bound to this motif, although the precise modes of interactions were slightly different. The enhancement of protein thermostability has been observed in the micellar phase for other GPCRs such as oxytocin receptor, which is a serotonin$_{1A}$ receptor containing a strict cholesterol consensus motif (Hanson et al. 2008). The interaction between GPCRs and cholesterol were postulated to lead to their preferential partition into the cholesterol-rich caveolae (Pucadyil and Chattopadhyay 2004; Song et al. 2014a).

1.6.4 Role of Cholesterol in Amyloid Precursor Proteins

While cholesterol binds to the "cleft" formed by TM helices of β2-AR and is locked by several bulky residues, another type of cholesterol binding site has been identified in the amyloid precursor protein (APP) (Barrett et al. 2012). APP is cleaved by an intramembrane protease, γ-secretase, which generates the C-terminal C99

TM domain and the amyloid-β (Aβ) peptide (Song et al. 2014a). Misfolding and accumulation of aggregated Aβ in the brain is the hallmark of Alzheimer's disease (Finder and Glockshuber 2007). The solution NMR structure of APP showed that the extracellular region of C99 is composed of an amphiphilic N-helix and a short N-loop connected to the TM domain (Barrett et al. 2012) (Fig. 1.13). The TM domain is highly bent and flexible, which may promote the cleavage of C99 by γ-secretase. The N-terminal half of the TM domain contains the GxxxGxxxG glycine zipper motif, which is frequently observed in the TM helix-helix interactions ((Kim et al. 2005); also see Chap. 4 by Morgado and Garvey and Chap. 10 by Leonova and Galzitskaya in this volume). Because of this motif, the C99 protein was thought to form dimers or higher-order oligomers. However, titration of cholesterol (up to 20 mol %) in bicelles, monitored by NMR, revealed a striking binding pattern of cholesterol to C99. Contrary to the expectation, two of the three Gly in the zipper motif, G700 and G704, are critically involved in cholesterol binding rather than oligomerization. In addition, the dynamic N-loop and the polar residues N698 and E693 also appear to contribute to the binding of cholesterol.

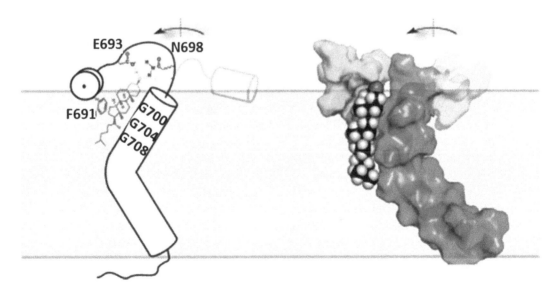

Fig. 1.13 Cholesterol binding to the C99 TM domain of amyloid precursor protein. Cholesterol binds to the flat surface created by glycine zipper motif (G700-xxx-G704- xxx-G708) and, additionally, by N-helix (F691) and N-loop (E693 and N698) (Figure adopted with permission from Song et al. 2014a)

Furthermore, the dynamic N-helix, which lies parallel to the membrane and perpendicular to the TM helix to which it is linked via the N-loop, effectively locks bound cholesterol by packing with the bulky aromatic residue, F690. Thus, the flat surface presented by the GxxxG motif, the flexible N-loop and the conformation of N-helix work together to accommodate bound cholesterol.

What is the physiological significance of cholesterol binding in the generation of Aβ? γ-secretases are known to preferentially partition into the cholesterol-rich membrane rafts (Lee et al. 1998). Cholesterol binding to APP is expected to favor partitioning into membrane rafts, and thereby co-localize APP with the γ-secretase. Cholesterol binding was also proposed to make APP a better substrate for γ-secretase (Osenkowski et al. 2008). Moreover, after cleavage, Aβ association with cholesterol may facilitate amyloid fibril formation (Yanagisawa 2005). These effects provide possible explanations for the link between the plasma levels of cholesterol and the development of Alzheimer's disease.

1.6.5 Role of PIP2 in the Gating Function of Inward Rectifier Potassium Channel

Phosphatidylinositol 4,5 diphosphate (PIP_2) is a minor lipid component in the plasma membrane that is critical in a number of signaling pathways as a secondary messenger and a targeting signal of soluble proteins on the plasma membrane (Tisi et al. 2004; Cho and Stahelin 2005). PIP_2 is also of particular interest as a direct modulator of several ion channels (Suh and Hille 2008; Suh et al. 2006, 2010). The crystal structures of inward rectifier potassium channel (Kir2.2) in apo- and in PIP_2 – bound forms solved by (Hansen et al. 2011) demonstrated the specific role of PIP_2 in the regulation of resting cell membrane potential (Fig. 1.14). In the apo form, the TM domain and the cytoplasmic domain are disengaged. Binding of PIP_2 at the interface between the two domains induces a substantial conformational change. The

PIP_2 head group binds to the conserved highly positively charged PIP_2 – binding motif located in the flexible interdomain linker (KxxRPKK motif that binds inositol 4,5 diphosphate) and to the interfacial region of the TM domain (RWR or KWR motif that binds phosphoester). Binding of the acyl chain of PIP_2 to the TM domain occurs in a non-specific manner. This PIP_2 binding, which induces a coil-to-helix transition of the interdomain linker, pulls away the inner helical gating region of the TM domain in the direction of the membrane plane, docks the cytoplasmic G-loop into the channel lumen, and opens up the channel.

1.7 Concluding Remarks

This chapter describes the active role of lipids in the folding, structure and function of integral membrane proteins. The results suggest that lipids in cell membranes not only serve as a medium for stabilizing membrane proteins but also act as critical functional and structural ligands for many of these proteins. Therefore, at the origin of life, the quasi two-dimensional film that constitutes cell membranes may have started as a simple permeability barrier, but evolved to fit to the biological needs in cellular processes by incorporating lipids with various chemical and physical properties and by utilizing them as ligands and stabilizers for membrane proteins.

Structural studies provided the molecular details of lipid-protein interactions for several membrane proteins. Still, it remains a far-reaching goal to quantify the energetics and dynamics of these interactions and elucidate their relationship to the conformational equilibria of membrane proteins which enable their function. In this context, quantitative studies on the energetics of the mechanosensitive channel gating (Perozo et al. 2002a, b; Wiggins and Phillips 2004) and NMR studies of the amyloid precursor protein revealing the dynamic features of this disease-causing protein, are truly intriguing (Barrett et al. 2012; Beel et al. 2010; Pester et al. 2013; Song et al. 2013, 2014b). For such efforts to be extended to other

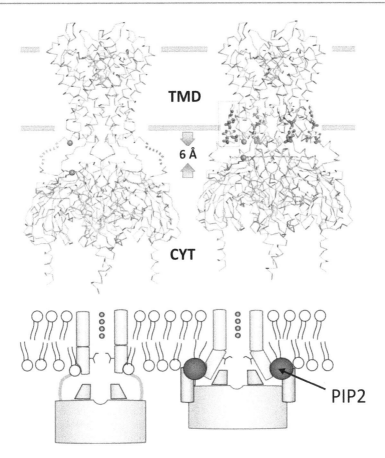

Fig. 1.14 Kir2.2 inward rectifier K⁺ channel gated by PIP₂ binding. Structures of Kir2.2 in apo- and PIP₂-bound forms (*top*). PIP₂ binding induces a conformational change that tightens the junction between the TM domain (TMD) and cytoplasmic domain (CYT) by 6 Å and opens the channel via a proposed mechanism shown in the *bottom*. PIP₂ is shown in *purple circles* (Figures adapted with permission from Hansen et al. 2011)

classes of membrane proteins, new experimental and theoretical frameworks will be necessary. A recent development of steric trapping method that allows the reversible control of the membrane protein folding will open new opportunities for studying the forces and mechanisms of folding of membrane proteins in their native bilayer background (Blois et al. 2009; Chang and Bowie 2014; Hong et al. 2010; Hong and Bowie 2011; Jefferson et al. 2013).

Acknowledgements This works was supported by start-up fund from Michigan State University and Hunt for Cure foundation cystic fibrosis research grant to H.H. The author is thankful for critical comments and suggestions by Hong lab members, anonymous reviewers and Dr. Olga Gursky, the volume editor.

References

Adair BD, Engelman DM (1994) Glycophorin A helical transmembrane domains dimerize in phospholipid bilayers: a resonance energy transfer study. Biochemistry 33(18):5539–5544

Adamian L, Liang J (2001) Helix-helix packing and interfacial pairwise interactions of residues in membrane proteins. J Mol Biol 311(4):891–907

Adamian L, Nanda V, Degrado WF, Liang J (2005) Empirical lipid propensities of amino acid residues in multispan alpha helical membrane proteins. Proteins 59(3):496–509

Anbazhagan V, Schneider D (2010) The membrane environment modulates self-association of the human GpA TM domain - implications for membrane protein folding and transmembrane signaling. Biochim Biophys Acta 1798(10):1899–1907

Ballesteros JA, Weinstein H (1995) Integrated methods for the construction of three-dimensional models and computational probing of structure-function relations in G protein-coupled receptors. Methods Neurosci 25(1):366–428

Barrera NP, Di Bartolo N, Booth PJ, Robinson CV (2008) Micelles protect membrane complexes from solution to vacuum. Science 321(5886):243–246

Barrett PJ, Song Y, Van Horn WD, Hustedt EJ, Schafer JM, Hadziselimovic A, Beel AJ, Sanders CR (2012) The amyloid precursor protein has a flexible transmembrane domain and binds cholesterol. Science 336(6085):1168–1171

Beel AJ, Sakakura M, Barrett PJ, Sanders CR (2010) Direct binding of cholesterol to the amyloid precursor protein: an important interaction in lipid-Alzheimer's disease relationships? Biochim Biophys Acta 1801(8):975–982

Bental N, Sitkoff D, Topol IA, Yang AS, Burt SK, Honig B (1997) Free energy of amide hydrogen bond formation in vacuum, in water, and in liquid alkane solution. J Phys Chem B 101(3):450–457

Bertero MG, Rothery RA, Palak M, Hou C, Lim D, Blasco F, Weiner JH, Strynadka NC (2003) Insights into the respiratory electron transfer pathway from the structure of nitrate reductase A. Nat Struct Biol 10(9):681–687

Blois TM, Hong H, Kim TH, Bowie JU (2009) Protein unfolding with a steric trap. J Am Chem Soc 131(39):13914–13915

Bogdanov M, Dowhan W (1999) Lipid-assisted protein folding. J Biol Chem 274(52):36827–36830

Bogdanov M, Xie J, Heacock P, Dowhan W (2008) To flip or not to flip: lipid-protein charge interactions are a determinant of final membrane protein topology. J Cell Biol 182(5):925–935

Bogdanov M, Heacock P, Guan Z, Dowhan W (2010) Plasticity of lipid-protein interactions in the function and topogenesis of the membrane protein lactose permease from Escherichia coli. Proc Natl Acad Sci U S A 107(34):15057–15062

Bogdanov M, Dowhan W, Vitrac H (2014) Lipids and topological rules governing membrane protein assembly. Biochim Biophys Acta Mol Cell Res 1843(8):1475–1488

Booth PJ, Riley ML, Flitsch SL, Templer RH, Farooq A, Curran AR, Chadborn N, Wright P (1997) Evidence that bilayer bending rigidity affects membrane protein folding. Biochemistry 36(1):197–203

Bowie JU (1997) Helix packing in membrane proteins. J Mol Biol 272(5):780–789

Bowie JU (2011) Membrane protein folding: how important are hydrogen bonds? Curr Opin Struct Biol 21(1):42–49

Burgess NK, Dao TP, Stanley AM, Fleming KG (2008) Beta-barrel proteins that reside in the Escherichia coli outer membrane in vivo demonstrate varied folding behavior in vitro. J Biol Chem 283(39):26748–26758

Burley SK, Petsko GA (1986) Amino-aromatic interactions in proteins. FEBS Lett 203(2):139–143

Bush MF, Hall Z, Giles K, Hoyes J, Robinson CV, Ruotolo BT (2010) Collision cross sections of proteins and their complexes: a calibration framework and database for gas-phase structural biology. Anal Chem 82(22):9557–9565

Cafiso DS (1994) Alamethicin: a peptide model for voltage gating and protein-membrane interactions. Annu Rev Biophys Biomol Struct 23:141–165

Cantor RS (1997) The lateral pressure profile in membranes: a physical mechanism of general anesthesia. Biochemistry 36(9):2339–2344

Cantor RS (1999) Lipid composition and the lateral pressure profile in bilayers. Biophys J 76(5):2625–2639

Chan RB, Oliveira TG, Cortes EP, Honig LS, Duff KE, Small SA, Wenk MR, Shui G, Di Paolo G (2012) Comparative lipidomic analysis of mouse and human brain with Alzheimer disease. J Biol Chem 287(4):2678–2688

Chang YC, Bowie JU (2014) Measuring membrane protein stability under native conditions. Proc Natl Acad Sci U S A 111(1):219–224

Cherezov V, Rosenbaum DM, Hanson MA, Rasmussen SG, Thian FS, Kobilka TS, Choi HJ, Kuhn P, Weis WI, Kobilka BK, Stevens RC (2007) High-resolution crystal structure of an engineered human beta2-adrenergic G protein-coupled receptor. Science 318(5854):1258–1265

Cho W, Stahelin RV (2005) Membrane-protein interactions in cell signaling and membrane trafficking. Annu Rev Biophys Biomol Struct 34:119–151

Chong SH, Ham S (2014) Interaction with the surrounding water plays a key role in determining the aggregation propensity of proteins. Angew Chem 53(15):3961–3964

Clark MR (2011) Flippin' lipids. Nat Immunol 12(5):373–375

Contreras FX, Ernst AM, Wieland F, Brugger B (2011) Specificity of intramembrane protein-lipid interactions. Cold Spring Harb Perspect Biol 3(6):1–18

Curran AR, Templer RH, Booth PJ (1999) Modulation of folding and assembly of the membrane protein bacteriorhodopsin by intermolecular forces within the lipid bilayer. Biochemistry 38(29):9328–9336

Dawidowicz EA (1987) Dynamics of membrane lipid metabolism and turnover. Annu Rev Biochem 56:43–61

Dechavigny A, Heacock PN, Dowhan W (1991) Sequence and inactivation of the pss gene of Escherichia coli. Phosphatidylethanolamine may not be essential for cell viability. J Biol Chem 266(8):5323–5332

Dill KA (1990) Dominant forces in protein folding. Biochemistry 29(31):7133–7155

Dowhan W, Bogdanov M (2009) Lipid-dependent membrane protein topogenesis. Annu Rev Biochem 78:515–540

Du Plessis DJ, Nouwen N, Driessen AJ (2011) The Sec translocase. Biochim Biophys Acta 1808(3):851–865

East JM, Melville D, Lee AG (1985) Exchange rates and numbers of annular lipids for the calcium and magnesium ion dependent adenosine triphosphatase. Biochemistry 24(11):2615–2623

Eilers M, Shekar SC, Shieh T, Smith SO, Fleming PJ (2000) Internal packing of helical membrane proteins. Proc Natl Acad Sci U S A 97(11):5796–5801

Engelman DM (2005) Membranes are more mosaic than fluid. Nature 438(7068):578–580

Engelman DM, Chen Y, Chin CN, Curran AR, Dixon AM, Dupuy AD, Lee AS, Lehnert U, Matthews EE, Reshetnyak YK, Senes A, Popot JL (2003) Membrane protein folding: beyond the two stage model. FEBS Lett 555(1):122–125

Escriba PV, Ozaita A, Ribas C, Miralles A, Fodor E, Farkas T, Garciasevilla JA (1997) Role of lipid polymorphism in G protein-membrane interactions: nonlamellar-prone phospholipids and peripheral protein binding to membranes. Proc Natl Acad Sci U S A 94(21):11375–11380

Finder VH, Glockshuber R (2007) Amyloid-beta aggregation. Neurodegener Dis 4(1):13–27

Fisher LE, Engelman DM, Sturgis JN (1999) Detergents modulate dimerization, but not helicity, of the glycophorin A transmembrane domain. J Mol Biol 293(3):639–651

Fisher LE, Engelman DM, Sturgis JN (2003) Effect of detergents on the association of the glycophorin a transmembrane helix. Biophys J 85(5):3097–3105

Fleming KG, Engelman DM (2001) Specificity in transmembrane helix-helix interactions can define a hierarchy of stability for sequence variants. Proc Natl Acad Sci U S A 98(25):14340–14344

Fleming KG, Ackerman AL, Engelman DM (1997) The effect of point mutations on the free energy of transmembrane alpha-helix dimerization. J Mol Biol 72(2):266–275

Fyfe PK, Jones MR (2005) Lipids in and around photosynthetic reaction centres. Biochem Soc Trans 33(Pt 5):924–930

Gallivan JP, Dougherty DA (1999) Cation-pi interactions in structural biology. Proc Natl Acad Sci U S A 96(17):9459–9464

Gessmann D, Chung YH, Danoff EJ, Plummer AM, Sandlin CW, Zaccai NR, Fleming KG (2014) Outer membrane beta-barrel protein folding is physically controlled by periplasmic lipid head groups and BamA. Proc Natl Acad Sci U S A 111(16):5878–5883

Gonen T, Cheng Y, Sliz P, Hiroaki Y, Fujiyoshi Y, Harrison SC, Walz T (2005) Lipid-protein interactions in double-layered two-dimensional AQP0 crystals. Nature 438(7068):633–638

Gray AN, Henderson-Frost JM, Boyd D, Sharafi S, Niki H, Goldberg MB (2011) Unbalanced charge distribution as a determinant for dependence of a subset of Escherichia coli membrane proteins on the membrane insertase YidC. MBio 2(6):1–10

Grigorieff N, Ceska TA, Downing KH, Baldwin JM, Henderson R (1996) Electron-crystallographic refinement of the structure of bacteriorhodopsin. J Mol Biol 259(3):393–421

Grove DE, Fan CY, Ren HY, Cyr DM (2011) The endoplasmic reticulum-associated Hsp40 DNAJB12 and Hsc70 cooperate to facilitate RMA1 E3-dependent degradation of nascent CFTR Delta F508. Mol Biol Cell 22(3):301–314

Gruner SM (1985) Intrinsic curvature hypothesis for biomembrane lipid-composition – a role for nonbilayer lipids. Proc Natl Acad Sci U S A 82(11):3665–3669

Gruner SM, Cullis PR, Hope MJ, Tilcock CP (1985) Lipid polymorphism: the molecular basis of nonbilayer phases. Annu Rev Biophys Biophys Chem 14:211–238

Gruner SM, Tate MW, Kirk GL, So PT, Turner DC, Keane DT, Tilcock CP, Cullis PR (1988) X-ray diffraction study of the polymorphic behavior of N-methylated dioleoyl phosphatidylethanolamine. Biochemistry 27(8):2853–2866

Guan L, Kaback HR (2006) Lessons from lactose permease. Annu Rev Biophys Biomol Struct 35:67–91

Guskov A, Kern J, Gabdulkhakov A, Broser M, Zouni A, Saenger W (2009) Cyanobacterial photosystem II at 2.9-A resolution and the role of quinones, lipids, channels and chloride. Nat Struct Mol Biol 16(3):334–342

Haltia T, Freire E (1995) Forces and factors that contribute to the structural stability of membrane-proteins. Biochim Biophys Acta Bioenerg 1228(1):1–27

Hampton RY (2002) ER-associated degradation in protein quality control and cellular regulation. Curr Opin Cell Biol 14(4):476–482

Hansen SB, Tao X, Mackinnon R (2011) Structural basis of PIP2 activation of the classical inward rectifier K+ channel Kir2.2. Nature 477(7365):495–498

Hanson MA, Cherezov V, Griffith MT, Roth CB, Jaakola VP, Chien EY, Velasquez J, Kuhn P, Stevens RC (2008) A specific cholesterol binding site is established by the 2.8 A structure of the human beta2-adrenergic receptor. Structure 16(6):897–905

He C, Klionsky DJ (2009) Regulation mechanisms and signaling pathways of autophagy. Annu Rev Genet 43:67–93

Hessa T, Kim H, Bihlmaier K, Lundin C, Boekel J, Andersson H, Nilsson I, White SH, Von Heijne G (2005) Recognition of transmembrane helices by the endoplasmic reticulum translocon. Nature 433(7024):377–381

Hessa T, Meindl-Beinker NM, Bernsel A, Kim H, Sato Y, Lerch-Bader M, Nilsson I, White SH, Von Heijne G (2007) Molecular code for transmembrane-helix recognition by the Sec61 translocon. Nature 450(7172):1026–1030

Hite RK, Li Z, Walz T (2010) Principles of membrane protein interactions with annular lipids deduced from aquaporin-0 2D crystals. EMBO J 29(10):1652–1628

Hong H, Bowie JU (2011) Dramatic destabilization of transmembrane helix interactions by features of natural membrane environments. J Am Chem Soc 133(29):11389–11398

Hong H, Tamm LK (2004) Elastic coupling of integral membrane protein stability to lipid bilayer forces. Proc Natl Acad Sci U S A 101(12):4065–4070

Hong H, Blois TM, Cao Z, Bowie JU (2010) Method to measure strong protein-protein interactions in lipid bilayers using a steric trap. Proc Natl Acad Sci U S A 107(46):19802–19807

Hong H, Chang YC, Bowie JU (2013) Measuring transmembrane helix interaction strengths in lipid bilayers using steric trapping. Methods Mol Biol 1063:37–56

Houck SA, Cyr DM (2012) Mechanisms for quality control of misfolded transmembrane proteins. Biochim Biophys Acta 1818(4):1108–1114

Hu CX, Van Der Heijden R, Wang M, Van Der Greef J, Hankemeier T, Xua GW (2009) Analytical strategies in lipidomics and applications in disease biomarker discovery. J Chromatogr B Anal Tech Biomed Life Sci 877(26):2836–2846

Huysmans GH, Baldwin SA, Brockwell DJ, Radford SE (2010) The transition state for folding of an outer membrane protein. Proc Natl Acad Sci U S A 107(9):4099–4104

Jaakola VP, Griffith MT, Hanson MA, Cherezov V, Chien EY, Lane JR, Ijzerman AP, Stevens RC (2008) The 2.6 angstrom crystal structure of a human A2A adenosine receptor bound to an antagonist. Science 322(5905):1211–1217

Jefferson RE, Blois TM, Bowie JU (2013) Membrane proteins can have high kinetic stability. J Am Chem Soc 135(40):15183–15190

Joh NH, Oberai A, Yang D, Whitelegge JP, Bowie JU (2009) Similar energetic contributions of packing in the core of membrane and water-soluble proteins. J Am Chem Soc 131(31):10846–10847

Keller SL, Bezrukov SM, Gruner SM, Tate MW, Vodyanoy I, Parsegian VA (1993) Probability of alamethicin conductance states varies with nonlamellar tendency of bilayer phospholipids. Biophys J 65(1):23–27

Kikkert M, Doolman R, Dai M, Avner R, Hassink G, Van Voorden S, Thanedar S, Roitelman J, Chau V, Wiertz E (2004) Human HRD1 is an E3 ubiquitin ligase involved in degradation of proteins from the endoplasmic reticulum. J Biol Chem 279(5):3525–3534

Kim S, Jeon TJ, Oberai A, Yang D, Schmidt JJ, Bowie JU (2005) Transmembrane glycine zippers: physiological and pathological roles in membrane proteins. Proc Natl Acad Sci U S A 102(40):14278–14283

Kleinschmidt JH, Tamm LK (1996) Folding intermediates of a beta-barrel membrane protein. Kinetic evidence for a multi-step membrane insertion mechanism. Biochemistry 35(40):12993–13000

Kleinschmidt JH, Tamm LK (2002) Secondary and tertiary structure formation of the beta-barrel membrane protein OmpA is synchronized and depends on membrane thickness. J Mol Biol 324(2):319–330

Knowles PF, Watts A, Marsh D (1979) Spin-label studies of lipid immobilization in dimyristoyl phosphatidylcholine-substituted cytochrome oxidase. Biochemistry 18(21):4480–4487

Kopito RR (1999) Biosynthesis and degradation of CFTR. Physiol Rev 79(1 Suppl):S167–S173

Krishnamachary N, Stephenson FA, Steggles AW, Holloway PW (1994) Stopped-flow fluorescence studies of the interaction of a mutant form of cytochrome b5 with lipid vesicles. J Fluoresc 4(3):227–233

Krogh A, Larsson B, Von Heijne G, Sonnhammer ELL (2001) Predicting transmembrane protein topology with a hidden Markov model: application to complete genomes. J Mol Biol 305(3):567–580

Kumar VV (1991) Complementary molecular shapes and additivity of the packing parameter of lipids. Proc Natl Acad Sci U S A 88(2):444–448

Kusumi A, Ike H, Nakada C, Murase K, Fujiwara T (2005) Single-molecule tracking of membrane molecules: plasma membrane compartmentalization and dynamic assembly of raft-philic signaling molecules. Semin Immunol 17(1):3–21

Kuznetsov G, Nigam SK (1998) Folding of secretory and membrane proteins. N Engl J Med 339(23):1688–1695

Laganowsky A, Reading E, Hopper JT, Robinson CV (2013) Mass spectrometry of intact membrane protein complexes. Nat Protoc 8(4):639–651

Lange C, Nett JH, Trumpower BL, Hunte C (2001) Specific roles of protein-phospholipid interactions in the yeast cytochrome bc1 complex structure. EMBO J 20(23):6591–6600

Lee AG (2011) Biological membranes: the importance of molecular detail. Trends Biochem Sci 36(9):493–500

Lee SJ, Liyanage U, Bickel PE, Xia W, Lansbury PT Jr, Kosik KS (1998) A detergent-insoluble membrane compartment contains A beta in vivo. Nat Med 4(6):730–734

Lehninger AL, Nelson DL, Cox MM (1993) Principles of biochemistry. Worth Publishers, New York

Lodish H, Berk A, Kaiser CA, Krieger T, Scott M, Pretscher A, Ploegh H, Matsudaira P (2000) Biomembrane structure (chapter 10). In: Molecular cell biology. W. H. Freeman, New York

London E, Feigenson GW (1981) Fluorescence quenching in model membranes. 1. Characterization of quenching caused by a spin-labeled phospholipid. Biochemistry 20(7):1932–1938

Luecke H, Schobert B, Richter HT, Cartailler JP, Lanyi JK (1999) Structure of bacteriorhodopsin at 1.55 A resolution. J Mol Biol 291(4):899–911

Lukacs GL, Verkman AS (2012) CFTR: folding, misfolding and correcting the DeltaF508 conformational defect. Trends Mol Med 18(2):81–91

Macgurn JA, Hsu PC, Emr SD (2012) Ubiquitin and membrane protein turnover: from cradle to grave. Annu Rev Biochem 81:231–259

Manglik A, Kruse AC, Kobilka TS, Thian FS, Mathiesen JM, Sunahara RK, Pardo L, Weis WI, Kobilka BK, Granier S (2012) Crystal structure of the micro-opioid receptor bound to a morphinan antagonist. Nature 485(7398):321–326

Marsh D (2008) Protein modulation of lipids, and vice-versa, in membranes. Biochim Biophys Acta 1778(7–8):1545–1575

Mendre C, Mouillac B (2010) Pharmacological chaperones: a potential therapeutic treatment for conformational diseases. M S Med Sci 26(6–7):627–635

Miller C (2010) CFTR: break a pump, make a channel. Proc Natl Acad Sci U S A 107(3):959–960

Milutinovic PS, Yang L, Cantor RS, Eger EI 2nd, Sonner JM (2007) Anesthetic-like modulation of a gamma-aminobutyric acid type A, strychnine-sensitive glycine, and N-methyl-d-aspartate receptors by coreleased neurotransmitters. Anesth Analg 105(2):386–392

Minton AP (2000) Implications of macromolecular crowding for protein assembly. Curr Opin Struct Biol 10(1):34–39

Mitra K, Ubarretxena-Belandia I, Taguchi T, Warren G, Engelman DM (2004) Modulation of the bilayer thickness of exocytic pathway membranes by membrane proteins rather than cholesterol. Proc Natl Acad Sci U S A 101(12):4083–4088

Moon CP, Fleming KG (2011) Side-chain hydrophobicity scale derived from transmembrane protein folding into lipid bilayers. Proc Natl Acad Sci U S A 108(25):10174–10177

Moon CP, Zaccai NR, Fleming PJ, Gessmann D, Fleming KG (2013) Membrane protein thermodynamic stability may serve as the energy sink for sorting in the periplasm. Proc Natl Acad Sci U S A 110(11):4285–4290

Morein S, Andersson A, Rilfors L, Lindblom G (1996) Wild-type Escherichia coli cells regulate the membrane lipid composition in a "window" between gel and non-lamellar structures. J Biol Chem 271(12):6801–6809

Morrison EA, Dekoster GT, Dutta S, Vafabakhsh R, Clarkson MW, Bahl A, Kern D, Ha T, Henzler-Wildman KA (2012) Antiparallel EmrE exports drugs by exchanging between asymmetric structures. Nature 481(7379):45–50

Navarro J, Toivio-Kinnucan M, Racker E (1984) Effect of lipid composition on the calcium/adenosine 5'-triphosphate coupling ratio of the Ca2+-ATPase of sarcoplasmic reticulum. Biochemistry 23(1):130–135

Nicolson GL (2014) The fluid-mosaic model of membrane structure: still relevant to understanding the structure, function and dynamics of biological membranes after more than 40 years. Biochem Biophys Acta Biomembr 1838(6):1451–1466

Noinaj N, Kuszak AJ, Gumbart JC, Lukacik P, Chang HS, Easley NC, Lithgow T, Buchanan SK (2013) Structural insight into the biogenesis of beta-barrel membrane proteins. Nature 501(7467):385–390

Noinaj N, Kuszak AJ, Balusek C, Gumbart JC, Buchanan SK (2014) Lateral opening and exit pore formation are required for BamA function. Structure 22(7):1055–1062

Oberai A, Joh NH, Pettit FK, Bowie JU (2009) Structural imperatives impose diverse evolutionary constraints on helical membrane proteins. Proc Natl Acad Sci U S A 106(42):17747–17750

Osenkowski P, Ye W, Wang R, Wolfe MS, Selkoe DJ (2008) Direct and potent regulation of gamma-secretase by its lipid microenvironment. J Biol Chem 283(33):22529–2240

Pareek S, Notterpek L, Snipes GJ, Naef R, Sossin W, Laliberte J, Iacampo S, Suter U, Shooter EM, Murphy RA (1997) Neurons promote the translocation of peripheral myelin protein 22 into myelin. J Neurosci 17(20):7754–7762

Park E, Rapoport TA (2012) Mechanisms of Sec61/SecY-mediated protein translocation across membranes. Annu Rev Biophys 41:21–40

Pasyk EA, Foskett JK (1995) Mutant (Delta-F508) Cystic-fibrosis transmembrane conductance regulator Cl- channel is functional when retained in endoplasmic-reticulum of mammalian cells. J Biol Chem 270(21):12347–12350

Patel GJ, Kleinschmidt JH (2013) The lipid bilayer-inserted membrane protein BamA of escherichia coli facilitates insertion and folding of outer membrane protein A from its complex with Skp. Biochemistry 52(23):3974–3986

Perozo E, Cortes DM, Sompornpisut P, Kloda A, Martinac B (2002a) Open channel structure of MscL and the gating mechanism of mechanosensitive channels. Nature 418(6901):942–948

Perozo E, Kloda A, Cortes DM, Martinac B (2002b) Physical principles underlying the transduction of bilayer deformation forces during mechanosensitive channel gating. Nat Struct Biol 9(9):696–703

Pester O, Barrett PJ, Hornburg D, Hornburg P, Probstle R, Widmaier S, Kutzner C, Durrbaum M, Kapurniotu A, Sanders CR, Scharnagl C, Langosch D (2013) The backbone dynamics of the amyloid precursor protein transmembrane helix provides a rationale for the sequential cleavage mechanism of gamma-secretase. J Am Chem Soc 135(4):1317–1329

Popot JL, Engelman DM (1990) Membrane protein folding and oligomerization: the two-stage model. Biochemistry 29(17):4031–4037

Popot JL, Engelman DM (2000) Helical membrane protein folding, stability, and evolution. Annu Rev Biochem 69:881–922

Powell GL, Knowles PF, Marsh D (1985) Association of spin-labelled cardiolipin with dimyristoyl phosphatidylcholine-substituted bovine heart cytochrome c oxidase. A generalized specificity increase rather than highly specific binding sites. Biochim Biophys Acta 816(1):191–194

Pucadyil TJ, Chattopadhyay A (2004) Cholesterol modulates ligand binding and G-protein coupling to serotonin (1A) receptors from bovine hippocampus. Biochim Biophys Acta 1663(1–2):188–200

Rajan RS, Kopito RR (2004) Role of protein misfolding in retinal degeneration caused by a mutation in rhodopsin. Abstr Am Chem Soc 227:U220

Rapoport TA (2007) Protein translocation across the eukaryotic endoplasmic reticulum and bacterial plasma membranes. Nature 450(7170):663–669

Rasmussen SG, Choi HJ, Rosenbaum DM, Kobilka TS, Thian FS, Edwards PC, Burghammer M, Ratnala VR, Sanishvili R, Fischetti RF, Schertler GF, Weis WI, Kobilka BK (2007) Crystal structure of the human beta2 adrenergic G-protein-coupled receptor. Nature 450(7168):383–387

Ravid T, Kreft SG, Hochstrasser M (2006) Membrane and soluble substrates of the Doa10 ubiquitin ligase are degraded by distinct pathways. EMBO J 25(3):533–543

Reinstein E, Ciechanover A (2006) Narrative review: protein degradation and human diseases: the ubiquitin connection. Ann Intern Med 145(9):676–684

Sakakura M, Hadziselimovic A, Wang Z, Schey KL, Sanders CR (2011) Structural basis for the trembler-J phenotype of Charcot-Marie-Tooth disease. Structure 19(8):1160–1169

Sanders CR, Myers JK (2004) Disease-related misassembly of membrane proteins. Annu Rev Biophys Biomol Struct 33:25–51

Sato BK, Schulz D, Do PH, Hampton RY (2009) Misfolded membrane proteins are specifically recognized by the transmembrane domain of the Hrd1p ubiquitin ligase. Mol Cell 34(2):212–222

Seddon JM, Templer RH (1995) Polymorphism of lipid-water systems (chapter 3). In: Lipowsky R, Sackmann E (eds) Handbook of biological physics, volume 1. Elsevier, Amsterdam

Sheppard DN, Welsh MJ (1999) Structure and function of the CFTR chloride channel. Physiol Rev 79(1 Suppl):S23–S45

Shimkets RA (1996) Hot papers – hypertension genetics – Liddle's syndrome: heritable human hypertension caused by mutations of the beta subunit of the epithelial sodium channel by R.A. Shimkets DG, Warnock CM, Bositis C, Nelson-Williams JH, Hansson M, Schambelan JR, Gill S, Ulick RV, Milora JW, Findling CM, Canessa BC, Rossier RP. Lifton – Comments. Scientist 10(22):14

Shinoda T, Ogawa H, Cornelius F, Toyoshima C (2009) Crystal structure of the sodium-potassium pump at 2.4 A resolution. Nature 459(7245):446–450

Shinzawa-Itoh K, Aoyama H, Muramoto K, Terada H, Kurauchi T, Tadehara Y, Yamasaki A, Sugimura T, Kurono S, Tsujimoto K, Mizushima T, Yamashita E, Tsukihara T, Yoshikawa S (2007) Structures and physiological roles of 13 integral lipids of bovine heart cytochrome c oxidase. EMBO J 26(6):1713–1725

Shyamsunder E, Gruner SM, Tate MW, Turner DC, So PT, Tilcock CP (1988) Observation of inverted cubic phase in hydrated dioleoylphosphatidylethanolamine membranes. Biochemistry 27(7):2332–2336

Siegel DP (1993) Energetics of intermediates in membrane fusion: comparison of stalk and inverted micellar intermediate mechanisms. Biophys J 65(5):2124–2140

Simons K, Sampaio JL (2011) Membrane organization and lipid rafts. Cold Spring Harb Perspect Biol 3(10):a004697

Singer SJ, Nicolson GL (1972) The fluid mosaic model of the structure of cell membranes. Science 175(4023):720–731

Song Y, Hustedt EJ, Brandon S, Sanders CR (2013) Competition between homodimerization and cholesterol binding to the C99 domain of the amyloid precursor protein. Biochemistry 52(30):5051–5064

Song Y, Kenworthy AK, Sanders CR (2014a) Cholesterol as a co-solvent and a ligand for membrane proteins. Protein Sci 23(1):1–22

Song Y, Mittendorf KF, Lu Z, Sanders CR (2014b) Impact of bilayer lipid composition on the structure and topology of the transmembrane amyloid precursor C99 protein. J Am Chem Soc 136(11):4093–4096

Spector AA, Yorek MA (1985) Membrane lipid composition and cellular function. J Lipid Res 26(9):1015–1035

Staub O, Gautschi I, Ishikawa T, Breitschopf K, Ciechanover A, Schild L, Rotin D (1997) Regulation of stability and function of the epithelial Na+channel (ENaC) by ubiquitination. EMBO J 16(21):6325–6336

Stenson PD, Ball E, Howells K, Phillips A, Mort M, Cooper DN (2008) Human gene mutation database: towards a comprehensive central mutation database. J Med Genet 45(2):124–126

Suh BC, Hille B (2008) PIP2 is a necessary cofactor for ion channel function: how and why? Annu Rev Biophys 37:175–195

Suh BC, Inoue T, Meyer T, Hille B (2006) Rapid chemically induced changes of PtdIns(4,5)P2 gate KCNQ ion channels. Science 314(5804):1454–1457

Suh BC, Leal K, Hille B (2010) Modulation of high-voltage activated Ca(2+) channels by membrane phosphatidylinositol 4,5-bisphosphate. Neuron 67(2):224–238

Sun J, Wu J, Carrasco N, Kaback HR (1996) Identification of the epitope for monoclonal antibody 4B1 which uncouples lactose and proton translocation in the lactose permease of Escherichia coli. Biochemistry 35(3):990–998

Sun F, Zhang RL, Gong XY, Geng XH, Drain PF, Frizzell RA (2006) Derlin-1 promotes the efficient degradation of the cystic fibrosis transmembrane conductance regulator (CFTR) and CFTR folding mutants. J Biol Chem 281(48):36856–36863

Szule JA, Fuller NL, Rand RP (2002) The effects of acyl chain length and saturation of diacylglycerols and phosphatidylcholines on membrane monolayer curvature. Biophys J 83(2):977–984

Takeda K, Sato H, Hino T, Kono M, Fukuda K, Sakurai I, Okada T, Kouyama T (1998) A novel three-dimensional crystal of bacteriorhodopsin obtained by successive fusion of the vesicular assemblies. J Mol Biol 283(2):463–474

Tamm LK, Hong H, Liang BY (2004) Folding and assembly of beta-barrel membrane proteins. Biochem Biophys Acta Biomembr 1666(1–2):250–263

Tate MW, Gruner SM (1987) Lipid polymorphism of mixtures of dioleoylphosphatidylethanolamine and saturated and monounsaturated phosphatidylcholines of various chain lengths. Biochemistry 26(1):231–236

Tatsuta T, Langer T (2008) Quality control of mitochondria: protection against neurodegeneration and ageing. EMBO J 27(2):306–314

Tieleman DP, Hess B, Sansom MS (2002) Analysis and evaluation of channel models: simulations of alamethicin. Biophys J 83(5):2393–2407

Tisi R, Belotti F, Wera S, Winderickx J, Thevelein JM, Martegani E (2004) Evidence for inositol triphosphate as a second messenger for glucose-induced calcium signalling in budding yeast. Curr Genet 45(2):83–89

Valiyaveetil FI, Zhou Y, Mackinnon R (2002) Lipids in the structure, folding, and function of the KcsA K+ channel. Biochemistry 41(35):10771–10777

Van Den Brink-Van Der Laan E, Chupin V, Killian JA, De Kruijff B (2004a) Stability of KcsA tetramer depends on membrane lateral pressure. Biochemistry 43(14):4240–4250

Van Den Brink-Van Der Laan E, Killian JA, De Kruijff B (2004b) Nonbilayer lipids affect peripheral and integral membrane proteins via changes in the lateral pressure profile. Biochim Biophys Acta 1666(1–2):275–288

Van Meer G, Voelker DR, Feigenson GW (2008) Membrane lipids: where they are and how they behave. Nat Rev Mol Cell Biol 9(2):112–124

Von Heijne G (1992) Membrane protein structure prediction. Hydrophobicity analysis and the positive-inside rule. J Mol Biol 225(2):487–494

Von Heijne G (2006) Membrane-protein topology. Nat Rev Mol Cell Biol 7(12):909–918

Wang C, Jiang Y, Ma J, Wu H, Wacker D, Katritch V, Han GW, Liu W, Huang XP, Vardy E, Mccorvy JD, Gao X, Zhou XE, Melcher K, Zhang C, Bai F, Yang H, Yang L, Jiang H, Roth BL, Cherezov V, Stevens RC, Xu HE (2013) Structural basis for molecular recognition at serotonin receptors. Science 340(6132):610–614

White SH, Von Heijne G (2008) How translocons select transmembrane helices. Annu Rev Biophys 37:23–42

White SH, Wimley WC (1999) Membrane protein folding and stability: physical principles. Annu Rev Biophys Biomol Struct 28:319–365

Wickner S, Maurizi MR, Gottesman S (1999) Posttranslational quality control: folding, refolding, and degrading proteins. Science 286(5446):1888–18893

Wiggins P, Phillips R (2004) Analytic models for mechanotransduction: gating a mechanosensitive channel. Proc Natl Acad Sci U S A 101(12):4071–4076

Wimley WC (2003) The versatile beta-barrel membrane protein. Curr Opin Struct Biol 13(4):404–411

Wimley WC, White SH (1992) Partitioning of tryptophan side-chain analogs between water and cyclohexane. Biochemistry 31(51):12813–12818

Wimley WC, White SH (1996) Experimentally determined hydrophobicity scale for proteins at membrane interfaces. Nat Struct Biol 3(10):842–848

Winckler B, Forscher P, Mellman I (1999) A diffusion barrier maintains distribution of membrane proteins in polarized neurons. Nature 397(6721):698–701

Wu CC, Maccoss MJ, Howell KE, Yates JR 3rd (2003) A method for the comprehensive proteomic analysis of membrane proteins. Nat Biotechnol 21(5):532–538

Yanagisawa K (2005) Cholesterol and amyloid beta fibrillogenesis. Subcell Biochem 38:179–202

Yang L, Huang HW (2002) Observation of a membrane fusion intermediate structure. Science 297(5588):1877–1879

Zhang W, Bogdanov M, Pi J, Pittard AJ, Dowhan W (2003) Reversible topological organization within a polytopic membrane protein is governed by a change in membrane phospholipid composition. J Biol Chem 278(50):50128–50135

Zhang W, Campbell HA, King SC, Dowhan W (2005) Phospholipids as determinants of membrane protein topology – phosphatidylethanolamine is required for the proper topological organization of the gamma-aminobutyric acid permease (GabP) of Escherichia coli. J Biol Chem 280(28):26032–26038

Zhou Y, Morais-Cabral JH, Kaufman A, Mackinnon R (2001) Chemistry of ion coordination and hydration revealed by a K+ channel-Fab complex at 2.0 A resolution. Nature 414(6859):43–48

Protein Misfolding in Lipid-Mimetic Environments

2

Vladimir N. Uversky

Abstract

Among various cellular factors contributing to protein misfolding and subsequent aggregation, membranes occupy a special position due to the two-way relations between the aggregating proteins and cell membranes. On one hand, the unstable, toxic pre-fibrillar aggregates may interact with cell membranes, impairing their functions, altering ion distribution across the membranes, and possibly forming non-specific membrane pores. On the other hand, membranes, too, can modify structures of many proteins and affect the misfolding and aggregation of amyloidogenic proteins. The effects of membranes on protein structure and aggregation can be described in terms of the "membrane field" that takes into account both the negative electrostatic potential of the membrane surface and the local decrease in the dielectric constant. Water-alcohol (or other organic solvent) mixtures at moderately low pH are used as model systems to study the joint action of the local decrease of pH and dielectric constant near the membrane surface on the structure and aggregation of proteins. This chapter describes general mechanisms of structural changes of proteins in such model environments and provides examples of various proteins aggregating in the "membrane field" or in lipid-mimetic environments.

V.N. Uversky (✉)
Department of Molecular Medicine and USF Health
Byrd Alzheimer's Research Institute, Morsani
College of Medicine, University of South Florida,
12901 Bruce B. Downs Blvd. MDC07, Tampa,
FL 33612, USA

Biology Department, Faculty of Science, King
Abdulaziz University, P.O. Box 80203,
Jeddah 21589, Kingdom of Saudi Arabia

Institute for Biological Instrumentation,
Russian Academy of Sciences, 142290 Pushchino,
Moscow Region, Russia

Laboratory of Structural Dynamics, Stability and
Folding of Proteins, Institute of Cytology, Russian
Academy of Sciences, St. Petersburg, Russia
e-mail: vuversky@health.usf.edu

© Springer International Publishing Switzerland 2015
O. Gursky (ed.), *Lipids in Protein Misfolding*, Advances in Experimental
Medicine and Biology 855, DOI 10.1007/978-3-319-17344-3_2

Keywords

Membrane field • Lipid mimetic • Intrinsically disordered protein • Protein misfolding • Protein aggregation • Protein-membrane interaction

Abbreviations

Aβ	Amyloid-β
AFM	Atomic force microscopy
ANS	Anilino-8-napthalene sulfonate
CD	Circular dichroism
DMPC	Dimyristoyl phosphatidylcholine
DMPG	1,2-dimyristoyl-*sn*-glycero-3-[phospho-rac-(1-glycerol)] sodium salt
DMPS	Dimyristoyl phosphatidylserine
EtOH	Ethanol
FTIR	Fourier transform infrared spectroscopy
HFiP	1,1,1,3,3,3-hexafluoro-2-propanol
HSA	Human serum albumin
IAPP	Islet amyloid polypeptide
IDP	Intrinsically disordered protein
IDPR	Intrinsically disordered protein region
LUV	Large unilamellar vesicle
MeOH	Methanol
nFGF-1	Newt acidic fibroblast growth factor
PA	1,2-dipalmitoyl-*sn*-glycero-3-phosphate
PC	1,2-dipalmitoyl-*sn*-glycero-3-phospho-choline
PE	1-palmitoyl-2-oleoyl-phosphoethanolamine
PG	1,2-dipalmitoyl-*sn*-glycero-3-phospho-RAC-(1-glycerol)
PI	Phosphatidylinositol
PrOH	Propanol
PS	1-palmitoyl-2-oleoyl-phosphatidylserine
SUV	Small unilamellar vesicle
TFE	2,2,2-trifluoroethanol
ThT	Thioflavin T

2.1 Introduction

2.1.1 Molecular Mechanisms of Protein Misfolding Diseases

Proteins are the major components of the living cell which play a crucial role in the maintenance of life. Protein dysfunctions cause various pathological conditions. Numerous human diseases, known as conformational or protein misfolding diseases, arise from the failure of a specific peptide or protein to adopt its native functional conformation. The obvious consequences of misfolding are protein aggregation and/or fibril formation, loss of function, and gain of toxic function. Some proteins have an intrinsic propensity to acquire a pathologic conformation, which becomes evident with aging or at persistently high concentrations. Interactions (or impaired interactions) with some endogenous factors (e.g., membranes, chaperones, intracellular or extracellular matrixes, other proteins, small molecules) can change protein conformation and influence its propensity to misfold. Misfolding can originate from point mutations or from an exposure to internal or external toxins, impaired posttranslational modifications (phosphorylation, advanced glycation, deamidation, racemization, etc.), increased degradation, impaired trafficking, lost binding partners or oxidative damage. All these factors can act independently or in concert with one another.

Misfolding diseases can affect a single organ or multiple tissues. Numerous neurodegenerative disorders and amyloidoses originate from the conversion of a specific protein from its soluble functional state into stable, highly ordered, filamentous aggregates, termed amyloid fibrils, that can deposit in various organs and tissues. In each disease, a specific protein or protein fragments deposit as fibrils (Kelly 1998; Dobson 1999; Bellotti et al. 1999; Uversky et al. 1999a, b; Rochet and Lansbury 2000; Uversky and Fink 2004). Amyloid fibrils display many common properties including a core cross-β-sheet structure in which β-strands run perpendicular to the fibril axis (Sunde et al. 1997). Morphologically, amyloid fibrils typically consist of two to six unbranched protofilaments 2–5 nm in diameter associated laterally or twisted together to form fibrils with 4–13 nm diameter (Shirahama et al. 1973; Shirahama and Cohen 1967; Jimenez et al. 1999; also see Gorbenko et al., Chap. 6 and Singh et al., Chap. 4 in this volume). Although the amyloid fibrils from different diseases have common structural and morphological characteristics, the polypeptides causing these diseases are extremely diverse and, prior to fibrillation, may be rich in β-sheet, α-helix, or be intrinsically disordered (Uversky and Fink 2004).

For years it has been assumed that amyloid fibril formation is limited to the disease-related proteins that possess specific sequence motifs encoding the unique structure of the amyloid core. However, later studies showed that many disease-unrelated proteins can also form fibrils (Dobson 1999; Uversky 2003b; Uversky and Fink 2004; Chiti et al. 1999). The current concept is that virtually any protein can form amyloid under appropriate conditions (Dobson 1999; Uversky 2003b; Uversky and Fink 2004). The structural diversity of amyloidogenic proteins and close similarity of the resultant fibrils imply that considerable structural rearrangements must occur in order for fibrils to form. In a well-folded globular protein, such rearrangements cannot take place due to the constraints of the tertiary structure. Therefore, it has been proposed that fibrillation of a globular protein requires the destabilization of its native structure, leading to the formation of a partially unfolded conformation (Fink 1998; Kelly 1998; Dobson 1999, 2001; Bellotti et al. 1999; Uversky et al. 1999a, b; Rochet and Lansbury 2000; Uversky and Fink 2004; Lansbury 1999; Zerovnik 2002).

2.1.2 Brief Introduction to Intrinsically Disordered Proteins

Recent years showed an increasing appreciation of proteins that lack unique 3D structure under physiological conditions in vitro, existing instead as dynamic ensembles of interconverting structures. These naturally flexible proteins are called intrinsically disordered (Dunker et al. 2001) among other terms; they have many similarities to non-native states of "normal" globular proteins. These intrinsically disordered proteins (IDPs) and IDP regions (IDPRs) under physiological conditions in vitro may contain regions of collapsed disorder (i.e. molten globules) or extended disorder (i.e. random coil or pre-molten globule) (Daughdrill et al. 2005; Dunker et al. 2001; Uversky 2003b).

IDPs and IDPRs differ from structured globular proteins and domains in many respects, including amino acid composition, sequence complexity (which is a numerical measure of repetitiveness of sequences), hydrophobicity, charge, flexibility, and type and rate of amino acid substitutions over evolutionary time, with IDPs/IDPRs typically possessing higher mutation rates. Typically, IDPs are significantly depleted in "order-promoting" residues, such as bulky aliphatic (Ile, Leu, and Val) and aromatic amino acids (Trp, Tyr, and Phe), which would normally form the hydrophobic core of a folded globular protein, and Cys and Asn residues. On the other hand, IDPs tend to be substantially enriched in Ala and in polar "disorder-promoting" amino acids: Arg, Gly, Gln, Ser, Pro, Glu, and Lys (Dunker et al. 2001; Romero et al. 2001; Williams et al. 2001; Radivojac et al. 2007). Many of these differences were utilized to develop numerous disorder prediction algorithms, including PONDR® (Predictor of Naturally Disordered Regions) (Li et al. 1999; Romero et al. 2001), charge-hydropathy plots

(Uversky et al. 2000), NORSp (Liu and Rost 2003), GlobPlot (Linding et al. 2003a, b), FoldIndex© (Prilusky et al. 2005), IUPred (Dosztanyi et al. 2005), and DisoPred (Jones and Ward 2003; Ward et al. 2004a, b), to name a few.

Despite lacking stable tertiary structure, IDPs are involved in a wide variety of crucial biological functions (Wright and Dyson 1999; Tompa 2002, 2005; Tompa and Csermely 2004; Dyson and Wright 2005; Xie et al. 2007b, a; Vucetic et al. 2007; Dunker et al. 2005 and references therein). The functional diversity provided by disordered regions was proposed to complement functions of ordered regions (Xie et al. 2007a, b; Vucetic et al. 2007). IDPs/IDPRs are commonly involved in regulation, signaling, and control pathways, in which interactions with multiple partners and high-specificity/low-affinity binding are often requisite (Dunker et al. 2005; Uversky et al. 2005). These proteins are highly abundant in nature, with eukaryotic proteomes being more enriched in IDPRs relative to bacterial and archaean proteomes (Dunker et al. 2000; Oldfield et al. 2005; Ward et al. 2004b). Another important feature of some IDPs/IDPRs is their unique ability to fold under a variety of conditions, e.g. as a result of interactions with other proteins, nucleic acids, membranes, or small molecules, and upon changes in the protein environment (Uversky 2011a, b, 2013a, b, c). The resulting conformations could be either relatively non-compact (i.e., remain substantially disordered) or tightly folded.

2.1.3 Intrinsically Disordered Proteins in Misfolding Diseases

As IDPs and IDPRs are devoid of ordered structure, the primary step of their fibrillogenesis requires the stabilization of a partially folded conformation, i.e., partial folding rather than unfolding (Uversky 2003b; Uversky and Fink 2004). Therefore, a general hypothesis of fibrillogenesis states: structural transformation of a polypeptide chain into a partially folded conformation is a prerequisite for protein fibrillation

(Uversky and Fink 2004). Such partially folded conformations facilitate specific intermolecular interactions, including electrostatic attraction, hydrogen bonding and hydrophobic contacts, which are necessary for oligomerization and fibrillation. These partially folded aggregation-prone intermediates are expected to be structurally different for different proteins. Furthermore, such intermediates may contain different amounts of ordered structure even for the same protein in different aggregation processes. The precursor of soluble aggregates is believed to be the most structured, whereas amyloid fibrils are formed from the least ordered conformation (Khurana et al. 2001). Furthermore, variations in the amount of the ordered structure in the amyloidogenic precursor may be responsible for the formation of fibrils with distinct morphologies (Smith et al. 2003).

Peculiarities of misfolding and fibrillogenesis of IDPs with numerous illustrative examples were covered in several reviews (Uversky 2008b, 2009a; Breydo and Uversky 2011). This chapter analyzes the effects of membrane-mimetic environments on structural and aggregation properties of ordered and intrinsically disordered proteins. The effects of artificial membranes and "membrane field" on structural properties of proteins are also briefly outlined.

2.1.4 Effect of Membranes and Membrane Field on Protein Structure

2.1.4.1 Membranes and IDPs

2.1.4.1.1 Interactions of α-Synuclein with Membranes

Many IDPs can bind efficiently to the artificial and natural membranes, which can be accompanied by a dramatic increase in the α-helical content. Various membrane mimics are utilized in biophysical studies of membrane-interacting proteins in vitro. A commonly used model utilizes lipids containing various hydrophilic head groups and hydrophobic fatty acyl chains. In an aqueous environment, these amphipathic molecules self-assemble in variously shaped macromolecular assemblies to sequester the acyl chains. Such assemblies depend on the molecular shape and chemical composition of lipids (see Chap. 1 by Hong in this volume). For example, lysolipids which have one acyl chain per headgroup form micelles, whereas phospholipids with two acyl chains form planar bilayers or liposome vesicles (Pfefferkorn et al. 2012).

Let us consider the behavior of α-synuclein as an example of an extended IDP in the presence of membranes (Perrin et al. 2000, 2001; Jo et al. 2000, 2002; Volles et al. 2001; Zhu et al. 2003; Furukawa et al. 2006; Georgieva et al. 2008; Jao et al. 2004, 2008; Madine et al. 2004; McLean et al. 2000; Mihajlovic and Lazaridis 2008; Nuscher et al. 2004; Tamamizu-Kato et al. 2006; van Rooijen et al. 2008). α-Synuclein is a 140 amino acid protein that deposits as amyloid in Lewy bodies of dopaminergic neurons in Parkinson's disease and is implicated in the pathogenesis of this and several other neurodegenerative diseases collectively known as synucleinopathies. Since major structural features of α-synuclein were first characterized (Weinreb et al. 1996), great progress has been achieved towards understanding its physicochemical properties (Uversky 2003a, 2007, 2008a; Uversky and Eliezer 2009; Uversky et al. 2002b). α-Synuclein is highly expressed in various regions of the brain, predominantly in the *substantia nigra pars compacta* where it is found in the pre-synaptic regions, bound to synaptic vesicles as well as in the cytosol (Maroteaux et al. 1988; Jakes et al. 1994). α-Synuclein is highly conserved across species and contains an imperfect consensus repeat, xxKTKEGVxxx, that is probably important for function (Maroteaux et al. 1988).

α-Synuclein binding to lipid membranes is probably important for its function (Jenco et al. 1998). Membrane-bound α-synuclein adopts an amphipathic α-helical structure that promotes interactions between lipid vesicles (Bonini and Giasson 2005). In addition, membrane-bound α-synuclein inhibits phospholipase D$_2$, which may contribute to vesicle regulation (Jenco et al. 1998; Mezey et al. 1998), as this enzyme hydrolyses PC to generate phosphatidic acid, which is important in regulating vesicle transport and changes in cell morphology (Chen et al. 1996; Mezey et al. 1998).

Although most α-synuclein is found in free cytosolic fraction, the membrane-bound protein was proposed to be important in fibril formation (Lee et al. 2002). Depending on their nature, membranes may promote or inhibit fibrillation and aggregation of human α-synuclein (Zhu and Fink 2003; Zhu et al. 2003). Therefore, α-synuclein was proposed to exist in two distinct forms in vivo: a helix-rich, membrane-bound form and a disordered, cytosolic form, with the membrane-bound protein generating nuclei that seed the aggregation of the more abundant cytosolic form (Lee et al. 2002). Different affinity for specific phospholipids is probably responsible for the location of the protein and may modulate its aggregation and fibril formation.

Interactions of α-synuclein with lipid vesicles of different size and composition have been analyzed in detail (Zhu et al. 2003; Zhu and Fink 2003). To this end, characteristic changes in far-UV circular dichroism (CD) spectra were used to analyze protein conformation on small and large unilamellar vesicles (SUVs and LUVs) comprised of different zwitterionic (1,2-dipalmitoyl-*sn*-glycero-3-phosphocholine, PC) or anionic phospholipids (1,2-dipalmitoyl-*sn*-glycero-3-phosphate, PA; 1,2-dipalmitoyl-*sn*-glycero-3-phospho-RAC-(1-glycerol), PG) and mixtures thereof. Since PC contains a polar headgroup with zero net charge, whereas PA and PG have positively charged headgroups, the surface properties of the resulting vesicles are very different. The SUVs used in these studies were relatively homogeneous, with the size similar to that of the synaptic vesicles (20–25 nm) (Zhu and Fink 2003).

Although the far-UV CD spectrum of α-synuclein at pH 7.5 was typical of an unfolded protein, addition of PA/PC (molar ratio 1:1) at a protein to lipid mass ratio of 5:1 led to a decrease in the negative molar ellipticity at 198 nm, indicating an increase in ordered secondary structure (Fig. 2.1a). When the protein to lipid mass ratio decreased to 1:1, the CD changes indicated α-helix formation. When the protein to lipid mass ratio fur-

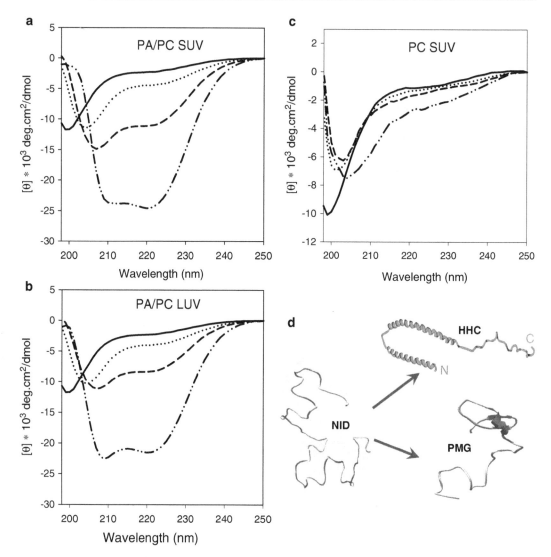

Fig. 2.1 Increased lipid concentration induces helical structure in α-synuclein. Far-UV CD spectra of α-synuclein in the presence of various types of lipid vesicles: α-synuclein alone (*solid line*); PA/PC vesicles, protein:lipid mass ratio 5:1 (*dotted line*), 1:1 (*dashed line*) and 1:5 (*dash-and-dot line*). (**a**) Data for PA/PC SUV. (**b**) Data for PA/PC LUV. (**c**) Data for PC SUV (Adapted from Zhu and Fink 2003). (**d**) A model illustrating structural changes induced in α-synuclein by high concentration of organic solvents (highly α-helical conformation determined by NMR, PDB ID 1XQ8), SUVs and lipid membranes, metals, pesticides or other factors stabilizing partially folded pre-molten globule-like conformation. Structures of the "native" intrinsically disordered conformation and pre-molten globule-like partially folded intermediate are hypothetical

ther decreased to 1:5, further CD changes indicated an increase in helical content from 32 % to 73 % upon increasing lipid concentration (Fig. 2.1a).

Importantly, Fig. 2.1a, b shows that helix was induced on binding to the acidic phospholipid vesicles of various sizes (SUV or LUV).

However, less helical structure was induced by LUVs than by the corresponding SUVs (Fig. 2.1a, b). If one uses lipid-induced increase in the helical structure as a reflection of binding efficiency, these data suggest that the protein preferentially binds to SUV as compared to

LUV. For the same amount of lipid, the available contact area (determined by the surface to volume ratio) in SUV is larger than that in LUV, which may contribute to higher helical content. Alternatively, the increased binding of α-synuclein to smaller vesicles could reflect better accommodation of the protein α-helices on more curved surfaces (Zhu and Fink 2003).

To understand whether the tendency to induce α-helix is a general property of acidic phospholipids, vesicles of PS/PG, PG/PC, PA/PG or PC alone were prepared, and far-UV CD spectra of α-synuclein in the presence of these vesicles were collected. When SUVs or LUVs were formed from the negatively charged PG/PC, PS/PC or PA/PC mixtures, the binding of α-synuclein to these vesicles was accompanied by similar increase in the α-helical content. However, PC vesicles only slightly decreased the ellipticity at 198 nm (Fig. 2.1c) and did not induce any measurable helical structure (Zhu and Fink 2003). Thus, the nature of the α-synuclein interaction with vesicles depends on the phospholipid charge, the protein to phospholipid ratio, and the vesicle size. The most extensive α-helical structure was induced by the small negatively charged vesicles (Zhu and Fink 2003).

The lipid-induced α-helical structure, which was detected by far-UV CD and Fourier transformed infrared spectroscopy (FTIR), was attributed to the formation of two curved α-helices, Val3-Val37 and Lys45-Thr92 (Fig. 2.1d) connected by a well-ordered extended linker (Ulmer and Bax 2005; Ulmer et al. 2005). The acidic, glutamate-rich C-terminal tail, Asp98-Ala140, remained unstructured even in the presence of membranes (Fig. 2.1d) (Jao et al. 2004; Ulmer and Bax 2005; Ulmer et al. 2005); this tail could attain a protease-insensitive conformation in the micelle-bound α-synuclein in the presence of calcium (de Laureto et al. 2006).

Numerous studies showed that α-synuclein binds preferentially to SUVs or LUVs containing anionic but not zwitterionic phospholipids. This binding preference suggests the electrostatic attraction between the negatively charged membrane surface and multiple positively charged lysines from the N-terminal region of α-synuclein.

However, electrostatics is not the sole driving force of this interaction, as the protein maintains substantial helical structure even in high ionic strength solution (up to 500 mM) where such interactions are effectively screened (Davidson et al. 1998). Moreover, α-synuclein can also interact with some zwitterionic lipids, with the interaction efficiency dependent on the lipid composition (Pfefferkorn et al. 2012). For example, zwitterionic 1-palmitoyl-2-oleoyl-phospho-ethanolamine (PE) enhances interaction of α-synuclein with membrane (Jo et al. 2000). Furthermore, α-synuclein shows clear preference for PA and phosphatidylinositol (PI) over 1-palmitoyl-2-oleoyl-phosphatidylserine (PS) and PG (Jo et al. 2000; Rhoades et al. 2006; van Rooijen et al. 2009; Middleton and Rhoades 2010; Shvadchak et al. 2011; Pfefferkorn et al. 2012), suggesting specific protein interactions with these anionic lipids.

2.1.4.1.2 Interaction of Other IDPs with Membranes

NMR analysis of the β- and γ-synucleins revealed that these IDPs acquired helical conformations upon binding to SDS micelles; all three synucleins showed extensive helical structure in their N-terminal lipid-binding domains and the relatively disordered acidic C-terminal tails (Sung and Eliezer 2006). There were noticeable structural differences among the micelle-bound conformational ensembles of α-, β-, and γ-synucleins, which were directly related to variations in their amino acid sequences (Sung and Eliezer 2006). In fact, an entire segment of the micelle-bound helical structure of β-synuclein was destabilized upon 11-residue deletion in the lipid-binding protein domain (Sung and Eliezer 2006). In micelle-bound γ-synuclein, the acidic C-terminal tail which has distinctly different primary structure was more disordered than those in α- and β-synucleins (Sung and Eliezer 2006).

Helical folding in the presence of membrane mimetics is not unique for synucleins. For example, SDS micelles induce partial helical formation in the intracellular region of human Jagged-1 (J1_tmic) (Popovic et al. 2007). This IDP, which is one of the five ligands of human Notch receptors, mediates protein-protein interactions through the C-terminal PDZ binding motif, and is involved in receptor-mediated endocytosis triggered by mono-ubiquitination. Similarly to α-synuclein, the increase in the helical content of J1_tmic was induced by lipid vesicles comprised of anionic phospholipids (1,2-dimyristoyl-*sn*-glycero-3-[phospho-rac-(1-glycerol)] sodium salt, DMPG, or dimyristoyl phosphatidylserine (DMPS)), whereas zwitterionic phospholipids (dimyristoyl phosphatidylcholine, DMPC) did not promote structure formation (Popovic et al. 2007). Notably, anionic phospholipids are prevalent components of the inner leaflet of the eukaryotic plasma membrane.

Other examples of lipid-induced helical structure include apolipoproteins and related lipid surface-binding proteins such as serum amyloid A. These IDPs/IDPRs can acquire amphipathic α-helical structure upon interactions with cell membranes as well as with anionic (SDS) or zwitterionic (DPC) lipid mimetics, and these interactions importantly influence amyloid formation by these proteins (see Ryan et al., Chap. 7, Gorbenko et al., Chap. 6 and Das and Gursky, Chap. 8 in this volume).

2.1.4.2 Effect of the Membrane Field on Ordered Proteins

The membrane surface is one of the factors that modify protein structure in the living cell. Some ordered proteins lose their native tertiary structure upon interaction with membranes, whereas IDPs typically gain structure. Examples of ordered proteins denatured by the membrane are various toxins undergoing a transition into the molten globule-like intermediate before inserting into the membrane, or transport proteins that must denature upon the transfer of ligands to the target cells (Ptitsyn et al. 1995). Partial denaturation of some ordered proteins is due to the negative electrostatic potential of the membrane surface (Endo and Schatz 1988). This negative potential attracts solution protons, resulting in a local decrease in pH that does not exceed two units in salt-free solutions and is even less pronounced at physiologic salt concentrations. Such "acidification" is insufficient for the pH-induced

denaturation of most globular proteins, so it cannot provide the sole denaturing factor of the membrane surface.

Another factor is the dielectric constant that is nearly tenfold lower at the water-hydrophobic interface than in the bulk water ($\varepsilon = 80$). Such a local decrease in dielectric is an additional structure-modifying factor of the membrane, which may cause partial denaturation of ordered globular proteins (Ptitsyn et al. 1995) or promote partial folding of extended IDPs (Uversky 2009b).

Water-alcohol mixtures at moderately acidic pH are used as model systems to study the joint action of local decrease of pH and dielectric permeability on the structure of ordered proteins (Ptitsyn et al. 1995; Bychkova et al. 1996; Uversky et al. 1997). This model system was validated by the observation that, at moderately low pH, an increase in methanol concentration transforms cytochrome *c* (Bychkova et al. 1996) and α-fetoprotein into a molten globule state (Narizhneva and Uversky 1997). Moreover, the molten globule-like intermediate can be induced by the low dielectric in the ordered protein β-lactoglobulin (Uversky et al. 1997).

Generally, alcohols promote secondary structure, especially α-helix formation, but at the same time act as destabilizers of tertiary and quaternary interactions in ordered proteins, leading to the formation of partially folded molten globule-like conformations (Buck 1998). Typically, alcohol-induced denaturation of globular proteins is accompanied by an increase in their α-helical content (Ptitsyn et al. 1995; Bychkova et al. 1996; Uversky et al. 1997; Narizhneva and Uversky 1997; Tanford 1968; Arakawa and Goddette 1985; Wilkinson and Mayer 1986; Jackson and Mantsch 1992; Buck et al. 1993; Dufour et al. 1993, 1994; Fan et al. 1993; Thomas and Dill 1993; Alexandrescu et al. 1994; Hamada et al. 1995). Precise mechanisms of alcohol-protein interactions are not well understood. Since alcohols have lower dielectric constant and are much weaker hydrogen bond acceptors than water, some of the proposed mechanisms include: (a) decrease in hydrophobic interactions (Thomas and Dill 1993); (b) changes in the protein hydration shell (Kentsis and Sosnick 1998; Walgers et al. 1998); (c) strengthening of intra-protein hydrogen bonding and decreased shielding of electrostatic interactions (Hirota et al. 1998); (d) the formation of alcohol clusters that provide local regions of low polarity (Hong et al. 1999); (e) locally assisting the secondary structures formation by breaking down the peptide hydration shell and providing a solvent matrix for side chain – side chain interactions (Reiersen and Rees 2000); and (f) preferential binding of alcohols to protein molecules with simultaneous protein dehydration (Gruber and Low 1988; Fioroni et al. 2002).

2.1.4.3 Effect of the Membrane Field on IDPs: The Example of α-Synuclein

Although, compared to globular proteins, much less is currently known about the behavior of IDPs in water/alcohol mixtures, one would expect similar effects. Studies of α-synuclein showed that α-helical structure is induced in aqueous solutions containing fluorinated alcohols, detergent micelles (Weinreb et al. 1996; Munishkina et al. 2003), or small unilamellar vesicles (Davidson et al. 1998; Bussell et al. 2005; Eliezer et al. 2001; Georgieva et al. 2008; Rhoades et al. 2006; Sung and Eliezer 2006).

In a quest for an experimental model to study the effects of the membrane field on human α-synuclein, the structure and aggregation/fibrillation properties of this protein have been systematically analyzed in the mixtures of water with simple alcohols differing in the aliphatic chain length (methanol, ethanol, propanol) and with fluoroalcohols (2,2,2-trifluoroethanol (TFE) and 1,1,1,3,3,3-hexafluoro-2-propanol (HFiP)) (Munishkina et al. 2003). Figure 2.2 shows far-UV CD spectra of α-synuclein in solutions containing different concentrations of these alcohols. The addition of any alcohol increased the content of ordered secondary structure manifested by a decrease in the negative ellipticity at 196 nm and an increase at 222 nm. Interestingly, the shape and the intensity of far-UV CD spectra measured at low alcohol concentrations were close to those reported for the partially folded intermediate

induced by low pH, high temperatures (Munishkina et al. 2003; Uversky et al. 2001a), certain metal cations (Uversky et al. 2001b), pesticides and herbicides (Uversky et al. 2001c, 2002a; Manning-Bog et al. 2002), or in moderate concentrations of a protein-folding osmolyte trimethylamine-N-oxide (Uversky et al. 2001d). Importantly, these initial structural changes observed by CD were completely reversible and were independent of the protein concentration in the range explored (0.1–2.5 mg/ml), indicating that the helix formation in α-synuclein at low alcohol concentrations was an intramolecular process and not self-association (Munishkina et al. 2003). This partially folded intermediate was associated with the fibrillation of α-synuclein (Uversky et al. 2001a) described below.

Intriguingly, structural transformations in α-synuclein at higher alcohol concentrations depended on the type of alcohol (Munishkina et al. 2003). For example, high concentrations of simple alcohols induced a β-sheet-enriched conformation, as shown for ethanol in Fig. 2.2a. This conformation is manifested by the pronounced CD minimum near 218 nm, which is typical of extensive β-sheet structure. These β-rich species involved oligomeric forms of α-synuclein (Munishkina et al. 2003).

Curiously, Fig. 2.2a shows a well-defined isosbestic point where the spectra intercept. This suggests that the structural changes in α-synuclein induced by ethanol (as well as other simple alcohols) can be described by two major conformations, mostly disordered and β-rich. Data analysis using spectral "phase diagram" approach (according to which the dependence of $[\theta]\lambda_1$ vs. $[\theta]\lambda_2$ is linear for all-or-none transitions between two different conformations) revealed that spectral changes induced by simple alcohols can be described by two straight lines, suggesting at least two independent transitions separating three different conformations: the natively-unfolded state, the partially folded intermediate, and the β-structure-enriched species (Munishkina et al. 2003).

A different situation was observed in TFE. Figure 2.2b shows that low alcohol concentration initially induced formation of the partially

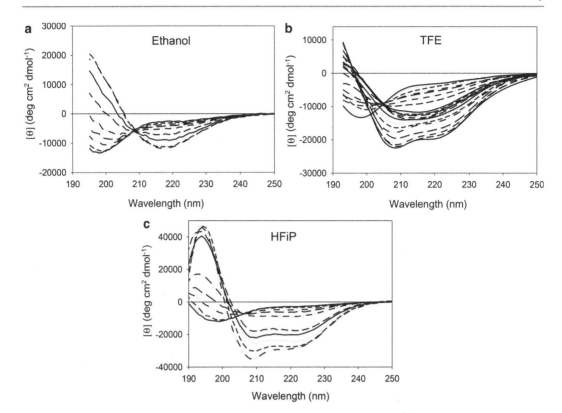

Fig. 2.2 The effect of alcohols on α-synuclein conformation. Far-UV CD spectra of α-synuclein in the presence of increasing concentrations of various alcohols. (**a**) Ethanol: The order of EtOH concentrations with increasing negative ellipticity at 220 nm is 0, 5, 10, 14, 18, 22, 28, 34, 40, 50 and 60 %. The data for methanol and propanol (not shown) were similar to that of ethanol. (**b**) TFE: The order of TFE concentrations with increasing negative ellipticity at 220 nm is 0, 2, 5, 6, 8, 10, 11, 12, 14, 15, 16, 20, 22, 24, 25, 28, 30, 32, 35, 40, 50, and 60 %. (**c**) HFiP: The order of HFiP concentrations with increasing negative ellipticity at 220 nm is 0, 1, 2, 2.5, 3, 4, 5, 7.5, 10, 15, 20, and 40 %. Measurements were carried out at 20 °C at a protein concentration of 0.5 mg/ml (Adapted with permission from Munishkina et al. 2003)

folded intermediate; at about 15 % TFE, the β-structure-enriched species was formed; finally, above 35 % TFE, the α-helical conformation was observed. The resulting protein species was initially monomeric but self-associated over longer incubation times (Munishkina et al. 2003). The complexity of this process is illustrated by the presence of a well-defined isosbestic point observed at low-to- moderate TFE concentrations, and another less well-defined point at high TFE concentrations, suggesting multiple conformational transitions. Spectroscopic phase diagram analysis supported multiple structural transitions: *I*, natively unfolded → partially folded intermediate (0–10 % TFE); *II*, partially folded intermediate → β-structure-enriched spe-

cies (10–15 % TFE); *III*, structural transformations within the β-structure species (15–25 % TFE); *IV*, β-structure-enriched conformation → α-helical state (25–35 % TFE); *V*, rearrangements within the α-helical species (35–60 % TFE) (Munishkina et al. 2003). Alcohol-induced stabilization of helical structure was also observed in many other proteins and peptides.

A third scenario was observed in HFiP. Figure 2.2c shows that the transition started from the partially folded intermediate but ended with the formation of a helix-rich species, as shown by the CD minima at 208 and 222 nm (Munishkina et al. 2003). The HFiP-induced changes in α-synuclein were complex, and the phase diagram indicated three transitions: *I*, natively unfolded → partially folded intermediate (0–3 % HFiP); *II*, partially folded intermediate → α-helical conformation (3–10 % HFiP); *III*, rearrangements within the α-helical species (10–20 % HFiP) (Munishkina et al. 2003).

Taken together, these studies showed that alcohols can cause intrinsically disordered α-synuclein to fold in a multistep manner. Although the mechanism of such folding was different for different solvents, all of them induced a common first stage, i.e. formation of a partially folded intermediate. The fate of this intermediate depended on the nature of the organic solvent. Simple alcohols and moderate concentrations of TFE induced β-structure-enriched oligomers, whereas high concentrations of fluoroalcohols gave rise to the α-helical conformation (Munishkina et al. 2003).

The structural transformations and oligomerization of α-synuclein in simple alcohols were proposed to be driven by the increase in the solvent hydrophobicity (Munishkina et al. 2003). This is suggested by Fig. 2.3a showing that the structure-promoting potential of simple alcohols correlates with the length of their aliphatic chains. Furthermore, the structure-forming potency of different simple alcohols, represented as a function of $[\theta]\lambda$ vs. dielectric constant, could be described by a "master curve" (Fig. 2.3b). This strongly supports the idea that the folding of α-synuclein in mixtures of water with simple alcohol is induced by the decrease in the dielectric constant (Munishkina et al. 2003). Similarly, the structural transformations induced in the globular protein β-lactoglobulin by various alcohols have been directly attributed to reduced solvent polarity (Dufour et al. 1993; Uversky et al.

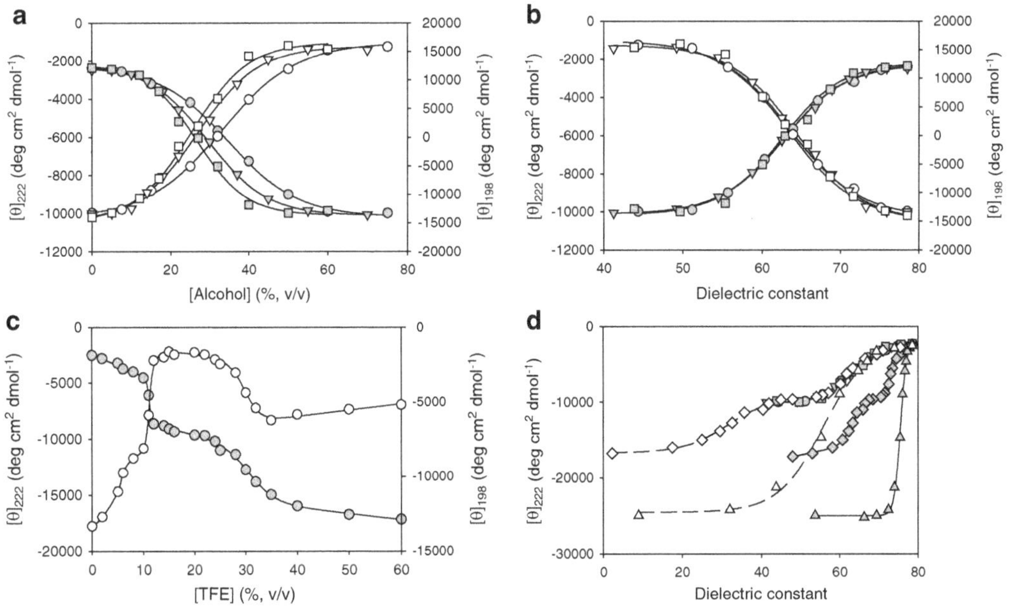

Fig. 2.3 Comparison of the effects of simple and fluorinated alcohols on far-UV CD of α-synuclein. Effects of the simple alcohols, MeOH (○), EtOH (▽), and PrOH (□), on the far-UV CD spectrum of α-synuclein: (**a**) Dependencies of $[\theta]_{222}$ (*gray symbols*) and $[\theta]_{198}$ (*open symbols*) on alcohol concentration. (**b**) Dependencies of $[\theta]_{222}$ (*gray symbols*) and $[\theta]_{198}$ (*open symbols*) on the dielectric constant of the media in the presence of the alcohols. The solvent dielectric constant is the main factor determining the effect of simple alcohols on α-synuclein conformation. (**c**) The effects of the TFE concentration on

far-UV CD spectra, $[\theta]_{222}$ (*gray circles*) and $[\theta]_{198}$ (*open circles*). (**d**) Comparison of the hydrophobic effect of simple and fluorinated alcohols on α-synuclein conformational transitions. The data are displayed as $[\theta]_{222}$ versus dielectric constant. MeOH, *gray circles*; EtOH, *gray inverted triangles*; PrOH, *gray squares*; TFE, *gray diamonds*; or HFiP, *gray triangles*. Data renormalized for the preferential solvation of the fluorinated alcohols (3 × [TFE] or 7 × [HFiP]) are shown by open symbols (see the text) (Adapted with permission from Munishkina et al. 2003)

1997). This excludes the specific protein-alcohol interactions, suggesting that water/alcohol mixtures may be used to model the effect of membrane field on both ordered and intrinsically disordered proteins.

Thus, studies of α-synuclein in water/alcohol mixtures revealed the effects of the membrane field on the protein prior to its specific interactions with the membrane. Overall, these studies suggested that, in the membrane field, α-synuclein first folds to the pre-molten globule-like conformation and then forms either fibrils or amorphous aggregates (Munishkina et al. 2003). The aggregation fate of α-synuclein probably depends upon the distance between the protein and the membrane. Protein molecules localized in close proximity to the membrane where the dielectric constant is lowest will most likely form amorphous aggregates, whereas the more distant protein molecules are more prone to form fibrils.

Fluorinated alcohols have more complex effects. The structure-forming effect of TFE involves at least three transitions separating four different conformations (Fig. 2.3c). The TFE- and HFiP-induced transitions are complex and cannot be attributed solely to the dependence of $[\theta]_{222}$ on dielectric constant (Fig. 2.3d). Consequently, these solvents interact specifically with α-synuclein and/or modify its solvation shell. This conclusion is in good agreement with studies showing that fluorinated alcohols preferentially bind to proteins and perturb their solvation shell (Diaz and Berger 2001; Gast et al. 2001; Diaz et al. 2002; Fioroni et al. 2002; Roccatano et al. 2002), whereas simple alcohols do not (Fioroni et al. 2002). In fact, the local concentration of TFE near a 14-residue peptide bombesin was shown to be nearly twice as high as the nominal TFE concentration in the bulk (Diaz et al. 2002). This effect may be substantially greater for larger proteins, and is definitely more profound for HFiP than for TFE. Thus, concentrated solutions of TFE and HFiP can be used to model the conformational changes in a protein resulting from its direct interactions with the membrane surface or even insertion into this surface. In other words, the α-helical conformation

induced in α-synuclein by high concentrations of TFE or HFiP probably corresponds to the membrane surface-bound or membrane-inserted protein forms (Munishkina et al. 2003), even though the details of protein interactions with fluoroalcohols and membranes are expected to be different.

Similar to simple alcohols, fluorinated alcohols decrease water polarity and thus, can modify the protein structure via the decreased dielectric constant. Therefore, the structural dependence on alcohol concentration should be renormalized to account for the preferential partitioning of fluorinated alcohols in the vicinity of a protein molecule. Such renormalization is illustrated in Fig. 2.3d. Here, it is assumed that the local concentrations of TFE and HFiP around the α-synuclein molecule were three and sevenfold higher than the bulk concentration (Munishkina et al. 2003). Figure 2.3d shows perfect correspondence between the results obtained for simple alcohols and the initial part of the renormalized TFE data, which represents the partially folded intermediate and β-structure formation. A similar agreement was observed for the part of the renormalized HFiP data describing the transition of α-synuclein into the partially folded conformation. Figure 2.3d also shows that the transition into the α-helical conformation induced by high concentrations of HFiP is unique and probably could not be attributed to changes in the dielectric. Thus, fluorinated alcohols cause dual effect on the protein and, depending on their concentration, can be used to model at least two different protein forms, one located in a membrane field (low alcohol concentrations) and another directly interacting with the membrane or embedded into it (high alcohol concentrations). The first form is a partially folded intermediate, which shows high propensity to fibrillate or to form amorphous aggregates, whereas the second form is highly α-helical and does not aggregate or form fibrils (Munishkina et al. 2003).

In addition to α-synuclein, many other IDPs also undergo large conformational changes in the presence of fluorinated alcohols. Examples include: PET domain of the Prickle protein, a key member of the planar cell polarity pathway, which is required for fetal tissue morphogenesis and for the maintenance of adult tissues in animals (Sweede et al. 2008); myelin basic protein (Libich and Harauz 2008); protein 1 from the model legume Lotus japonicus, LjIDP1 (Haaning et al. 2008), and many others. Comparative analysis of the intrinsically disordered J1_tmic in various structure-promoting environments revealed that TFE is a more potent promoter of the helical structure than SDS micelles or phospholipid vesicles (Popovic et al. 2007). Since TFE does not interact strongly with the hydrophobic moieties of the polypeptide chain, it drives the secondary structure formation by reducing the exposure of the protein backbone to the aqueous solvent, thereby favoring intramolecular hydrogen bonding (Popovic et al. 2007; Roccatano et al. 2002). Hence, the structure formed in the presence of fluorinated alcohols reflects the secondary structure propensity of an intrinsically disordered polypeptide. Since disordered regions with increased folding propensity often form primary sites of protein-protein interactions, treatment of IDPs with fluoroalcohols can be used as a tool for the analysis of local intrinsic propensity to adopt helical structure that is formed during binding-induced folding. Although this approach is well-suited for the analysis of IDPs with largely helical propensity, its application to other proteins comes with a caveat, as fluorinated alcohols can induce non-native helical structure in some proteins (described below).

2.2 Aggregation of Proteins in Lipid-Mimetic Environments

2.2.1 IDPs in Lipid-Mimetic Milieu

2.2.1.1 Aggregation of α-Synuclein in Various Alcohols

Structurally, human α-synuclein is a typical IDP (Uversky et al. 2001a; Uversky 2003a, 2007, 2008a; Wu et al. 2008; Uversky and Eliezer 2009; Wu and Baum 2010; Breydo et al. 2012; Dikiy and Eliezer 2012; Binolfi et al. 2012; Fauvet et al.

2012a, b; Moriarty et al. 2013). In the presence of several divalent and trivalent metal ions (Uversky et al. 2001b, 2002a; Breydo and Uversky 2011) and/or several common pesticides (Uversky 2004; Uversky et al. 2001c, 2002a; Silva et al. 2013), α-synuclein can adopt a partially folded conformation, which is critical for fibrillation, suggesting that these metals and pesticides can serve as potential promoters of Parkinson's disease.

The binding of α-synuclein to some synthetic and natural membranes is accompanied by an increase in α-helical content described above. The peculiarities of α-synuclein aggregation in the presence of lipids, fatty acids, and detergents have been carefully analyzed in the recent review (Dikiy and Eliezer 2012). This analysis indicated that a critical factor is the ratio of protein to lipid or detergent, with higher ratios (low concentrations of lipid or detergent) driving protein aggregation, and lower ratios preventing it (Dikiy and Eliezer 2012). This observation suggests that one mechanism by which lipids or lipid-like molecules may facilitate α-synuclein aggregation is by a two-dimensional condensation of a protein on a surface of a vesicle or a micelle, and thereby driving aggregation through mass action (Dikiy and Eliezer 2012).

In vitro α-synuclein readily assembles into fibrils whose morphologies and staining characteristics are similar to the fibrils extracted from the disease-affected brain. Fibrillation occurs via a nucleation-dependent mechanism, with the critical primary stage involving a partially folded intermediate (Uversky et al. 2001a). α-Synuclein aggregation and fibrillation is accelerated by interactions with the long-chain fatty acids and lipid droplets (Lee et al. 2002).

The membrane-bound protein represents only ~15 % of the total cellular α-synuclein (Lee et al. 2002) suggesting only transient association with the neuronal membranes. To analyze the effects of the membrane field on α-synuclein aggregation, water-alcohol mixtures were used to study joint action of the local decrease in pH and dielectric constant. Low concentrations of alcohols induced a partially folded protein intermediate with a strong propensity to fibrillate. However, protein fibrillation was completely abolished at high alcohol concentrations, with different structural effects produced by different alcohols.

Kinetics of the Alcohol-Induced Fibril Formation The histological dye Thioflavin T (ThT) was used to monitor amyloid fibril formation. To this end, the time-dependent changes in the ThT fluorescence during incubation of α-synuclein at 37 °C were analyzed as a function of alcohol concentration. Fibril formation in aqueous solution was accelerated in low concentrations of all alcohols studied, but was completely inhibited in high concentrations (Munishkina et al. 2003). The former effect was due to the stabilization of the partially folded intermediate at the critical early step of α-synuclein fibrillogenesis. The fact that, in the presence of 40 % EtOH or 20 % TFE, α-synuclein self-associates to form β-structure-enriched oligomers but does not fibrillate, suggests that these stable oligomers are not on the fibrillation pathway. This conclusion also applies to the α-helical conformation of α-synuclein induced by high concentrations of fluoroalcohols (Munishkina et al. 2003; Uversky 2003a).

Morphology of the Alcohol-Induced α-Synuclein Aggregates Electron microscopy (EM) revealed a strong correlation between the amount of fibrils determined from the EM images and the ThT fluorescence intensity. EM analysis also showed that α-synuclein aggregates formed in the absence and in the presence of various alcohols had similar overall morphology and formed amyloid-like fibrils (Fig. 2.4a–c). However, not all these fibrils were identical. Fibrils formed in the presence of 10 % PrOH (Fig. 2.4b) looked more smooth and elongated compared to those formed in the presence of 5 % TFE (Fig. 2.4c); the latter were morphologically similar to fibrils grown in the absence of alcohol (Fig. 2.4a). Furthermore, after incubation of α-synuclein for several days under conditions stabilizing the β-sheet- and α-helix-enriched conformations, most aggregated material was amorphous, with very few short fibrils. This confirms that alcohols inhibit fibrillation of the β-sheet- and α-helix-enriched conformations (Munishkina et al. 2003; Uversky 2003a). Finally, in agreement

with the kinetics analysis, Fig. 2.4d shows that in the presence of 40 % EtOH or 20 % TFE, α-synuclein forms oligomers and does not fibrillate.

Based on the effects of various alcohols on the structure and aggregation properties of α-synuclein, it was proposed that, in addition to the α-helical membrane-bound form, this protein may also have a membrane-related form induced by the membrane field which involves the partially folded intermediate with high propensity to fibrillate. Depending on the protein's affinity to the particular type of membrane, one of the two membrane-related forms was proposed to dominate, giving rise to two different aggregation scenarios: lack of fibrils in case of the helical conformation or aggregation into fibrils in case of the partially folded intermediate. This model can account for the apparently contradictory reports of acceleration or inhibition of α-synuclein fibrillation by lipid membranes (Munishkina et al. 2003).

Such a complex aggregation behavior of α-synuclein is not an exception. Sections below describe other proteins whose complex aggregation behavior in organic solvents has been reported.

2.2.1.2 TFE-Induced Aggregation of Aβ

Amyloid-β (Aβ) occupies a special position among amyloidogenic proteins, because this short 36–43 residue polypeptide was one of the first proteins identified as a potential cause of a neurodegenerative disease. Thirty years ago Aβ was purified from the vascular amyloid (Glenner and Wong 1984a, b; Glenner et al. 1984) and from the plaques in cerebral cortex (Masters et al. 1985) of the patients afflicted with Alzheimer's disease. Aβ is produced through successive endoproteolytic cleavage at specific sites inside or near the transmembrane α-helical domain of the amyloid-β precursor protein (reviewed by Morgado and Garvey, Chap. 3, this volume). Similar to α-synuclein, Aβ is an intrinsically disordered polypeptide whose fibrillation in vitro involves the conversion of a mostly disordered conformer into a stable β-sheet-rich assembly (Teplow 1998). Based on the detailed structural analysis of Aβ fibrillation, it was proposed that fibrillogenesis involved transient formation of an α-helix-containing intermediate (Walsh et al.

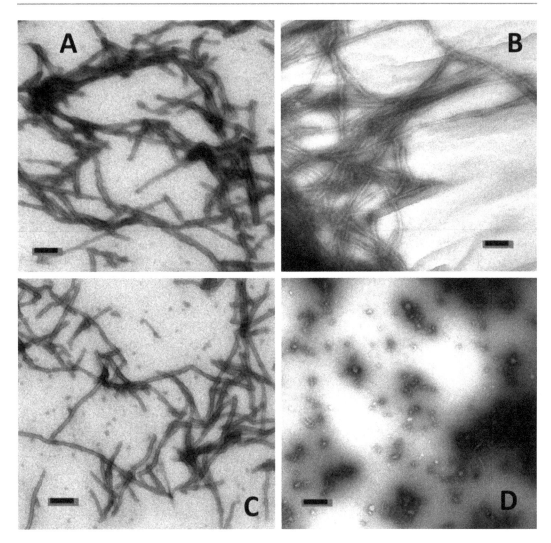

Fig. 2.4 Negatively stained transmission electron micrographs of various α-synuclein aggregates induced by different alcohols. Fibrils prepared from α-synuclein in the absence of organic solvents (**a**). Fibrils formed from the partially folded intermediate induced by 10 % PrOH (**b**) or 5 % TFE (**c**). Soluble oligomers formed in 20 % TFE (**d**). The scale bars are 100 nm (Adapted with permission from Munishkina et al. 2003)

1997, 1999; Kirkitadze et al. 2001). This study, together with similar observations for other amyloidogenic proteins containing α-helices in segments that are predicted to form β-strands led to a hypothesis that these discordant α-helix/β-strand sequences can be used to predict protein propensity to form amyloid (Kallberg et al. 2001). In agreement with this hypothesis, a comprehensive experimental analysis of a fibrillation process of some of these discordant helix-containing proteins revealed that they can form typical β-sheet-containing fibrils (Kallberg et al. 2001). These findings suggested that stabilization of α-helical conformation in amyloid-prone regions can block fibril formation (Kallberg et al. 2001). In agreement with this hypothesis, Aβ$_{1-40}$, Aβ$_{1-42}$ and α-synuclein exhibited predominantly α-helical conformations and did not fibrillate in the presence of high concentrations of TFE (>40 %) (Barrow et al. 1992; Sticht et al. 1995; Coles et al. 1998; Shao et al. 1999; Munishkina et al. 2003). These findings indicated that, although dynamic helical structure might be transiently formed during the fibrillogenesis of some extended IDPs, stable helical structure efficiently inhibited fibrillation.

Careful far-UV CD analysis revealed a steep transition between ~5 and 30 % TFE, from the mostly disordered (~70 % random coil) to the highly helical structure (~80 % α-helix at high TFE concentrations) in both $A\beta_{1-40}$ and $A\beta_{1-42}$ peptides (Fezoui and Teplow 2002). This transition occurred prior to peptide aggregation. However, both peptides rapidly formed β-sheet-containing assemblies at intermediate concentrations (15–20 % TFE) (Fezoui and Teplow 2002). Also, an inverse relationship was reported between the β-sheet and α-helix content, where formation of the β-sheet observed in 20 % TFE was paralleled by a comparable loss of the α-helix (Fezoui and Teplow 2002).

The reason for the rapid β-sheet folding in 20 % TFE was formation of amyloid fibrils, as evident from the morphological analysis of Aβ assemblies. Although the rates of the β-sheet and fibril formation by the more hydrophobic $A\beta_{1-42}$ were significantly higher than those by $A\beta_{1-40}$, the qualitative effects of TFE on these rates were similar, with the maximal fibrillation rates near 20 % TFE for both peptides (Fezoui and Teplow 2002). EM analysis showed that the fibril morphology was dramatically affected by TFE, suggesting that TFE alters the interchain packing (Fezoui and Teplow 2002). Also, fibril formation occurred at much lower peptide concentrations in the presence of TFE (Fezoui and Teplow 2002). Since at low TFE concentrations noticeable helical content was induced in $A\beta_{1-40}$ and $A\beta_{1-42}$, and since both peptides formed fibrils faster than in the absence of TFE, the partially helical conformer was proposed to be important for the fibril formation, as it facilitates correct intermolecular packing within the Aβ oligomers at the early fibrillation stages (Fezoui and Teplow 2002).

2.2.1.3 Aggregation of Amylin in Alcohol-Heparin Mixtures

Islet amyloid polypeptide (IAPP, a.k.a. amylin) is a 37-residue β-cell polypeptide co-stored and co-released with insulin (reviewed by Singh et al., Chap. 4, this volume). Human IAPP adopts a β-structure-enriched conformation that easily oligomerizes and forms insoluble fibrils. Amyloid deposition by IAPP is linked to pathological events responsible for cell death and organ dysfunctions in type 2 diabetes (Clark and Nilsson

2004). Deposition of IAPP amyloid in human pancreatic islet is a common pathological feature of type 2 diabetes (Clark and Nilsson 2004; Marzban et al. 2003). Natural proline substitutions in the region 24–29 of rodent IAPP (e.g., A25P, S28P, and S29P in rat and mouse IAPPs, or G24P and L27P in hamster IAPP) prevent fibrillation (Clark and Nilsson 2004).

In type 2 diabetes and other amyloidosis, the fibrillar deposits in apoptotic tissues often contain heparan sulfate proteoglycans, lipids, apolipoproteins and other macromolecules (Rochet and Lansbury 2000; Ancsin 2003; Marzban et al. 2003; Clark and Nilsson 2004). To model the effects of natural polyanions and non-polar cofactors on the aggregation of IAPP, the peptide was incubated in the presence of heparin and various alcohols (Konno et al. 2007). Based on the changes in ThT fluorescence and the morphological analysis by EM, the authors concluded that heparin and alcohols act synergistically to promote IAPP fibrillation. In the presence of both cofactors, IAPP readily formed amyloid-like fibrils. Without TFE, heparin alone even at 100 µg/ml was unable to promote fibrillation, yet in the presence of TFE as little as 0.5 µg/ml heparin was sufficient for fibril formation. Similarly, in the absence of heparin, up to 15 % TFE was an ineffective promoter of fibrillation, whereas 5 % or less of TFE with a trace amount of heparin induced fibrillation (Konno et al. 2007). Other alcohols could also promote IAPP fibrillation in the presence of 5 µg/ml heparin. Curiously, the order of the fibrillation promotion efficiency was HFiP > TFE > iPrOH > EtOH (Konno et al. 2007). Since a similar order was reported for many other proteins and peptides (e.g. α-synuclein), similar molecular mechanisms are potentially at play in modulating aggregation of various proteins and peptides by alcohols.

2.2.2 Ordered Proteins in Lipid-Mimetic Environments

2.2.2.1 Aggregation of Insulin in Ethanol and TFE

Similar to many other ordered proteins involved in amyloidosis, insulin fibrillates (Jimenez et al.

2002) under moderately denaturing conditions, such as heating at low pH (Nielsen et al. 2001), agitation in the presence of hydrophobic surfaces (Sluzky et al. 1991), and low concentrations of guanidine hydrochloride (Ahmad et al. 2003) or urea (Ahmad et al. 2004). Insulin is a small 51-residue pancreatic hormone with α/β structure comprised of two polypeptide chains linked by two interchain and one intrachain disulfide bonds. Although insulin is stored in the pancreas as a Zn^{2+}-containing hexamer, it circulates in plasma in monomeric or dimeric forms and binds the receptor as a monomer. In typical in vitro conditions, insulin forms a complex mixture of hexamers, dimers, and monomers (Grudzielanek et al. 2005). In contrast, at low pH and low temperatures, monomers and dimers represent the majority of pre-aggregated, native-like insulin (Nielsen et al. 2001). A possible explanation is that insulin hexamer is stabilized by Zn ions whose coordinating histidines get protonated at low pH and lose their affinity for Zn. Curiously, mostly monomeric insulin was found in the presence of 20 % of EtOH or TFE, showing that these alcohols can dissociate higher-order oligomers while preserving some secondary and tertiary structure (Grudzielanek et al. 2005). However, the structure was markedly distorted, since far-UV CD spectra showed that the addition of as low as 5 % of EtOH or TFE induced substantial β-sheet unfolding. At higher alcohol concentrations, the protein became increasingly disordered and completely lost its β-structure while retaining a substantial amount of α-helix (Grudzielanek et al. 2005).

Similar to the alcohol-induced fibrillation of α-synuclein and Aβ, insulin fibril formation is favored at low and disfavored at high concentrations of EtOH and TFE (Grudzielanek et al. 2005). Atomic force microscopy revealed that the morphology of the aggregated insulin was affected by the nature and the concentration of alcohols, with low concentrations typically favoring straight (TFE) or bent and circular fibrils (EtOH), and high concentrations promoting amorphous aggregation (Grudzielanek et al. 2005).

2.2.2.2 Aggregation of β_2-Microglobulin and Its Fragment in Fluorinated Alcohols

Aggregation of the light chain of the class-I major histocompatibility complex, β_2-microglobulin, which is a major complication of dialysis, involves accumulation of β_2-microglobulin-enriched amyloid in the skeletal joints. In its native state, this 12 kDa protein adopts an immunoglobulin fold, with a β-sandwich formed by seven antiparallel β-strands, two of which are connected by a disulfide bond. An array of biophysical techniques such as electrospray ionization-mass spectrometry (ESI-MS), ion mobility-mass spectrometry (IM-MS), fluorescence and CD spectroscopy were used to characterize the conformational properties of β_2-microglobulin (Santambrogio et al. 2011). Under conditions promoting fibrillation (neutral pH and 20 % TFE), β_2-microglobulin was shown to populate α-helical conformational ensembles with a minor fraction of a partially folded state. The helical structure disappeared upon heating, leading to a more disordered conformation (Santambrogio et al. 2011). Aggregation of K3 peptide, which is a 22-residue fragment of intact β_2-microglobulin, was analyzed in high concentrations of TFE and HFiP in 10 mM HCl (pH~2) (Yamaguchi et al. 2006). The K3 peptide formed partly α-helical conformation that was converted over time into a highly ordered β-sheet-enriched aggregates with fibrillar morphology, as seen by atomic force microscopy (Yamaguchi et al. 2006). Similar to many other proteins and peptides, the TFE and HFiP-induced fibrillation of K3 was concentration dependent, with a maximum near 20 % TFE and 10 % HFiP (Yamaguchi et al. 2006). This is well below the concentrations at which TFE and HFiP form dynamic clusters. In fact, earlier small-angle X-ray scattering analysis of HFiP-water and TFE-water mixtures revealed micelle-like highly dynamic clusters of alcohols, reaching maximum at about 30 % TFE and 40 % HFiP (v/v) (Kuprin et al. 1995; Hong et al. 1999). Hence, K3 fibrillation in these studies was not affected by the alcohol clustering in solution.

The fibrils formed in different fluorinated alcohols had distinct morphology and secondary structure. Thin and long untwisted fibrils

1.4–1.7 nm in height and with a CD minimum at 210 nm (f210 fibrils) were formed in 20 % (v/v) TFE; short thick fibrils with a CD minimum at 218 nm (f218 fibrils) were formed in 10 % HFiP (Yamaguchi et al. 2006). Tertiary structure analysis by near-UV CD showed that the α-helical species stabilized by high concentrations of TFE or HFiP did not have a unique tertiary structure. However, the f210 fibrils formed in 20 % TFE and the f218 fibrils formed in 10 % HFiP had distinct near-UV CD spectra indicating different aromatic packing in these two types of fibers (Yamaguchi et al. 2006).

2.2.2.3 Accelerated Aggregation of Serum Albumin in the Presence of Ethanol and Copper (II)

In line with the aforementioned synergistic action of alcohols and polyanions in fiber formation is a study of the copper-promoted aggregation of human serum albumin (HSA) in ethanol (Pandey et al. 2010). HSA is a complex α-helical protein (>60 % helix) with 17 disulfide bridges, which contains three domains, each comprised of two subdomains. Fibrillation of this protein requires destabilization of its native helical structure upon lowering the pH, increasing the temperature, as well as adding salts, denaturants, metal ions, and 40–60 % ethanol (Taboada et al. 2006). HAS fibrillation was greatly accelerated when HSA was incubated with 60 % ethanol and Cu(II) (Pandey et al. 2010). Since Cu(II) plays an important role in HSA transport and metabolism in vivo (Appleton and Sarkar 1971), and since the N-terminal metal binding site of HSA was reported to have picomolar affinity to Cu(II) (Rozga et al. 2007), the presence of Cu(II) was expected to stabilize the protein. Therefore, the enhanced aggregation of HSA under denaturing conditions in the presence of Cu(II) was rather unexpected. A potential explanation of this phenomenon lies in the modulation of the electrostatic interactions by Cu(II), which shifts the balance of forces between protein molecules, leading to changes in protein aggregation (Juarez et al. 2009).

2.2.2.4 Aggregation of Acidic Fibroblast Growth Factor in TFE

Many sequence-unrelated globular proteins and IDPs can aggregate into fibrils with characteristic structural and histological features of amyloid (Dobson 1999; Uversky 2003b; Uversky and Fink 2004; Chiti et al. 1999). To better understand the structural properties of the "amyloid-prone" partially folded intermediates, the newt acidic fibroblast growth factor (nFGF-1) was investigated. nFGF-1 can form amyloid-like fibrils that were studied by an array of biophysical techniques, such as Congo red and ThT binding, transmission EM, x-ray fiber diffraction, intrinsic fluorescence, far-UV CD, anilino-8-napthalene sulfonate (ANS) binding, multidimensional NMR, and FTIR (Srisailam et al. 2003). These studies revealed that the onset of the nFGF-1 fibrillation was preceded by the formation of a partially structured intermediate.

In its native state, this 15-kDa all-β-sheet protein with no disulfide bonds has an atypical far-UV CD spectrum with a positive band at 228 nm and an intense minimum at around 205 nm. Similarly shaped far-UV CD spectra were described for other β-sheet proteins containing aromatic clusters, e.g., carbonic anhydrase (Rodionova et al. 1989). Consistent with this idea, the intrinsic fluorescence of the lone tryptophan on nFGF-1, Trp 121, was quenched in the native state by the nearby imidazole and pyrrole groups (Arunkumar et al. 2002). In 20 % TFE, the positive CD band at 228 nm disappeared and a negative CD band appeared near 218 nm. This suggested TFE-induced disruption of the native 3D structure and formation of an intermediate with extended β-sheet structure, which at high TFE concentrations was transformed into a helical conformation that was obviously a nonnative misfolded form (Srisailam et al. 2003). In 70 % TFE, the non-native helix estimated by far-UV CD was about 20 %, with over 60 % random coil (Srisailam et al. 2003). Substantial unfolding of nFGF-1 by TFE was supported by the analysis of the intrinsic tryptophan fluorescence showing maximal unfolding in 15 % TFE (Srisailam et al. 2003). Also, ANS, a hydrophobic fluorescent dye whose emission increases upon binding to solvent-exposed hydrophobic surfaces in partially folded states (Semisotnov et al. 1991), showed maximal emission at 15 % TFE (Srisailam et al. 2003). The loss of the native β-barrel architecture in 15 % TFE was further supported by the FTIR analysis of nFGF-1 showing that the amide I doublet bands at 1,618 and 1,639 cm^{-1} characterizing the β-barrel structure disappeared in the presence of alcohol (Srisailam et al. 2003).

Denaturation of nFGF-1 in TFE was accompanied by protein aggregation and β-sheet formation. The partially structured intermediate showed a high tendency to aggregate and form intermolecular β-sheet. The resultant aggregates could bind diagnostic dyes Congo red and ThT and possessed all other characteristics of amyloid fibrils. EM analysis revealed straight unbranched fibrils with diameters of 10–15 nm (Srisailam et al. 2003). The presence of specific cross-β structure in the nFGF-1 aggregates was confirmed by X-ray fiber diffraction showing a dominant meridional reflection at 4.7 Å and an equatorial reflection at 10.3 Å, which are hallmarks of amyloid.

The authors proposed a sequential mechanism of the TFE-induced amyloid fibril formation by nFGF-1. Here, the first conformational transition induced in >10 % TFE was attributed to the disruption of the hydrophobic contacts in the native β-barrel structure, leading to the formation of a sticky partially structured intermediate. At the next stage, the amyloid-like fibrils were formed upon rearrangement and annealing of the extended β-sheets through intermolecular hydrogen bonding. Formation of β-sheets was demonstrated by the FTIR analysis of the TFE-induced amyloid-like fibrils (Srisailam et al. 2003). Similar to other amyloidogenic proteins, aggregation of nFGF-1 was inhibited by high concentrations of TFE (>50 %), possibly due to weakened hydrophobic interactions between the β-sheets (Srisailam et al. 2003).

2.2.2.5 Cytotoxic Aggregates of SH3 Domain from Bovine Phosphatidylinositol 3-Kinase and the N-Terminal Domain of the E. coli HypF Protein in TFE

The SH3 domain from bovine phosphatidyl-inositol-3′-kinase (PI3-SH3) is a ~60-residue globular β-barrel protein that can form fibrils in vitro (Guijarro et al. 1998; Bucciantini et al. 2002). These fibrils have all characteristics of amyloid as evidenced by electron microscopy, Congo red and ThT binding, and X-ray diffraction (Bucciantini et al. 2002). Incubation of PI3-SH3 in aqueous solution containing 25 % TFE (pH 5.5) or HCl (pH 2.0) resulted in the formation of granular or fibrillar aggregates, respectively. Both types of aggregates showed enhanced ThT fluorescence suggesting amyloid-like features (Bucciantini et al. 2002). Cytotoxicity studies revealed that the highly structured PI3-SH3 fibrils were not cytotoxic even at the highest concentration tested, whereas the granular aggregates significantly reduced cell viability (Bucciantini et al. 2002). Similarly, fibrils of many other proteins are much less toxic to cells as compared to low-order protein aggregates (see Gorbenko et al., Chap. 6 and Singh et al., Chap. 4 in this volume).

The N-terminal "acylphosphatase-like" domain of the E. coli HypF protein (HypF-N) can also form various types of aggregates in 30 % TFE (pH 5.5, 22 °C). This TFE-induced aggregation is a sequential process, where the initial aggregates formed within minutes were non-fibrillar and non-granular species characterized by high β-structure content and noticeable ThT fluorescence. After 48 h of incubation in 30 % TFE, fibril-like species were formed. These short (25–60 nm) aggregates were 4–8 nm in width, similar to the protofibrils observed in other proteins at relatively early stages of fibrillation (Bucciantini et al. 2002). After 20 days of incubation, only long unbranched fibrils with widths of 3–5 nm or 7–9 nm were seen (Bucciantini et al. 2002). Different aggregated forms of HypF-N showed different cytotoxicity: the aggregates formed after 48 h of incubation in 30 % TFE were most toxic while mature fibrils were largely non-toxic (Bucciantini et al. 2002).

These results suggest that protein aggregates that are not associated with disease can be inherently cytotoxic and cause impairment of cellular function or even cell death. These cytotoxic effects were proposed to result from combinatorial display of amino acids on the surface of disordered aggregates, enabling these aggregates to interact inappropriately with a wide range of cellular components (Bucciantini et al. 2002). Therefore, not only the ability to form amyloid fibrils, but also cytotoxicity of aggregated proteins could be a general phenomenon. This suggests that the pathogenicity of the protein-deposition diseases could be primarily related to the structural properties of the aggregates, rather than to the specific protein sequences (Bucciantini et al. 2002).

2.2.2.6 α-Chymotrypsin Aggregation in TFE

α-Chymotrypsin, a serine protease with an all-β fold, readily forms amyloid in intermediate TFE concentrations (Pallares et al. 2004). This complex protein is comprised of three chains connected by five inter- and intra-chain disulfide bonds; the native structure contains two antiparallel β-barrel domains consisting of a Greek key motif followed by an antiparallel hairpin motif (Wright 1973). α-Chymotrypsin aggregated at intermediate TFE concentrations, with the maximum circa 35 % TFE (Pallares et al. 2004), and forms amyloid-like fibrils, as suggested by Congo red and ThT binding and by transmission EM. Intrinsic fluorescence, ANS binding, CD and FTIR spectroscopy suggested that the onset of aggregation was preceded by the formation of a partially structured intermediate (Pallares et al. 2004). This intermediate was characterized by the non-native extended β-sheet conformation with exposed hydrophobic surfaces (Pallares et al. 2004). Molecular dynamics simulations of the early events in the TFE-induced conformational changes prior to fibrillation revealed that TFE molecules rapidly replaced water molecules in the solvation shell of α-chymotrypsin (Rezaei-Ghaleh et al. 2008). As a result, the radius of gyration, total and hydrophobic solvent-accessible surface areas of α-chymotrypsin, and the extended β-conformation were significantly increased in the presence of TFE (Rezaei-Ghaleh et al. 2008). Other characteristics of the TFE-induced alterations in α-chymotrypsin structure were its molten-globule-like properties, distorted catalytic active site, distorted S1 binding pocket, and altered backbone flexibility (Rezaei-Ghaleh et al. 2008). Analysis of the TFE-induced aggre-

gation of α-chymotrypsin clearly showed that alcohols can promote fibrillation of both monomeric and oligomeric proteins.

2.2.2.7 TFE-Induced Fibrillation of an α-Helical Protein, the FF Domain of the URN1 Splicing Factor

FF domains are small protein-protein interaction modules containing 50–70 residues with two conserved Phe at the *N*- and *C*-termini. Native FF structure contains three α-helices arranged as an orthogonal bundle, with a 3_{10} helix connecting the second and the third helix (Bedford and Leder 1999; Gasch et al. 2006; Bonet et al. 2008). FF domains, which are often organized in tandem arrays, are found in various eukaryotic nuclear transcription and splicing factors and are commonly involved in RNA splicing, signal transduction and transcription pathways (Bedford and Leder 1999; Marinelli et al. 2013). A comprehensive structural analysis revealed that fibrillation of the FF domain of the URN1 splicing factor (URN1-FF) at acidic pH is modulated strongly by TFE (Marinelli et al. 2013).

At low pH, yeast URN1-FF domain forms a partially folded, molten globule-like intermediate that retains most of the native α-helical structure. Although this intermediate is devoid of any detectable β-structure, it can self-assemble into highly ordered amyloid fibrils (Castillo et al. 2013). Curiously, N-terminal helix 1 of the yeast URN1-FF domain had highest α-helical as well as amyloid-forming propensities, thereby controlling the transition between soluble and aggregated protein states (Castillo et al. 2013). To explain the increased aggregation propensity of the pH-induced α-helical intermediate, two alternative mechanisms were proposed: (i) tertiary structure opening and exposure of buried apolar residues promotes hydrophobic intermolecular contacts; or (ii) fluctuations of the helical secondary structure cause transient formation of disordered regions that form intermolecular backbone hydrogen bonds, leading to formation of β-sheet-rich oligomers (Castillo et al. 2013).

Far-UV CD analysis of this intermediate revealed that, at pH 2.5 in 15 % or 25 % TFE,

URN1-FF had enhanced helical structure (Castillo et al. 2013). The addition of TFE at pH 2.5 induced a more open conformation, as suggested by the red shift in the intrinsic Trp fluorescence. This TFE-induced conformation also had reduced affinity to bis-ANS, suggesting the decreased solvent-accessible hydrophobic surface (Castillo et al. 2013). Protein incubation for 7 days at acid pH in the absence or the presence of 15 and 25 % TFE (v/v) revealed that alcohol reinforced intrinsic α-helical structure and reduced aggregation (Castillo et al. 2013). This behavior seems atypical, since it contrasts with maximal aggregation observed at intermediate TFE concentrations for other proteins described in this chapter. However, in contrast to many other proteins, URN1-FF domain has highly stable helical structure at pH 2.5 in 15 % or 25 % TFE, and stable helical structure disfavors protein aggregation. Since fluorinated alcohols stabilize helical structure, and since proteins vary in their helical propensity, this propensity is expected to define the alcohol concentration at which the helical conformation is formed. Hence, proteins with relatively high helical propensity are expected to form highly helical species with low aggregation propensity at relatively low alcohol concentrations.

2.2.2.8 Amyloid Formation by Hen Egg Lysozyme in Concentrated Ethanol Solution

Hen egg white lysozyme forms amyloid fibrils in 80 % EtOH at 22 °C with constant agitation, as evident from biophysical studies using light scattering, ThT fluorescence, CD spectroscopy, seeding experiments, and AFM (Holley et al. 2008). Hen egg white lysozyme is a very stable α-helical protein; CD analysis showed that addition of 80 % EtOH did not alter its secondary structure and hence, EtOH alone is insufficient to induce significant partial unfolding of this protein (Holley et al. 2008). However, moderate agitation over the course of a few weeks resulted in destabilization, partial unfolding and fibrillation of lysozyme, and its α-helical structure was

lost and substituted by β-sheet (Holley et al. 2008).

Curiously, when higher concentrations of EtOH (90 %) were used in the presence of 10 mM NaCl, amyloid fibrils were formed after 1 week even in the absence of agitation (Goda et al. 2000). This accelerated fibrillation was attributed to the fact that the initial state of lysozyme in 90 % ethanol with 10 mM NaCl was a non-native conformation with increased β-sheet structure (Goda et al. 2000; Holley et al. 2008). These results suggest that partial unfolding is an obligatory early step in amyloid fibril formation by lysozyme in high ethanol concentrations (Holley et al. 2008).

2.2.2.9 Step-Wise Fibrillation of Acylphosphatase in the Presence of TFE

Human muscle acylphosphatase is an 98-residue α/β protein. Incubation for just a few seconds in the presence of 25 % (vol/vol) TFE, pH 5.5, 25 °C, converted the protein in a denatured state with extensive α-helical content indicated by far-UV CD (Chiti et al. 1999). Incubation in TFE-containing solutions for 2–3 h transformed this α-helical conformation into a β-sheet rich structure accompanied by formation granular aggregates, with no evidence for extended fibrils. Incubation for 14 days produced noticeable amounts of the fibrillar material, and after 1–2 months incubation, large bundles of protofibrils were found to form rope-like structures of various diameters (Chiti et al. 1999). At this stage, more than 50 % of the protein was present in fibrillar form (Chiti et al. 1999). Acylphosphatase does not have unusually high propensity to form β-sheet structure, and a partially folded state that rapidly formed in the presence of 25 % (vol/vol) TFE, pH 5.5, 25 °C contained substantial amount of α-helix but little, if any, β-sheet. The authors concluded that a prerequisite for fibril formation was not the conformation of the partially folded state per se, but the partially denaturing conditions that destabilize the native structure (Chiti et al. 1999).

2.2.3 Aggregation of Intrinsic Membrane Proteins in Alcohols

Analysis of two detergent-solubilized intrinsic membrane proteins from *E. coli*, the 176-residue four-transmembrane helix disulfide-exchanging protein DsbB and the 388-residue sodium–hydrogen antiporter NhaA, in the increasing alcohol concentrations revealed similar conformational trends as those observed in globular proteins (Otzen et al. 2007). Multiparametric conformational analysis of these membrane proteins revealed that methanol (MeOH), ethanol (EtOH), n-propyl (PrOH), isopropyl (iPrOH), and hexafluoroisopropyl (HFiP) produced comparable effects: α-helical structure was not affected by low alcohol concentrations; conformationally destabilized species with high aggregation propensities were formed at intermediate alcohol concentrations; and misfolded helically enriched species, which are soluble irreversibly denatured forms with higher than native α-helical content, were found in high alcohol concentrations (Otzen et al. 2007). When the structural changes induced by different alcohols were plotted against the relative dielectric constant, a convergence but not complete coalescence of the data was observed, suggesting that reduced solvent polarity could not entirely explain the effects of alcohols. FTIR analysis revealed that in solutions of moderate alcohol concentrations, the β-sheet structure dominated at the expense of an α-helix (Otzen et al. 2007). The alcohol-aggregated forms of DsbB and NhaA had high affinity to ThT but did not bind Congo red. AFM analysis of protein deposits showed layer-like and spherical aggregates, which were interpreted as early fibrillation species trapped by strong hydrophobic contacts (Otzen et al. 2007). These observations suggested that, similar to soluble proteins, alcohols can induce conformational changes in membrane proteins, leading to protein destabilization and precipitation (Otzen et al. 2007).

It remains unclear whether intrinsic membrane proteins can form mature amyloid fibrils in aqueous or mixed solutions. However, these proteins have high hydrophobicity that often correlates with amyloid-forming propensity. Hence, release of membrane proteins or their fragments from the membrane is expected to lead to aggregation and potential amyloid formation. Aβ peptide that originates, in part, from the membrane-spanning helix of the amyloid precursor protein, exemplifies this effect. It is tempting to hypothesize that since the sequence propensity of intrinsic membrane proteins to form amyloid fibrils (or, more generally, to aggregate in aqueous solutions) is very high, the only thing that stops them from doing so is their strong interactions within the membrane.

2.3 Summary

There is no universal rule describing the effects of alcohols on protein aggregation. Simple and fluorinated alcohols generally induce structural changes in both globular and disordered proteins and, at moderate concentrations, promote protein aggregation. These effects were observed in a wide range of proteins differing in size and native structure. Table 2.1 summarizes basic information on all proteins described in this chapter. Moreover, similar effects were observed in several transmembrane proteins. However, there are exceptions from this rule. For example, the effect of TFE on the aggregation of URN1-FF domain is different from the general trend, since the fibrillation efficiency of this protein is significantly reduced in the presence of moderate TFE concentrations (Castillo et al. 2013). More work is clearly needed to better understand molecular mechanisms underlying protein aggregation in the presence of alcohols and its relevance to protein aggregation at the membrane surface.

Table 2.1 Basic structural information on proteins described in this chapter

Protein	UniProt ID	Number of amino acids	Type	Native fold	Structure (if known)
α-Synuclein	P37840	140	IDP	Intrinsically disordered with residual α-helical structure	PDB ID: 2KKW; bound to micelles of the sodium lauroyl sarcosinate (SLAS) detergent
β-Synuclein	Q16143	134	IDP	Intrinsically disordered with residual α-helical structure	Model ID: e5ac3d66bfb3e5fdb168ad4d33e145ba_UP000054_1; based on template 2KKW
γ-Synuclein	O76070	127	IDP	Intrinsically disordered with residual α-helical structure	Modeled structure, based on template 2KKW
Aβ (672–713)	P05067	42	IDP	Intrinsically disordered	PDB ID: 1AML; in the presence of 40 % TFE / PDB ID: 1IYT; bound to SDS micelles
Islet amyloid polypeptide (residues 34–70)	P10997	37	IDP	Intrinsically disordered	PDB ID 2KJ7: Rat islet amyloid polypeptide in DPC micelles

(continued)

Table 2.1 (continued)

Protein	UniProt ID	Number of amino acids	Type	Native fold	Structure (if known)
Jagged-1 residues (1,086–1,218)	P78504	133	IDP	Intrinsically disordered with residual α-helical structure	
PET domain of the Prickle-like protein (14–122)	Q96MT3	109	IDP	Intrinsically disordered with residual α-helical and β-sheet structure	
Myelin basic protein	P02686	304	IDP, peripheral membrane	Intrinsically disordered with residual α-helical structure	
Intrinsically disordered protein 1, LjIDP1	C6ZFX3	94	IDP	Intrinsically disordered with residual α-helical structure	
Cytochrome c	P00004	105	Globular	All-α protein	PDB ID: 1AKK
α-Fetoprotein	P02771	591	Globular	All-α protein, serum albumin-like fold	Modeled structure, based on 1N5U

Protein	ID	Length	Shape	Fold	Structure
β-Lactoglobulin	P02754	160	Globular	All-β protein, lipocalin fold	 PDB ID: 1BEB
Insulin (residues chain A: 90–110; chain B: 25–54)	P01308	51	Globular	Small protein, all-α protein	 PDB ID: 1HLS
β₂-Microglobulin	P61769	99	Globular	All-β protein, immunoglobulin-like β-sandwich fold	 PDB ID: 1JNJ
Serum albumin	P02769	583	Globular	All-α protein, serum albumin-like fold	 PDB ID: 1N5U
Acidic fibroblast growth factor	P05230	140	Globular	All-β protein, β-trefoil fold	 PDB ID: 1DZC

(continued)

Table 2.1 (continued)

Protein	UniProt ID	Number of amino acids	Type	Native fold	Structure (if known)
α-Chymotrypsin	P00766	241	Globular	All-β protein, trypsin-like serine protease fold	PDB ID: 1AB9
SH3 domain (residues 3–79) of phosphatidyl-inositol 3-kinase	P23727	77	Globular	All-β protein, SH3-like barrel fold	PDB ID: 2PNI
N-terminal domain (residues 1–91) of *E. coli* HypF protein	P30131	91	Globular	α+β protein, ferredoxin-like fold	PDB ID:
FF domain (residues 212–266) of the URN1 splicing factor	Q06525	55	Globular	All-α protein	PDB ID: 2JUC

Hen egg while lysozyme	P00698	129	Globular	α+β protein, lysozyme-like fold	 PDB ID: 2LYZ
Human muscle acylphosphatase	P07311	99	Globular	α/β protein, 2-layer sandwich fold	 PDB ID: 2VH7
Disulfide-exchanging protein DsbB	P0A6M2	176	Membrane	All-α transmembrane protein	 PDB ID: 2 K73
Sodium–hydrogen antiporter NhaA	P13738	388	Membrane	Mainly α-helical protein, up-down bundle fold	 PDB ID 1ZCD

Acknowledgements This work was supported in part by a grant from Russian Science Foundation RSCF № 14-24-00131

References

Ahmad A, Millett IS, Doniach S, Uversky VN, Fink AL (2003) Partially folded intermediates in insulin fibrillation. Biochemistry 42(39):11404–11416

Ahmad A, Millett IS, Doniach S, Uversky VN, Fink AL (2004) Stimulation of insulin fibrillation by urea-induced intermediates. J Biol Chem 279(15):14999–15013

Alexandrescu AT, Ng YL, Dobson CM (1994) Characterization of a trifluoroethanol-induced partially folded state of alpha-lactalbumin. J Mol Biol 235(2):587–599

Ancsin JB (2003) Amyloidogenesis: historical and modern observations point to heparan sulfate proteoglycans as a major culprit. Amyloid 10(2):67–79

Appleton DW, Sarkar B (1971) The absence of specific copper (II)-binding site in dog albumin. A comparative study of human and dog albumins. J Biol Chem 246(16):5040–5046

Arakawa T, Goddette D (1985) The mechanism of helical transition of proteins by organic solvents. Arch Biochem Biophys 240(1):21–32

Arunkumar AI, Kumar TK, Kathir KM, Srisailam S, Wang HM, Leena PS, Chi YH, Chen HC, Wu CH, Wu RT, Chang GG, Chiu IM, Yu C (2002) Oligomerization of acidic fibroblast growth factor is not a prerequisite for its cell proliferation activity. Protein Sci 11(5):1050–1061

Barrow CJ, Yasuda A, Kenny PT, Zagorski MG (1992) Solution conformations and aggregational properties of synthetic amyloid beta-peptides of Alzheimer's disease. Analysis of circular dichroism spectra. J Mol Biol 225(4):1075–1093

Bedford MT, Leder P (1999) The FF domain: a novel motif that often accompanies WW domains. Trends Biochem Sci 24(7):264–265

Bellotti V, Mangione P, Stoppini M (1999) Biological activity and pathological implications of misfolded proteins. Cell Mol Life Sci 55(6–7):977–991

Binolfi A, Theillet FX, Selenko P (2012) Bacterial in-cell NMR of human alpha-synuclein: a disordered monomer by nature? Biochem Soc Trans 40(5):950–954

Bonet R, Ramirez-Espain X, Macias MJ (2008) Solution structure of the yeast URN1 splicing factor FF domain: comparative analysis of charge distributions in FF domain structures-FFs and SURPs, two domains with a similar fold. Proteins 73(4):1001–1009

Bonini NM, Giasson BI (2005) Snaring the function of alpha-synuclein. Cell 123(3):359–361

Breydo L, Uversky VN (2011) Role of metal ions in aggregation of intrinsically disordered proteins in neurodegenerative diseases. Metallomics 3(11):1163–1180

Breydo L, Wu JW, Uversky VN (2012) Alpha-synuclein misfolding and Parkinson's disease. Biochim Biophys Acta 2:261–285

Bucciantini M, Giannoni E, Chiti F, Baroni F, Formigli L, Zurdo J, Taddei N, Ramponi G, Dobson CM, Stefani M (2002) Inherent toxicity of aggregates implies a common mechanism for protein misfolding diseases. Nature 416(6880):507–511

Buck M (1998) Trifluoroethanol and colleagues: cosolvents come of age. Recent studies with peptides and proteins. Q Rev Biophys 31(3):297–355

Buck M, Radford SE, Dobson CM (1993) A partially folded state of hen egg white lysozyme in trifluoroethanol: structural characterization and implications for protein folding. Biochemistry 32(2):669–678

Bussell R Jr, Ramlall TF, Eliezer D (2005) Helix periodicity, topology, and dynamics of membrane-associated alpha-synuclein. Protein Sci 14(4):862–872

Bychkova VE, Dujsekina AE, Klenin SI, Tiktopulo EI, Uversky VN, Ptitsyn OB (1996) Molten globule-like state of cytochrome c under conditions simulating those near the membrane surface. Biochemistry 35(19):6058–6063

Castillo V, Chiti F, Ventura S (2013) The N-terminal helix controls the transition between the soluble and amyloid states of an FF domain. PLoS One 8(3):e58297

Chen YG, Siddhanta A, Austin CD, Hammond SM, Sung TC, Frohman MA, Morris AJ, Shields D (1996) Phospholipase D stimulates release of nascent secretory vesicles from trans-Golgi network. J Cell Biol 138(3):495–504

Chiti F, Webster P, Taddei N, Clark A, Stefani M, Ramponi G, Dobson CM (1999) Designing conditions for in vitro formation of amyloid protofilaments and fibrils. Proc Natl Acad Sci U S A 96(7):3590–3594

Clark A, Nilsson MR (2004) Islet amyloid: a complication of islet dysfunction or an aetiological factor in type 2 diabetes? Diabetologia 47(2):157–169

Coles M, Bicknell W, Watson AA, Fairlie DP, Craik DJ (1998) Solution structure of amyloid beta-peptide(1–40) in a water-micelle environment. Is the membrane-spanning domain where we think it is? Biochemistry 37(31):11064–11077

Daughdrill GW, Pielak GJ, Uversky VN, Cortese MS, Dunker AK (2005) Natively disordered proteins. In: Buchner J, Kiefhaber T (eds) Handbook of protein folding. Wiley-VCH, Weinheim, pp 271–353

Davidson WS, Jonas A, Clayton DF, George JM (1998) Stabilization of alpha-synuclein secondary structure upon binding to synthetic membranes. J Biol Chem 273(16):9443–9449

de Laureto PP, Tosatto L, Frare E, Marin O, Uversky VN, Fontana A (2006) Conformational properties of the SDS-bound state of alpha-synuclein probed by limited proteolysis: unexpected rigidity of the acidic C-terminal tail. Biochemistry 45(38):11523–11531

Diaz MD, Berger S (2001) Preferential solvation of a tetrapeptide by trifluoroethanol as studied by intermolecular NOE. Magn Reson Chem 39(7):369–373

Diaz MD, Fioroni M, Burger K, Berger S (2002) Evidence of complete hydrophobic coating of bombesin by tri-fluoroethanol in aqueous solution: an NMR spectroscopic and molecular dynamics study. Chemistry 8(7):1663–1669

Dikiy I, Eliezer D (2012) Folding and misfolding of alpha-synuclein on membranes. Biochim Biophys Acta 1818(4):1013–1018

Dobson CM (1999) Protein misfolding, evolution and disease. Trends Biochem Sci 24(9):329–332

Dobson CM (2001) The structural basis of protein folding and its links with human disease. Philos Trans R Soc Lond B Biol Sci 356(1406):133–145

Dosztanyi Z, Csizmok V, Tompa P, Simon I (2005) IUPred: web server for the prediction of intrinsically unstructured regions of proteins based on estimated energy content. Bioinformatics 21(16):3433–3434

Dufour E, Bertrand-Harb C, Haertle T (1993) Reversible effects of medium dielectric constant on structural transformation of beta-lactoglobulin and its retinol binding. Biopolymers 33(4):589–598

Dufour E, Robert P, Bertrand D, Haertle T (1994) Conformation changes of beta-lactoglobulin: an ATR infrared spectroscopic study of the effect of pH and ethanol. J Protein Chem 13(2):143–149

Dunker AK, Obradovic Z, Romero P, Garner EC, Brown CJ (2000) Intrinsic protein disorder in complete genomes. Genome Informatics Ser Workshop 11:161–171

Dunker AK, Lawson JD, Brown CJ, Williams RM, Romero P, Oh JS, Oldfield CJ, Campen AM, Ratliff CM, Hipps KW, Ausio J, Nissen MS, Reeves R, Kang C, Kissinger CR, Bailey RW, Griswold MD, Chiu W, Garner EC, Obradovic Z (2001) Intrinsically disordered protein. J Mol Graph Model 19(1):26–59

Dunker AK, Cortese MS, Romero P, Iakoucheva LM, Uversky VN (2005) Flexible nets. The roles of intrinsic disorder in protein interaction networks. FEBS J 272(20):5129–5148

Dyson HJ, Wright PE (2005) Intrinsically unstructured proteins and their functions. Nat Rev Mol Cell Biol 6(3):197–208

Eliezer D, Kutluay E, Bussell R Jr, Browne G (2001) Conformational properties of alpha-synuclein in its free and lipid-associated states. J Mol Biol 307(4):1061–1073

Endo T, Schatz G (1988) Latent membrane perturbation activity of a mitochondrial precursor protein is exposed by unfolding. EMBO J 7(4):1153–1158

Fan P, Bracken C, Baum J (1993) Structural characterization of monellin in the alcohol-denatured state by NMR: evidence for beta-sheet to alpha-helix conversion. Biochemistry 32(6):1573–1582

Fauvet B, Fares MB, Samuel F, Dikiy I, Tandon A, Eliezer D, Lashuel HA (2012a) Characterization of semisynthetic and naturally Nalpha-acetylated alpha-synuclein in vitro and in intact cells: implications for aggregation and cellular properties of alpha-synuclein. J Biol Chem 287(34):28243–28262

Fauvet B, Mbefo MK, Fares MB, Desobry C, Michael S, Ardah MT, Tsika E, Coune P, Prudent M, Lion N, Eliezer D, Moore DJ, Schneider B, Aebischer P, El-Agnaf OM, Masliah E, Lashuel HA (2012b) Alpha-synuclein in central nervous system and from erythrocytes, mammalian cells, and Escherichia coli exists predominantly as disordered monomer. J Biol Chem 287(19):15345–15364

Fezoui Y, Teplow DB (2002) Kinetic studies of amyloid beta-protein fibril assembly. Differential effects of alpha-helix stabilization. J Biol Chem 277(40):36948–36954

Fink AL (1998) Protein aggregation: folding aggregates, inclusion bodies and amyloid. Fold Des 3(1):R9–R23

Fioroni M, Diaz MD, Burger K, Berger S (2002) Solvation phenomena of a tetrapeptide in water/trifluoroethanol and water/ethanol mixtures: a diffusion NMR, intermolecular NOE, and molecular dynamics study. J Am Chem Soc 124(26):7737–7744

Furukawa K, Matsuzaki-Kobayashi M, Hasegawa T, Kikuchi A, Sugeno N, Itoyama Y, Wang Y, Yao PJ, Bushlin I, Takeda A (2006) Plasma membrane ion permeability induced by mutant alpha-synuclein contributes to the degeneration of neural cells. J Neurochem 97(4):1071–1077

Gasch A, Wiesner S, Martin-Malpartida P, Ramirez-Espain X, Ruiz L, Macias MJ (2006) The structure of Prp40 FF1 domain and its interaction with the Crn-TPR1 motif of Clf1 gives a new insight into the binding mode of FF domains. J Biol Chem 281(1):356–364

Gast K, Siemer A, Zirwer D, Damaschun G (2001) Fluoroalcohol-induced structural changes of proteins: some aspects of cosolvent-protein interactions. Eur Biophys J 30(4):273–283

Georgieva ER, Ramlall TF, Borbat PP, Freed JH, Eliezer D (2008) Membrane-bound alpha-synuclein forms an extended helix: long-distance pulsed ESR measurements using vesicles, bicelles, and rodlike micelles. J Am Chem Soc 130(39):12856–12857

Glenner GG, Wong CW (1984a) Alzheimer's disease and Down's syndrome: sharing of a unique cerebrovascular amyloid fibril protein. Biochem Biophys Res Commun 122(3):1131–1135

Glenner GG, Wong CW (1984b) Alzheimer's disease: initial report of the purification and characterization of a novel cerebrovascular amyloid protein. Biochem Biophys Res Commun 120(3):885–890

Glenner GG, Wong CW, Quaranta V, Eanes ED (1984) The amyloid deposits in Alzheimer's disease: their nature and pathogenesis. Appl Pathol 2(6):357–369

Goda S, Takano K, Yamagata Y, Nagata R, Akutsu H, Maki S, Namba K, Yutani K (2000) Amyloid protofilament formation of hen egg lysozyme in highly concentrated ethanol solution. Protein Sci 9(2):369–375

Gruber HJ, Low PS (1988) Interaction of amphiphiles with integral membrane proteins. I. Structural destabilization of the anion transport protein of the erythrocyte membrane by fatty acids, fatty alcohols, and fatty amines. Biochim Biophys Acta 944(3):414–424

Grudzielanek S, Jansen R, Winter R (2005) Solvational tuning of the unfolding, aggregation and amyloidogenesis of insulin. J Mol Biol 351(4):879–894

Guijarro JI, Sunde M, Jones JA, Campbell ID, Dobson CM (1998) Amyloid fibril formation by an SH3 domain. Proc Natl Acad Sci U S A 95(8):4224–4228

Haaning S, Radutoiu S, Hoffmann SV, Dittmer J, Giehm L, Otzen DE, Stougaard J (2008) An unusual intrinsically disordered protein from the model legume Lotus japonicus stabilizes proteins in vitro. J Biol Chem 283(45):31142–31152

Hamada D, Kuroda Y, Tanaka T, Goto Y (1995) High helical propensity of the peptide fragments derived from beta-lactoglobulin, a predominantly beta-sheet protein. J Mol Biol 254(4):737–746

Hirota N, Mizuno K, Goto Y (1998) Group additive contributions to the alcohol-induced alpha-helix formation of melittin: implication for the mechanism of the alcohol effects on proteins. J Mol Biol 275(2):365–378

Holley M, Eginton C, Schaefer D, Brown LR (2008) Characterization of amyloidogenesis of hen egg lysozyme in concentrated ethanol solution. Biochem Biophys Res Commun 373(1):164–168

Hong D-P, Hoshino M, Kuboi R, Goto Y (1999) Clustering of fluorine-substituted alcohols as a factor responsible for their marked effects on proteins and peptides. J Am Chem Soc 121:8427–8433

Jackson M, Mantsch HH (1992) Halogenated alcohols as solvents for proteins: FTIR spectroscopic studies. Biochim Biophys Acta 1118(2):139–143

Jakes R, Spillantini MG, Goedert M (1994) Identification of two distinct synucleins from human brain. FEBS Lett 345(1):27–32

Jao CC, Der-Sarkissian A, Chen J, Langen R (2004) Structure of membrane-bound alpha-synuclein studied by site-directed spin labeling. Proc Natl Acad Sci U S A 101(22):8331–8336

Jao CC, Hegde BG, Chen J, Haworth IS, Langen R (2008) Structure of membrane-bound alpha-synuclein from site-directed spin labeling and computational refinement. Proc Natl Acad Sci U S A 105(50):19666–19671

Jenco JM, Rawlingson A, Daniels B, Morris AJ (1998) Regulation of phospholipase D2: selective inhibition of mammalian phospholipase D isoenzymes by alpha- and beta-synucleins. Biochemistry 37(14):4901–4909

Jimenez JL, Guijarro JI, Orlova E, Zurdo J, Dobson CM, Sunde M, Saibil HR (1999) Cryo-electron microscopy structure of an SH3 amyloid fibril and model of the molecular packing. EMBO J 18(4):815–821

Jimenez JL, Nettleton EJ, Bouchard M, Robinson CV, Dobson CM, Saibil HR (2002) The protofilament structure of insulin amyloid fibrils. Proc Natl Acad Sci U S A 99(14):9196–9201

Jo E, McLaurin J, Yip CM, St George-Hyslop P, Fraser PE (2000) Alpha-synuclein membrane interactions and lipid specificity. J Biol Chem 275(44):34328–34334

Jo E, Fuller N, Rand RP, St George-Hyslop P, Fraser PE (2002) Defective membrane interactions of familial Parkinson's disease mutant A30P alpha-synuclein. J Mol Biol 315(4):799–807

Jones DT, Ward JJ (2003) Prediction of disordered regions in proteins from position specific score matrices. Proteins 53(Suppl 6):573–578

Juarez J, Lopez SG, Cambon A, Taboada P, Mosquera V (2009) Influence of electrostatic interactions on the fibrillation process of human serum albumin. J Phys Chem B 113(30):10521–10529

Kallberg Y, Gustafsson M, Persson B, Thyberg J, Johansson J (2001) Prediction of amyloid fibril-forming proteins. J Biol Chem 276(16):12945–12950

Kelly JW (1998) The alternative conformations of amyloidogenic proteins and their multi-step assembly pathways. Curr Opin Struct Biol 8(1):101–106

Kentsis A, Sosnick TR (1998) Trifluoroethanol promotes helix formation by destabilizing backbone exposure: desolvation rather than native hydrogen bonding defines the kinetic pathway of dimeric coiled coil folding. Biochemistry 37(41):14613–14622

Khurana R, Gillespie JR, Talapatra A, Minert LJ, Ionescu-Zanetti C, Millett I, Fink AL (2001) Partially folded intermediates as critical precursors of light chain amyloid fibrils and amorphous aggregates. Biochemistry 40(12):3525–3535

Kirkitadze MD, Condron MM, Teplow DB (2001) Identification and characterization of key kinetic intermediates in amyloid beta-protein fibrillogenesis. J Mol Biol 312(5):1103–1119

Konno T, Oiki S, Morii T (2007) Synergistic action of polyanionic and non-polar cofactors in fibrillation of human islet amyloid polypeptide. FEBS Lett 581(8):1635–1638

Kuprin S, Graslund A, Ehrenberg A, Koch MH (1995) Nonideality of water-hexafluoropropanol mixtures as studied by X-ray small angle scattering. Biochem Biophys Res Commun 217(3):1151–1156

Lansbury PT Jr (1999) Evolution of amyloid: what normal protein folding may tell us about fibrillogenesis and disease. Proc Natl Acad Sci U S A 96(7):3342–3344

Lee HJ, Choi C, Lee SJ (2002) Membrane-bound alpha-synuclein has a high aggregation propensity and the ability to seed the aggregation of the cytosolic form. J Biol Chem 277(1):671–678

Li X, Romero P, Rani M, Dunker AK, Obradovic Z (1999) Predicting protein disorder for N-, C-, and internal regions. Genome Informatics Ser Workshop 10:30–40

Libich DS, Harauz G (2008) Solution NMR and CD spectroscopy of an intrinsically disordered, peripheral membrane protein: evaluation of aqueous and membrane-mimetic solvent conditions for studying the conformational adaptability of the 18.5 kDa isoform of myelin basic protein (MBP). Eur Biophys J 37(6):1015–1029

Linding R, Jensen LJ, Diella F, Bork P, Gibson TJ, Russell RB (2003a) Protein disorder prediction: implications for structural proteomics. Structure 11(11):1453–1459

Linding R, Russell RB, Neduva V, Gibson TJ (2003b) GlobPlot: exploring protein sequences for globularity and disorder. Nucleic Acids Res 31(13):3701–3708

Liu J, Rost B (2003) NORSp: predictions of long regions without regular secondary structure. Nucleic Acids Res 31(13):3833–3835

Madine J, Doig AJ, Middleton DA (2004) The aggregation and membrane-binding properties of an alpha-synuclein peptide fragment. Biochem Soc Trans 32(Pt 6):1127–1129

Manning-Bog AB, McCormack AL, Li J, Uversky VN, Fink AL, Di Monte DA (2002) The herbicide paraquat causes up-regulation and aggregation of alpha-synuclein in mice: paraquat and alpha-synuclein. J Biol Chem 277(3):1641–1644

Marinelli P, Castillo V, Ventura S (2013) Trifluoroethanol modulates amyloid formation by the all alpha-helical URN1 FF domain. Int J Mol Sci 14(9):17830–17844

Maroteaux L, Campanelli JT, Scheller RH (1988) Synuclein: a neuron-specific protein localized to the nucleus and presynaptic nerve terminal. J Neurosci 8(8):2804–2815

Marzban L, Park K, Verchere CB (2003) Islet amyloid polypeptide and type 2 diabetes. Exp Gerontol 38(4):347–351

Masters CL, Simms G, Weinman NA, Multhaup G, McDonald BL, Beyreuther K (1985) Amyloid plaque core protein in Alzheimer disease and Down syndrome. Proc Natl Acad Sci U S A 82(12):4245–4249

McLean PJ, Kawamata H, Ribich S, Hyman BT (2000) Membrane association and protein conformation of alpha-synuclein in intact neurons. Effect of Parkinson's disease-linked mutations. J Biol Chem 275(12):8812–8816

Mezey E, Dehejia A, Harta G, Papp MI, Polymeropoulos MH, Brownstein MJ (1998) Alpha synuclein in neurodegenerative disorders: murderer or accomplice? Nat Med 4(7):755–757

Middleton ER, Rhoades E (2010) Effects of curvature and composition on alpha-synuclein binding to lipid vesicles. Biophys J 99(7):2279–2288

Mihajlovic M, Lazaridis T (2008) Membrane-bound structure and energetics of alpha-synuclein. Proteins 70(3):761–778

Moriarty GM, Janowska MK, Kang L, Baum J (2013) Exploring the accessible conformations of N-terminal acetylated alpha-synuclein. FEBS Lett 587(8):1128–1138

Munishkina LA, Phelan C, Uversky VN, Fink AL (2003) Conformational behavior and aggregation of alpha-synuclein in organic solvents: modeling the effects of membranes. Biochemistry 42(9):2720–2730

Narizhneva NV, Uversky VN (1997) Human alpha-fetoprotein is in the molten globule state under conditions modelling protein environment near the membrane surface. Protein Pept Lett 4(4):243–249

Nielsen L, Khurana R, Coats A, Frokjaer S, Brange J, Vyas S, Uversky VN, Fink AL (2001) Effect of environmental factors on the kinetics of insulin fibril formation: elucidation of the molecular mechanism. Biochemistry 40(20):6036–6046

Nuscher B, Kamp F, Mehnert T, Odoy S, Haass C, Kahle PJ, Beyer K (2004) Alpha-synuclein has a high affinity for packing defects in a bilayer membrane: a thermodynamics study. J Biol Chem 279(21):21966–21975

Oldfield CJ, Cheng Y, Cortese MS, Brown CJ, Uversky VN, Dunker AK (2005) Comparing and combining predictors of mostly disordered proteins. Biochemistry 44(6):1989–2000

Otzen DE, Sehgal P, Nesgaard LW (2007) Alternative membrane protein conformations in alcohols. Biochemistry 46(14):4348–4359

Pallares I, Vendrell J, Aviles FX, Ventura S (2004) Amyloid fibril formation by a partially structured intermediate state of alpha-chymotrypsin. J Mol Biol 342(1):321–331

Pandey NK, Ghosh S, Dasgupta S (2010) Fibrillation in human serum albumin is enhanced in the presence of copper(II). J Phys Chem B 114(31):10228–10233

Perrin RJ, Woods WS, Clayton DF, George JM (2000) Interaction of human alpha-synuclein and Parkinson's disease variants with phospholipids. Structural analysis using site-directed mutagenesis. J Biol Chem 275(44):34393–34398

Perrin RJ, Woods WS, Clayton DF, George JM (2001) Exposure to long chain polyunsaturated fatty acids triggers rapid multimerization of synucleins. J Biol Chem 276(45):41958–41962

Pfefferkorn CM, Jiang Z, Lee JC (2012) Biophysics of alpha-synuclein membrane interactions. Biochim Biophys Acta 1818(2):162–171

Popovic M, De Biasio A, Pintar A, Pongor S (2007) The intracellular region of the Notch ligand Jagged-1 gains partial structure upon binding to synthetic membranes. FEBS J 274(20):5325–5336

Prilusky J, Felder CE, Zeev-Ben-Mordehai T, Rydberg EH, Man O, Beckmann JS, Silman I, Sussman JL (2005) FoldIndex: a simple tool to predict whether a given protein sequence is intrinsically unfolded. Bioinformatics 21(16):3435–3438

Ptitsyn OB, Bychkova VE, Uversky VN (1995) Kinetic and equilibrium folding intermediates. Philos Trans R Soc Lond B Biol Sci 348(1323):35–41

Radivojac P, Iakoucheva LM, Oldfield CJ, Obradovic Z, Uversky VN, Dunker AK (2007) Intrinsic disorder and functional proteomics. Biophys J 92(5):1439–1456

Reiersen H, Rees AR (2000) Trifluoroethanol may form a solvent matrix for assisted hydrophobic interactions between peptide side chains. Protein Eng 13(11):739–743

Rezaei-Ghaleh N, Amininasab M, Nemat-Gorgani M (2008) Conformational changes of alpha-chymotrypsin in a fibrillation-promoting condition: a molecular dynamics study. Biophys J 95(9):4139–4147

Rhoades E, Ramlall TF, Webb WW, Eliezer D (2006) Quantification of alpha-synuclein binding to lipid vesicles using fluorescence correlation spectroscopy. Biophys J 90(12):4692–4700

Roccatano D, Colombo G, Fioroni M, Mark AE (2002) Mechanism by which 2,2,2-trifluoroethanol/water mixtures stabilize secondary-structure formation in peptides: a molecular dynamics study. Proc Natl Acad Sci U S A 99(19):12179–12184

Rochet JC, Lansbury PT Jr (2000) Amyloid fibrillogenesis: themes and variations. Curr Opin Struct Biol 10(1):60–68

Rodionova NA, Semisotnov GV, Kutyshenko VP, Uverskii VN, Bolotina IA (1989) Staged equilibrium of carbonic anhydrase unfolding in strong denaturants. Mol Biol 23(3):683–692

Romero P, Obradovic Z, Li X, Garner EC, Brown CJ, Dunker AK (2001) Sequence complexity of disordered protein. Proteins 42(1):38–48

Rozga M, Sokolowska M, Protas AM, Bal W (2007) Human serum albumin coordinates Cu(II) at its N-terminal binding site with 1 pM affinity. J Biol Inorg Chem 12(6):913–918

Santambrogio C, Ricagno S, Sobott F, Colombo M, Bolognesi M, Grandori R (2011) Characterization of beta2-microglobulin conformational intermediates associated to different fibrillation conditions. J Mass Spectrom 46(8):734–741

Semisotnov GV, Rodionova NA, Razgulyaev OI, Uversky VN, Gripas AF, Gilmanshin RI (1991) Study of the "molten globule" intermediate state in protein folding by a hydrophobic fluorescent probe. Biopolymers 31(1):119–128

Shao H, Jao S, Ma K, Zagorski MG (1999) Solution structures of micelle-bound amyloid beta-(1–40) and beta-(1–42) peptides of Alzheimer's disease. J Mol Biol 285(2):755–773

Shirahama T, Cohen AS (1967) High-resolution electron microscopic analysis of the amyloid fibril. J Cell Biol 33(3):679–708

Shirahama T, Benson MD, Cohen AS, Tanaka A (1973) Fibrillar assemblage of variable segments of immunoglobulin light chains: an electron microscopic study. J Immunol 110(1):21–30

Shvadchak VV, Falomir-Lockhart LJ, Yushchenko DA, Jovin TM (2011) Specificity and kinetics of alpha-synuclein binding to model membranes determined with fluorescent excited state intramolecular proton transfer (ESIPT) probe. J Biol Chem 286(15):13023–13032

Silva BA, Breydo L, Fink AL, Uversky VN (2013) Agrochemicals, alpha-synuclein, and Parkinson's disease. Mol Neurobiol 47(2):598–612

Sluzky V, Tamada JA, Klibanov AM, Langer R (1991) Kinetics of insulin aggregation in aqueous solutions upon agitation in the presence of hydrophobic surfaces. Proc Natl Acad Sci U S A 88(21):9377–9381

Smith DP, Jones S, Serpell LC, Sunde M, Radford SE (2003) A systematic investigation into the effect of protein destabilisation on beta 2-microglobulin amyloid formation. J Mol Biol 330(5):943–954

Srisailam S, Kumar TK, Rajalingam D, Kathir KM, Sheu HS, Jan FJ, Chao PC, Yu C (2003) Amyloid-like fibril formation in an all beta-barrel protein. Partially structured intermediate state(s) is a precursor for fibril formation. J Biol Chem 278(20):17701–17709

Sticht H, Bayer P, Willbold D, Dames S, Hilbich C, Beyreuther K, Frank RW, Rosch P (1995) Structure of amyloid A4-(1-40)-peptide of Alzheimer's disease. Eur J Biochem 233(1):293–298

Sunde M, Serpell LC, Bartlam M, Fraser PE, Pepys MB, Blake CC (1997) Common core structure of amyloid fibrils by synchrotron X-ray diffraction. J Mol Biol 273(3):729–739

Sung YH, Eliezer D (2006) Secondary structure and dynamics of micelle bound beta- and gamma-synuclein. Protein Sci 15(5):1162–1174

Sweede M, Ankem G, Chutvirasakul B, Azurmendi HF, Chbeir S, Watkins J, Helm RF, Finkielstein CV, Capelluto DG (2008) Structural and membrane binding properties of the prickle PET domain. Biochemistry 47(51):13524–13536

Taboada P, Barbosa S, Castro E, Mosquera V (2006) Amyloid fibril formation and other aggregate species formed by human serum albumin association. J Phys Chem B 110(42):20733–20736

Tamamizu-Kato S, Kosaraju MG, Kato H, Raussens V, Ruysschaert JM, Narayanaswami V (2006) Calcium-triggered membrane interaction of the alpha-synuclein acidic tail. Biochemistry 45(36):10947–10956

Tanford C (1968) Protein denaturation. Adv Protein Chem 23:121–282

Teplow DB (1998) Structural and kinetic features of amyloid beta-protein fibrillogenesis. Amyloid 5(2):121–142

Thomas PD, Dill KA (1993) Local and nonlocal interactions in globular proteins and mechanisms of alcohol denaturation. Protein Sci 2(12):2050–2065

Tompa P (2002) Intrinsically unstructured proteins. Trends Biochem Sci 27(10):527–533

Tompa P (2005) The interplay between structure and function in intrinsically unstructured proteins. FEBS Lett 579(15):3346–3354

Tompa P, Csermely P (2004) The role of structural disorder in the function of RNA and protein chaperones. FASEB J 18(11):1169–1175

Ulmer TS, Bax A (2005) Comparison of structure and dynamics of micelle-bound human alpha-synuclein and Parkinson disease variants. J Biol Chem 280(52):43179–43187

Ulmer TS, Bax A, Cole NB, Nussbaum RL (2005) Structure and dynamics of micelle-bound human alpha-synuclein. J Biol Chem 280(10):9595–9603

Uversky VN (2003a) A protein-chameleon: conformational plasticity of alpha-synuclein, a disordered protein involved in neurodegenerative disorders. J Biomol Struct Dyn 21(2):211–234

Uversky VN (2003b) Protein folding revisited. A polypeptide chain at the folding-misfolding-nonfolding crossroads: which way to go? Cell Mol Life Sci 60(9):1852–1871

Uversky VN (2004) Neurotoxicant-induced animal models of Parkinson's disease: understanding the role of rotenone, maneb and paraquat in neurodegeneration. Cell Tissue Res 318(1):225–241

Uversky VN (2007) Neuropathology, biochemistry, and biophysics of alpha-synuclein aggregation. J Neurochem 103(1):17–37

Uversky VN (2008a) Alpha-synuclein misfolding and neurodegenerative diseases. Curr Protein Pept Sci 9(5):507–540

Uversky VN (2008b) Amyloidogenesis of natively unfolded proteins. Curr Alzheimer Res 5(3): 260–287

Uversky VN (2009a) Intrinsic disorder in proteins associated with neurodegenerative diseases. Front Biosci 14:5188–5238

Uversky VN (2009b) Intrinsically disordered proteins and their environment: effects of strong denaturants, temperature, pH, counter ions, membranes, binding partners, osmolytes, and macromolecular crowding. Protein J 28(7–8):305–325

Uversky VN (2011a) Intrinsically disordered proteins from A to Z. Int J Biochem Cell Biol 43(8):1090–1103

Uversky VN (2011b) Multitude of binding modes attainable by intrinsically disordered proteins: a portrait gallery of disorder-based complexes. Chem Soc Rev 40(3):1623–1634

Uversky VN (2013a) A decade and a half of protein intrinsic disorder: biology still waits for physics. Protein Sci 22(6):693–724

Uversky VN (2013b) Intrinsic disorder-based protein interactions and their modulators. Curr Pharm Des 19(23):4191–4213

Uversky VN (2013c) Unusual biophysics of intrinsically disordered proteins. Biochim Biophys Acta 1834(5):932–951

Uversky VN, Eliezer D (2009) Biophysics of Parkinson's disease: structure and aggregation of alpha-synuclein. Curr Protein Pept Sci 10(5):483–499

Uversky VN, Fink AL (2004) Conformational constraints for amyloid fibrillation: the importance of being unfolded. Biochim Biophys Acta 1698(2):131–153

Uversky VN, Narizhneva NV, Kirschstein SO, Winter S, Lober G (1997) Conformational transitions provoked by organic solvents in beta-lactoglobulin: can a molten globule like intermediate be induced by the decrease in dielectric constant? Fold Des 2(3):163–172

Uversky VN, Talapatra A, Gillespie JR, Fink AL (1999a) Protein deposits as the molecular basis of amyloidosis. I. Systemic amyloidosis. Med Sci Monit 5:1001–1012

Uversky VN, Talapatra A, Gillespie JR, Fink AL (1999b) Protein deposits as the molecular basis of amyloidosis. II. Localized amyloidosis and neurodegenerative disorders. Med Sci Monit 5:1238–1254

Uversky VN, Gillespie JR, Fink AL (2000) Why are "natively unfolded" proteins unstructured under physiologic conditions? Proteins 41(3):415–427

Uversky VN, Li J, Fink AL (2001a) Evidence for a partially folded intermediate in alpha-synuclein fibril formation. J Biol Chem 276(14):10737–10744

Uversky VN, Li J, Fink AL (2001b) Metal-triggered structural transformations, aggregation, and fibrillation of human alpha-synuclein. A possible molecular link between Parkinson's disease and heavy metal exposure. J Biol Chem 276(47):44284–44296

Uversky VN, Li J, Fink AL (2001c) Pesticides directly accelerate the rate of alpha-synuclein fibril formation:

a possible factor in Parkinson's disease. FEBS Lett 500(3):105–108

Uversky VN, Li J, Fink AL (2001d) Trimethylamine-N-oxide-induced folding of alpha-synuclein. FEBS Lett 509(1):31–35

Uversky VN, Li J, Bower K, Fink AL (2002a) Synergistic effects of pesticides and metals on the fibrillation of alpha-synuclein: implications for Parkinson's disease. Neurotoxicology 23(4–5):527–536

Uversky VN, Li J, Souillac P, Millett IS, Doniach S, Jakes R, Goedert M, Fink AL (2002b) Biophysical properties of the synucleins and their propensities to fibrillate: inhibition of alpha-synuclein assembly by beta- and gamma-synucleins. J Biol Chem 277(14):11970–11978

Uversky VN, Oldfield CJ, Dunker AK (2005) Showing your ID: intrinsic disorder as an ID for recognition, regulation and cell signaling. J Mol Recognit 18(5):343–384

van Rooijen BD, Claessens MM, Subramaniam V (2008) Membrane binding of oligomeric alpha-synuclein depends on bilayer charge and packing. FEBS Lett 582(27):3788–3792

van Rooijen BD, Claessens MM, Subramaniam V (2009) Lipid bilayer disruption by oligomeric alpha-synuclein depends on bilayer charge and accessibility of the hydrophobic core. Biochim Biophys Acta 1788(6):1271–1278

Volles MJ, Lee SJ, Rochet JC, Shtilerman MD, Ding TT, Kessler JC, Lansbury PT Jr (2001) Vesicle permeabilization by protofibrillar alpha-synuclein: implications for the pathogenesis and treatment of Parkinson's disease. Biochemistry 40(26):7812–7819

Vucetic S, Xie H, Iakoucheva LM, Oldfield CJ, Dunker AK, Obradovic Z, Uversky VN (2007) Functional anthology of intrinsic disorder 2. Cellular components, domains, technical terms, developmental processes, and coding sequence diversities correlated with long disordered regions. J Proteome Res 6(5):1899–1916

Walgers R, Lee TC, Cammers-Goodwin A (1998) An indirect chaotropic mechanism of the stabilization of helix conformation of peptides in aqueous trifluoroethanol and hexafluoro-2-propanol. J Am Chem Soc 120:5073–5079

Walsh DM, Lomakin A, Benedek GB, Condron MM, Teplow DB (1997) Amyloid beta-protein fibrillogenesis. Detection of a protofibrillar intermediate. J Biol Chem 272(35):22364–22372

Walsh DM, Hartley DM, Kusumoto Y, Fezoui Y, Condron MM, Lomakin A, Benedek GB, Selkoe DJ, Teplow DB (1999) Amyloid beta-protein fibrillogenesis. Structure and biological activity of protofibrillar intermediates. J Biol Chem 274(36):25945–25952

Ward JJ, McGuffin LJ, Bryson K, Buxton BF, Jones DT (2004a) The DISOPRED server for the prediction of protein disorder. Bioinformatics 20(13):2138–2139

Ward JJ, Sodhi JS, McGuffin LJ, Buxton BF, Jones DT (2004b) Prediction and functional analysis of native disorder in proteins from the three kingdoms of life. J Mol Biol 337(3):635–645

Weinreb PH, Zhen W, Poon AW, Conway KA, Lansbury PT Jr (1996) NACP, a protein implicated in Alzheimer's disease and learning, is natively unfolded. Biochemistry 35(43):13709–13715

Wilkinson KD, Mayer AN (1986) Alcohol-induced conformational changes of ubiquitin. Arch Biochem Biophys 250(2):390–399

Williams RM, Obradovi Z, Mathura V, Braun W, Garner EC, Young J, Takayama S, Brown CJ, Dunker AK (2001) The protein non-folding problem: amino acid determinants of intrinsic order and disorder. Pac Symp Biocomput 89–100

Wright HT (1973) Comparison of the crystal structures of chymotrypsinogen-A and alpha-chymotrypsin. J Mol Biol 79(1):1–11

Wright PE, Dyson HJ (1999) Intrinsically unstructured proteins: re-assessing the protein structure-function paradigm. J Mol Biol 293(2):321–331

Wu KP, Baum J (2010) Detection of transient interchain interactions in the intrinsically disordered protein alpha-synuclein by NMR paramagnetic relaxation enhancement. J Am Chem Soc 132(16):5546–5547

Wu KP, Kim S, Fela DA, Baum J (2008) Characterization of conformational and dynamic properties of natively unfolded human and mouse alpha-synuclein ensem-

bles by NMR: implication for aggregation. J Mol Biol 378(5):1104–1115

Xie H, Vucetic S, Iakoucheva LM, Oldfield CJ, Dunker AK, Obradovic Z, Uversky VN (2007a) Functional anthology of intrinsic disorder 3. Ligands, post-translational modifications, and diseases associated with intrinsically disordered proteins. J Proteome Res 6(5):1917–1932

Xie H, Vucetic S, Iakoucheva LM, Oldfield CJ, Dunker AK, Uversky VN, Obradovic Z (2007b) Functional anthology of intrinsic disorder. 1. Biological processes and functions of proteins with long disordered regions. J Proteome Res 6(5):1882–1898

Yamaguchi K, Naiki H, Goto Y (2006) Mechanism by which the amyloid-like fibrils of a beta 2-microglobulin fragment are induced by fluorine-substituted alcohols. J Mol Biol 363(1):279–288

Zerovnik E (2002) Amyloid-fibril formation. Proposed mechanisms and relevance to conformational disease. Eur J Biochem 269(14):3362–3371

Zhu M, Fink AL (2003) Lipid binding inhibits alpha-synuclein fibril formation. J Biol Chem 278(19):16873–16877

Zhu M, Li J, Fink AL (2003) The association of alpha-synuclein with membranes affects bilayer structure, stability, and fibril formation. J Biol Chem 278(41):40186–40197

Lipids in Amyloid-β Processing, Aggregation, and Toxicity

3

Isabel Morgado and Megan Garvey

Abstract

Aggregation of amyloid-beta (Aβ) peptide is the major event underlying neuronal damage in Alzheimer's disease (AD). Specific lipids and their homeostasis play important roles in this and other neurodegenerative disorders. The complex interplay between the lipids and the generation, clearance or deposition of Aβ has been intensively investigated and is reviewed in this chapter. Membrane lipids can have an important influence on the biogenesis of Aβ from its precursor protein. In particular, increased cholesterol in the plasma membrane augments Aβ generation and shows a strong positive correlation with AD progression. Furthermore, apolipoprotein E, which transports cholesterol in the cerebrospinal fluid and is known to interact with Aβ or compete with it for the lipoprotein receptor binding, significantly influences Aβ clearance in an isoform-specific manner and is the major genetic risk factor for AD. Aβ is an amphiphilic peptide that interacts with various lipids, proteins and their assemblies, which can lead to variation in Aβ aggregation in vitro and in vivo. Upon interaction with the lipid raft components, such as cholesterol, gangliosides and phospholipids, Aβ can aggregate on the cell membrane and thereby disrupt it, perhaps by forming channel-like pores. This leads to perturbed cellular calcium homeostasis, suggesting that Aβ-lipid interactions at the cell membrane probably trigger the neurotoxic cascade in AD. Here, we overview the roles of specific lipids, lipid assemblies and apolipoprotein E in Aβ processing, clearance and aggregation, and discuss the contribution of these factors to the neurotoxicity in AD.

I. Morgado (✉)
Centre of Marine Sciences, University of Algarve,
Faro, Portugal

Present Affiliation: Department of Physiology
and Biophysics, Boston University School
of Medicine, Boston, MA, USA
e-mail: imorgado@bu.edu

M. Garvey
School of Medicine, Deakin University,
Waurn Ponds, VIC, Australia

© Springer International Publishing Switzerland 2015
O. Gursky (ed.), *Lipids in Protein Misfolding*, Advances in Experimental
Medicine and Biology 855, DOI 10.1007/978-3-319-17344-3_3

Keywords

Amyloid-β peptide • Alzheimer's disease • Cellular membranes and lipid rafts • Cholesterol • Gangliosides • Peptide oligomers, protofibrils and fibrils • Apolipoprotein E • Amyloid precursor protein

Abbreviations

AD	Alzheimer's disease
APP	Amyloid precursor protein
Aβ	Amyloid-β peptide
CMC	Critical micelle concentration
DHPC	1,2-dihexanoyl-sn-glycero-3-phosphocholine
GM1	Monosialotetrahexosyl ganglioside
HDL	High-density lipoprotein
LRP1	Low-density lipoprotein receptor-related protein
VLDLR	Very low-density lipoprotein receptor

3.1   Introduction

3.1.1   Role of Lipids in Alzheimer's Disease

Alzheimer's disease (AD) is the most common neurodegenerative disorder and the leading cause of dementia in the Western world. AD currently affects 36 million people worldwide, and the numbers are continuously increasing (Alzheimer's Disease International, http://www.alzheimers.net/resources/alzheimers-statistics/). Clinically, AD is characterized by a progressive, irreversible loss of memory and cognitive functions, eventually leading to death. In late-onset AD, which is the most common form affecting over 90 % of patients, the symptoms first appear after the age of 65. Late-onset AD is sporadic with a large hereditary component. In the early-onset AD, which is a rare genetic form of AD, the symptoms can appear decades earlier. The major neuropathological hallmark of AD is the accumulation of extracellular amyloid plaques in the brain, followed by formation of intracellular tangles, loss of synaptic function and neuronal death. According to the widely accepted "amyloid cascade hypothesis", plaque deposition is the crucial causative event in AD pathology (Hardy and Higgins 1992; Hardy and Selkoe 2002). Currently, there is no cure for AD and the treatment options are limited, which makes AD a major public health challenge as well as the focus of the extensive search for pharmacologic interventions (Golde et al. 2010; Karran et al. 2011).

A possible relation between AD and aberrant lipid homeostasis was first suggested over 100 years ago by Alois Alzheimer, who found adipose inclusions and altered lipid composition in the brain tissues of AD patients (Alzheimer 1911; Foley 2010). The connection was firmly established decades later when the ε4 allele of apolipoprotein E (apoE) gene was identified as the strongest genetic risk factor in AD (Corder et al. 1993). ApoE, which is the major lipid transporter in the central nervous system, can significantly influence Aβ clearance and aggregation in AD (Kim et al. 2009). Cholesterol, gangliosides and a wide range of other lipids are involved in various ways with the key pathogenic processes in AD (Fig. 3.1). Thus, lipids are implicated in the activity and processing of the proteins that are crucial to AD, in particular, the amyloid-β precursor protein (APP) and the transmembrane proteases, such as γ-secretase, which process it. The critical role of APP processing is underscored by the findings that familial AD is caused by mutations either in APP near the γ-secretase cleavage site, or in presenilin 1 or 2 (Cruts and Van Broeckhoven 1998; Brouwers et al. 2008), which form the catalytic domain in the γ-secretase complexes (Krishnaswamy et al. 2009).

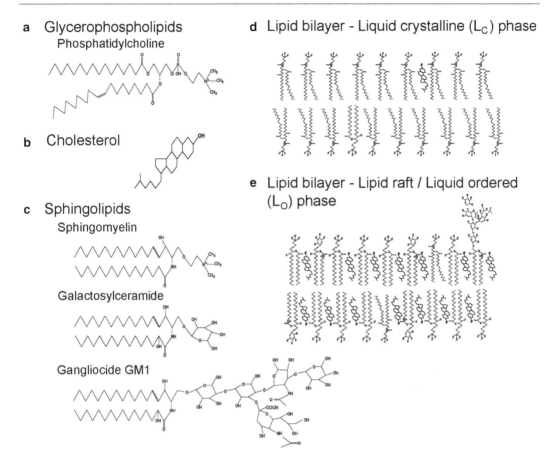

a Glycerophospholipids
  Phosphatidylcholine

b Cholesterol

c Sphingolipids
  Sphingomyelin

  Galactosylceramide

  Gangliocide GM1

d Lipid bilayer - Liquid crystalline (L_C) phase

e Lipid bilayer - Lipid raft / Liquid ordered (L_O) phase

Fig. 3.1 Membrane lipids and lipid rafts implicated in AD. (**a**) Phosphatidylcholine is the most abundant membrane phospholipid. (**b**) Cholesterol, which is essential for lipid rafts formation, is critical for AD pathogenesis. (**c**) Sphingolipids, such as sphingomyelin and galactoslyceramide, are essential components of lipid rafts. Gangliosides such as GM1 cluster in lipid rafts and are proposed to form platforms for Aβ aggregation. (**d**) Cartoon illustrating lipid bilayer in liquid crystalline (Lc) phase. (**e**) Cartoon illustrating lipid bilayer showing liquid-ordered (Lo) phase, also known as a lipid raft, with increased acyl chain order and reduced fluidity (Figures are modified from Fantini et al. 2002)

Cholesterol is an important modulator of APP processing and Aβ generation (Bodovitz and Klein 1996; Barrett et al. 2012). In addition, lipids modulate the pathogenic aggregation of Aβ peptide and Tau protein, which are believed to be the major players in AD. Furthermore, cytotoxicity of the aggregated Aβ in AD is mostly exerted via interference with cellular membranes, and lipids and apoE are ubiquitous constituents of amyloid plaques. This chapter reviews the role of lipids and their transport protein, apoE, in modulating the biogenesis, clearance and aggregation of Aβ.

3.1.2 Aggregation of Amyloid-β Peptide

The major pathological hallmark of AD is accumulation of extracellular amyloid plaques and intracellular tangles in the brain. The plaques are comprised of insoluble misfolded proteins, mainly of fibrillar Aβ (Roher et al. 1993), along with carbohydrates such as glycosaminoglycans and lipids. Intensive in vivo and in vitro research, including biochemical, cell biological, genetic, structural, and animal model studies, has provided

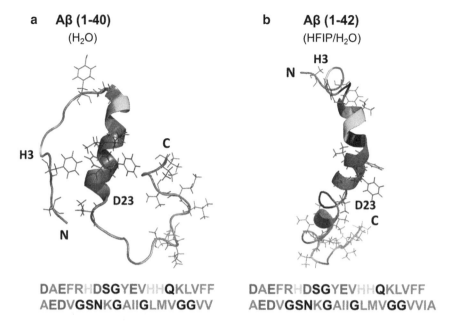

a Aβ (1-40)
(H₂O)

b Aβ (1-42)
(HFIP/H₂O)

DAEFRHDSGYEVHHQKLVFF
AEDVGSNKGAIIGLMVGGVV

DAEFRHDSGYEVHHQKLVFF
AEDVGSNKGAIIGLMVGGVVIA

Fig. 3.2 Solution structure of Aβ monomer and its amino acid sequence. Residues are color coded: negatively charged (*red*), positively charged (*blue*), ionizable (*teal*), polar (*black*) and apolar (*grey*). (**a**) NMR structure of Aβ(1–40) in aqueous environment (PDB ID: 2LFM). The central hydrophobic region forms an α-helix (residues H13-D23) and the N- and C-termini collapse upon clustering of hydrophobic residues (Vivekanandan et al. 2011). (**b**) NMR structure of Aβ(1–42) in aqueous solution containing 30 % hexafluoroisopropanol that induces α-helical structure (PDB:1Z0Q) (Tomaselli et al. 2006)

strong evidence for the causative role of Aβ aggregation and fibril formation in AD pathology (Lacor et al. 2007; Reed et al. 2011; Matsuzaki 2011; Gilbert 2013; Bloom 2014; Hong et al. 2014).

Aβ is a 39- to 43-residue peptide that is a proteolytic product of APP. This amphipathic peptide has a relatively polar N-terminal region that originates from the extracellular domain of APP, and a highly hydrophobic C-terminal region that originates from the transmembrane helical domain of APP (Figs. 3.2 and 3.3). The major physiologically relevant peptide species are Aβ(1–40) and Aβ(1–42), the latter having two additional C-terminal hydrophobic residues, I41 and A42, leading to much lower solubility and greater fibril-forming propensity. Monomeric Aβ has limited solubility and is mostly unstructured in aqueous solution (Zhang et al. 2000), except for short segments that can form ordered secondary structure such as α-helices (Tomaselli et al. 2006;

Vivekanandan et al. 2011) (Fig. 3.2). In response to the environmental factors, including certain lipid surfaces, Aβ can acquire additional helical structure that can undergo an α-helix to β-sheet transition (Fezoui and Teplow 2002; Morgado and Faendrich 2011; Hou et al. 2004; Coles et al. 1998). At elevated concentrations, Aβ can self-associate into transient β-sheet-rich species including dimers, oligomers and protofibrils, culminating in the formation of mature amyloid fibrils such as those found in AD plaques (Benseny-Cases et al. 2007). Such fibrils, which have been characterized extensively by various biophysical methods, present a common overall architecture comprised of intermolecular in-register β-sheets running perpendicular to the fibril axis in a highly stable "cross-β" conformation (Sunde et al. 1997; Petkova et al. 2002; Luhrs et al. 2005; Sachse et al. 2008; Lu et al. 2013). However, even for the same peptide the

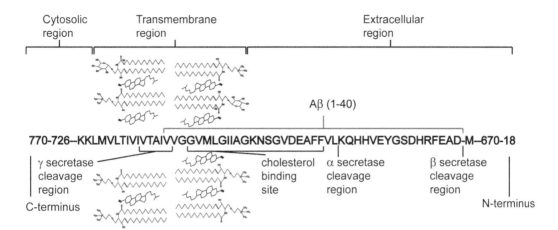

Fig. 3.3 Amyloid-β precursor protein and its processing by α-, β- and γ-secretases. APP and its processing secretases are integral membrane proteins that preferentially partition into lipid rafts. The steps involved in APP processing at specific sites and the effects of cholesterol on these steps are described in the text

fibrillar morphology can vary, suggesting the underlying variations in the peptide conformation and packing (Petkova et al. 2005; Meinhardt et al. 2009 and references therein).

In vitro, Aβ aggregation in solution and formation of amyloid fibrils has been well-characterized (reviewed in Morgado and Faendrich 2011) as a nucleated-growth process with two major kinetic phases: the initial lag phase involves amyloid nucleation, and the following growth phase involves fibril polymerization and elongation (Harper and Lansbury 1997; Wetzel 2006; Teplow et al. 2006; Pellarin and Caflisch 2006; Xue et al. 2008; Esler et al. 2000). Such kinetics is illustrated in Fig. 3.4 below, and the process is described in detail by Dovidchenko and Galzitskaya in Chap. 9 of this volume. Notably, Aβ aggregation occurs through a secondary nucleation mechanism that can be described as an exponential growth of protofibrils that involves branching (bifurcation) and formation of secondary nuclei (Cohen et al. 2013; Meisl et al. 2014; Dovidchenko et al. 2014). Theoretically, only the exponential growth is compatible with the existence of a pronounced lag phase that can be much longer than the following protofibril growth phase (Dovidchenko et al. 2014). In vivo, the process of Aβ aggregation that leads to the formation of toxic species is more complex and can be influenced by a wide range of factors. Some of these factors include membrane surfaces containing anionic lipids or lipid rafts, which may serve as templates for Aβ amyloid nucleation and/or modulate fibril growth (see Sects. 3.4 and 3.5 of this chapter).

Although it was initially believed that the amyloid fibrils are responsible for AD pathology, numerous studies (including in vivo mouse models, in vitro binding assays, electrophysiology, cell toxicity and synaptotoxicity assays, to name a few) have established that pre-fibrillar intermediates such as Aβ oligomers or protofibrils are highly cytotoxic and probably represent the major pathogenic species (Kayed et al. 2003; Cleary et al. 2005; Demuro et al. 2005; Haass and Selkoe 2007; Alberdi et al. 2010; Zhao et al. 2012; Hong et al. 2014). For example, studies of transgenic mice overexpressing human APP reported that AD neuropathology is linked to formation of Aβ oligomers (Gandy et al. 2010). Oligomers have also been shown to interfere with synapses, disrupt neuronal function, and induce cell death through an as yet unclear process (Walsh et al. 2002; Walsh and Selkoe 2004; Zhao et al. 2012). The increased toxicity of Aβ oligomers can be attributed, in part, to their large exposed hydrophobic surface that can interact with cellular membranes. Such interactions are believed to be crucial for Aβ aggregation and toxicity, as detailed in Sects. 3.4 and 3.5 below.

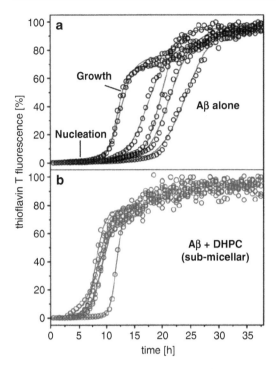

Fig. 3.4 Effects of DHPC on the nucleation and growth of Aβ amyloid. Aggregation kinetics of Aβ alone (**a**) or in the presence of sub-micellar DHPC (**b**) was monitored by using a diagnostic dye, thioflavin-T, that shows enhanced fluorescence upon binding to amyloid-like structures. DHPC concentration was 1.9 mM, well below its CMC of ~13 mM. Six replicates of each experiment illustrate variability of kinetic measurements. Aβ amyloid nucleation was accelerated in the presence of DHPC, as evident from the shorter lag phase. This effect was observed at 0.4–2.0 mM DHPC and above 9 mM DHPC, indicating that individual lipid molecules as well as the lipid micelles accelerate amyloid nucleation. The lesser variability during the nucleation phase, which was observed in the presence of DHPC, is attributed to the lipid molecules acting as nucleation templates for Aβ (Adopted with permission from Dhase et al. 2010)

Two main isoforms, Aβ(1–40) and Aβ(1–42), have been linked to AD. While Aβ(1–42) is the most aggregation-prone amyloidogenic isoform, Aβ(1–40) is most abundant in amyloid plaques as well as in the plasma, cerebrospinal fluid, and the brain interstitial fluid, where it is found mainly in the soluble form (Ghiso and Frangione 2002). The levels of Aβ are elevated in the brain of AD patients. Increased total levels of Aβ as well as the increased ratio of Aβ(1–42) to Aβ(1–40) have

been implicated in triggering amyloid formation and AD pathology (Pauwels et al. 2012).

In vivo studies of AD patients suggest that accumulation of Aβ in the brain arises from an imbalance between the peptide production and its clearance (Bateman et al. 2006; Mawuenyega et al. 2010; Wildsmith et al. 2013) via mechanisms which have been under intensive scrutiny. Lipids, particularly lipid raft components such as cholesterol, play a crucial role in Aβ production, whereas apoE, which is the major lipid carrier in the brain, is important for the clearance of Aβ via the lipoprotein receptor-related mechanisms described in Sect. 3.3 of this chapter.

3.2 Lipids Influence the Biogenesis of Aβ

3.2.1 Generation of Aβ from Its Precursor Protein, APP

Aβ is generated at the plasma membrane upon sequential cleavage of APP by β- and γ-secretases. First, the extracellular part of APP is cleaved at the N-terminus by β-secretase to produce the transmembrane β-carboxy-terminal fragment, termed C99, comprised of C-terminal 99 residues 672–770 of APP (Fig. 3.3). C99 is further cleaved at the C-terminus by γ-secretase at variable loci inside the membrane, giving rise to various Aβ isoforms containing 39–43 residues (Takami et al. 2009; Kukar et al. 2011). Alternatively, cleavage of C99 by α-secretase before residue 17 of Aβ precludes Aβ production (Fig. 3.3). This non-amyloidogenic cleavage is predominant in vivo and has been proposed to serve as a protection mechanism (Haass and Selkoe 2007).

The fact that APP and its processing enzymes are integral membrane proteins suggests that the lipid environment is important for Aβ production. Moreover, APP and its processing secretases have been found in lipid rafts (Lee et al. 1998; Golde and Eckman 2001; Riddell et al. 2001; Hooper et al. 2002). Lipid rafts, which form membrane microdomains enriched in cholesterol and sphingolipids such as gangliosides, have been implicated in signal transduction, protein trafficking, and proteolytic processing (Simons and Ikonen 1997; Brown and London 2000). Specifically, lipid rafts are believed to provide platforms for the amyloidogenic processing of APP. Importantly, APP processing that occurs outside lipid rafts predominantly follows the non-amyloidogenic α-secretase pathway (Hartmann et al. 2007; Vetrivel and Thinakaran 2010) which is stimulated by cholesteryl esters (Puglielli et al. 2001). On the other hand, in vitro and in vivo studies show that cellular cholesterol modulates the amyloidogenic processing by γ-secretase and production of Aβ (Simons et al. 1998; Frears et al. 1999; Refolo et al. 2000; Fassbender et al. 2001; Chauhan 2003; Zha et al. 2004). Together, these and other findings support the role of cholesterol in Aβ production.

The proteolytic processing of Aβ has been tested as a possible target to develop drugs against AD. Clinical trials have been performed with compounds that block γ-secretase activity (Eriksen et al. 2003; Doody et al. 2013). Unfortunately, these trials were discontinued as the drugs failed to improve cognitive status and showed adverse side effects (Doody et al. 2013, http://investor.myriad.com/releasedetail.cfm?releaseid=325471). These and other attempts to develop immunotherapy-based approaches are aimed at targeting Aβ generation, blocking its aggregation or promoting plaque clearance (Kukar et al. 2008; Lannfelt et al. 2014). Alternative strategies to develop treatments for AD have been focused on preventing the formation of Aβ plaques, the aggregation of Tau, and also trying to limit brain dysfunction (https://www.alz.washington.edu/NONMEMBER/SPR12/Ryan.pdf). Attempts to treat AD by targeting cholesterol levels are described in the next section.

numerous in vitro and in vivo studies (Kuo et al. 1998; Kivipelto et al. 2001; also see Chap. 10 by Leonova and Galzitskaya in this volume). Animal model studies reported that a cholesterol-rich diet promotes accumulation of intracellular Aβ and increases its cerebral load (Refolo et al. 2000; Shie et al. 2002). Cholesterol levels in the human brain were reported to increase in the early stages of AD (Wood et al. 2002; Cutler et al. 2004). Some (Kivipelto et al. 2001) but not all clinical studies (Reitz et al. 2004) showed a positive correlation between high levels of serum cholesterol and increased risk of late-onset AD in humans. Although the connection between elevated serum cholesterol and late-onset AD is the subject of debate (Proitsi et al. 2014), it raises a possibility that cholesterol-lowering drugs may help to prevent or alleviate AD.

Cholesterol-lowering drugs, such as statins, were reported to reduce intra- and extracellular levels of Aβ(1–40) and Aβ(1–42) in cultured neurons (Simons et al. 1998) as well as in the cerebrospinal fluid and the brain of guinea pigs (Fassbender et al. 2001). These and other cell and animal model studies led to a hypothesis that statins may benefit AD patients. In fact, early non-randomised population-based studies reported that statin treatment decreased the risk of AD (Wolozin et al. 2000; Rockwood et al. 2002) or improved cognitive function in AD patients (Simons et al. 2002; Sparks et al. 2005). However, later randomised controlled trials failed to show any clear beneficial effects of statins in the prevention or treatment of AD (McGuinness and Passmore 2010; Richardson et al. 2013). The authors concluded that "larger and better-designed studies are needed to draw unequivocal conclusions about the effect of statins on cognition" in AD.

3.2.2 Role of Cholesterol in AD Pathogenesis

Multiple lines of evidence suggest strongly that cholesterol plays an essential role in AD pathogenesis. Elevated plasma levels of cholesterol are an established risk factor for AD, as suggested by

3.2.3 Role of Cholesterol in APP Processing

The exact mechanism of cholesterol action in AD is far from clear and may involve multiple pathways (see Chap. 10 by Leonova and Galzitskaya in this volume). Structural studies have provided important information on the conformation of APP C-terminal C99 domain and demonstrated its interaction with cholesterol (Beel et al. 2008; Barrett et al. 2012). One mechanistic explanation for the effect of cholesterol on Aβ generation was proposed by Sanders and colleagues (Barrett et al. 2012). The NMR structure of the C99 domain of APP was determined in lipid micelles containing lyso-myristoyl phosphatidylglycerol. These structural studies, coupled with cholesterol titration studies, revealed a novel cholesterol binding site in APP that was proposed to modulate its amyloidogenic processing (Barrett et al. 2012).

C99 contains a single-span transmembrane α-helix flanked by two membrane surface-associated N- and C-helices. The transmembrane portion of C99 contains two tandem GxxxG motifs. Such motifs are often involved in dimerization of transmembrane helices by forming a glycine zipper (Kim et al. 2005). The NMR studies by Sanders team revealed that the GxxxGxxxG motif in APP has an additional role as a cholesterol binding site in C99. Cholesterol binding at this site is probably driven by hydrogen bonds involving N698 and E693, and by van der Waals interactions between the flat helical surface formed by the adjacent glycines (G700, G704, G708) and the cholesterol molecule (Barrett et al. 2012; also see Chap. 1 by Hong in this volume). The cholesterol binding site is located in the Aβ domain between the cleavage sites of α- and γ-secretase (Fig. 3.3). This location suggests that cholesterol binding can interfere with the APP processing. This idea was previously proposed in cell culture studies revealing that cholesterol interferes with the interactions between APP and its processing proteases (Bodovitz and Klein 1996). Moreover, direct binding of cholesterol can promote APP partitioning into lipid rafts and thereby co-localize APP with β- and γ-secretases, which is expected to promote the generation of Aβ (Barrett et al. 2012).

The fact that APP cleavage by β- and γ-secretases takes place in a healthy state suggests a normal physiological function for Aβ peptides (Haass et al. 1992; Seubert et al. 1992). The normal functions of Aβ and their underlying mechanisms are not entirely clear. Although some early studies suggested that Aβ deposition in the brain at normal physiological concentrations could be neuroprotective and act as a hemorrhagic sealant (reviewed in (Atwood et al. 2003)), other studies attribute the beneficial effects of Aβ to its soluble form (see Chap. 10 for detail). In vitro studies have revealed several important roles for Aβ, including antioxidant activity that is due to its ability to capture redox metals (Baruch-Suchodolsky and Fischer 2009; Kepp 2012), anti-microbial activity (Soscia et al. 2010), and action as a transcription factor (Bailey et al. 2011). Furthermore, in vitro and cell culture studies have implicated Aβ in control of cholesterol transport and homeostasis (Yao and Papadopoulos 2002; Igbavboa et al. 2009; Barrett et al. 2012), which is important for synaptic function.

3.3 Clearance of Aβ Is Influenced by Apolipoprotein E

3.3.1 Aβ Clearance Pathways

According to the "amyloid cascade hypothesis", Aβ accumulation in the brain and AD pathology result from an imbalance between the generation of Aβ and its degradation or clearance (Hardy and Selkoe 2002). While the rare genetic forms of AD involve increased generation of Aβ upon mutations in APP or presenilin proteins, the common sporadic form of AD has been linked mainly to defective clearance of Aβ (reviewed in Wildsmith et al. 2013).

Strong in vivo evidence supports this idea. For example, mouse model studies using microdialysis showed that Aβ deposition and plaque formation are promoted by an age-dependent decrease in Aβ clearance from the brain and the consequent increase in Aβ levels in the interstitial fluid (Yan et al. 2009). Measurements of the biosynthesis and clearance of Aβ in the human brain by isotope labelling revealed that even a small imbalance in Aβ homeostasis augments plaque formation (Bateman et al. 2006). These and other studies indicate that defective clearance of Aβ is a major triggering event in late-onset AD (Wildsmith et al. 2013).

Clearance of Aβ occurs either through enzymatic degradation in the brain cells or through the efflux across the blood-brain barrier (described below). These processes are mediated by two lipoprotein receptors, low-density lipoprotein receptor-related protein (LRP1) and very low-density lipoprotein receptor (VLDLR), and are importantly modulated by apoE that binds these receptors. The critical role of apoE in AD emerged when the ε4 allele of the APOE gene was found to provide the major genetic risk factor in both early- and late-onset AD in humans (Corder et al. 1993; Saunders et al. 1993; Strittmatter et al. 1993). This finding was supported by the observation that apoE4-positive AD patients have increased levels of Aβ and increased density of amyloid plaques in their brains (Schmechel et al. 1993; William Rebeck et al. 1993; Tiraboschi et al. 2004). Later studies suggested that apoE has complex effects on Aβ homeostasis. These effects were proposed to involve direct or indirect interactions of apoE with Aβ, which may depend on the degree of apoE lipidation and Aβ aggregation (Tokuda et al. 2000; Bell et al. 2007). An alternative mechanism that was proposed to explain the effects of apoE on Aβ accumulation in vivo is the competition between apoE and Aβ for binding to LRP1 that mediates Aβ clearance from the central nervous system (Holtzman 2001; Verghese et al. 2011). In sum, although the exact mechanisms underlying the effects of apoE in AD are subjects of debate, the current consensus is that the primary role of apoE in AD pathology is through modulating the Aβ clearance from the brain (Castellano et al. 2011).

3.3.2 ApoE Structure, Function and Isoform-Specific Effects

ApoE (299 amino acids) is an important lipid transport protein in plasma and the major lipid transporter in the central nervous system where it circulates mainly on high-density lipoproteins (HDL). ApoE in the brain is produced by astrocytes and microglia as a free protein that rapidly accrues lipids to generate nascent and, ultimately, mature HDL, which are the sole lipoproteins in the central nervous system (Pitas et al. 1987). ApoE on HDL forms an important functional ligand that binds to the lipoprotein receptors LRP1 and VLDLR in the brain. ApoE-containing HDL is the major carrier of cholesterol in the brain, which supports synaptic formation and plasticity, as well as the maintenance of myelin and neuronal membranes (Mahley 1988; Saher et al. 2005). In addition to its lipid transport function, apoE is also important in synaptic formation and neuronal signalling, which is essential for learning and memory (Grootendorst et al. 2005; Korwek et al. 2009).

Structurally, apoE is a two-domain protein comprised of a 22 kDa N-terminal receptor-binding domain and a hydrophobic 10 kDa C-terminal domain that forms the primary lipid binding site (Weisgraber 1994; Zhong and Weisgraber 2009). In solution, the N-domain forms a four-helix bundle, while the C-domain adopts a dynamic conformation ((Chen et al. 2011) and references therein). Lipid binding induces a stable α-helical conformation in the C-domain; this is followed by the N-domain opening and conformational reorganization of apoE that forms the structural scaffold on HDL surface (Phillips 2013). The receptor binding site in the N-domain is obscured in lipid-free apoE but acquires receptor-competent conformation on HDL (Chen et al. 2011).

In addition to lipid binding, the C-domain residues 209–255 can also bind Aβ in vitro (Tamamizu-Kato et al. 2008; Hauser and Ryan 2013). Although Aβ can bind both lipid-free apoE and HDL in vitro, the presence of Aβ reportedly reduces the lipoprotein-binding ability of apoE (Tamamizu-Kato et al. 2008), suggesting that Aβ potentially interferes with cholesterol transport via HDL.

There are three major isoforms of human apoE differing in Cys/Arg substitutions in positions 112 and 158 of the N-domain: apoE2 (Cys112, Cys158), apoE3 (Cys112, Arg158) and apoE4 (Arg158, Arg158) (Zannis et al. 1993; Cedazo-Mínguez and Cowburn 2001; Phillips

2014). ApoE3 is the most common human iso-form, while apoE4 and apoE2 are found in a small percentage of the population. Compared to apoE3, the presence of one and, particularly, two apoE4 alleles increases the risk of AD in humans by more than threefold and tenfold, respectively (Roses 1996), making apoE4 the major genetic risk factor for AD (Corder et al. 1993; Saunders et al. 1993; Strittmatter et al. 1993). In contrast, apoE2 is more protective against AD as com-pared to apoE3 and apoE4 (Rebeck et al. 2002; Conejero-Goldberg et al. 2014).

Although the molecular basis for the isoform-specific effects of apoE is subject of debate, it is generally attributed to altered inter-domain inter-actions (Mahley et al. 2006). The Arg/Cys substi-tution sites in the three apoE isoforms are located in the globular N-domain, yet the effects of these substitutions can propagate to the C-domain which, in turn, influences the interactions of the N- and C-domains with each other, with lipids, and with other ligands such as the lipoprotein receptors and Aβ (Zhong and Weisgraber 2009). These effects can lead to subtle changes in the structural stability of the apoE isoforms, as well as in their biological functions such as lipid bind-ing, receptor recognition, or Aβ binding (Verghese et al. 2011; Suri et al. 2013). The rank order of the structural stability of apoE isoforms in solu-tion, E4<E3<E2, correlates inversely with Aβ deposition in AD. One possible explanation is that enhanced degradation of the less stable apoE4 isoform reduces the total levels of apoE in the brain and thereby decelerates Aβ clearance. Similarly, apoE affinity for HDL goes in order E4<E3<E2 (Acharya et al. 2002). If apoE affinity for HDL decreases, a larger fraction of the protein will dissociate from HDL in the labile lipid-free state, contributing to a general decrease in apoE4 levels relative to apoE2 or apoE3. This reasoning is in line with increased Aβ deposition in apoE4 carriers (Bales et al. 2009). Similarly, interactions of Aβ with apoE-containing HDL were found to be isoform-specific; the rank order of the binding preference, E4<<E3<E2 (LaDu et al. 1994; Aleshkov et al. 1997; Tokuda et al. 2000), is inverse to Aβ amyloid deposition in AD (Holtzman et al. 2000; Bales et al. 2009). This inverse correlation

is consistent with the idea that apoE mediates Aβ clearance from the brain, perhaps via the direct interactions with Aβ.

3.3.3 Role of ApoE in the Clearance of Aβ

The two predominant modes through which apoE is thought to influence Aβ removal from the brain involve Aβ degradation in astrocytes and microglia and apoE-mediated efflux across the blood-brain barrier. In the first pathway, microglia, which are the macrophages of the brain, take up soluble and fibrillar Aβ via an endocytosis pathway that is facilitated by apoE. In vitro cell culture studies have shown that apoE can promote internaliza-tion and subsequent degradation of Aβ by the brain cells such as microglia and astrocytes in a isoform-specific manner (Cole and Ard 2000; Yamauchi et al. 2002; Koistinaho et al. 2004). In the second pathway, which is the major clearance mechanism of Aβ, transcytosis of soluble Aβ from brain to blood is mediated by LRP1 recep-tor (Shibata et al. 2000; Zlokovic 2008). Aβ binds directly to LRP1 on the cell surface (Deane et al. 2004) to initiate brain-to-blood transport. In vivo studies in mice have shown that Aβ transport is greatly decelerated or impaired upon complex formation between Aβ and apoE (Bell et al. 2007; Deane et al. 2008). This effect can be explained by the shift from LRP1, which transports free Aβ, to VLDLR, which transports Aβ bound to apoE-containing HDL. VLDLR has much slower endocytosis rates than LRP1, resulting in poor clearance of Aβ-apoE complexes from the brain (Li et al. 2001).

Another variable to consider is the aggregation level of Aβ, which can influence Aβ degradation, clearance and interactions with apoE. In one recent study, astrocytes and microglia were exposed to oligomeric and fibrillar Aβ. In the presence of apolipoproteins such as apoE, the uptake of Aβ oligomers was reduced in astrocytes but not in microglia; an opposite effect was observed for Aβ fibrils (Mulder et al. 2014). This suggests that different brain cells have distinct roles in Aβ clearance and are differently affected by the apolipoproteins depending on the degree of aggregation of Aβ. Moreover, apoE may influence Aβ aggregation. ApoE co-deposits with Aβ in amyloid plaques and was proposed to promote amyloid formation by Aβ in AD (Wisniewski et al. 1994; Hatters et al. 2006), although the mechanism underlying this effect is unclear (see Chap. 8 by Das and Gursky in this volume). Thus, the mutual interplay between the interactions of Aβ with apoE and the Aβ aggregation and clearance adds an additional layer of complexity to this already complex issue.

Part of the difficulty in dissecting the precise role of apoE in Aβ homeostasis lies in the fact that, although apoE and Aβ interact with each other and with a wide range of other ligands in vitro, the relevance of these interactions to in vivo conditions is often unclear. Interestingly, recent studies by Holtzman and colleagues revealed only minimal direct association between apoE and soluble Aβ in vivo in human cerebrospinal fluid (Verghese et al. 2013). To explain the isoform-specific effects of apoE on Aβ clearance in the absence of direct binding, the authors proposed an alternative mechanism whereby apoE and Aβ compete for binding to LRP1 receptor that mediates Aβ efflux from the brain.

3.4 Effects of Lipids on Aβ Aggregation and Amyloid Formation

3.4.1 Interactions of Aβ with Lipids Surfaces: Electrostatic Effects

The self-assembly of Aβ from soluble monomers into oligomers and, ultimately, amyloid fibrils can be influenced not only by the concentrations of Aβ peptides but also by their biochemical modifications such as mutations, truncations and pyroglutamate modifications (Morgado and Faendrich 2011; Wittnam et al. 2012), as well as by solvent composition, ionic conditions (Klement et al. 2007; Garvey et al. 2011) and a wide range of other factors including lipids (Terzi et al. 1995, 1997; Butterfield and Lashuel 2010). The combination of the hydrophilic N-terminal

segment and a highly hydrophobic C-terminal tail renders amphipathic properties to Aβ and confers its limited aqueous solubility, high aggregation propensity, and ability to interact with various lipids and their assemblies.

Numerous biophysical studies have demonstrated that lipids and lipid-mimicking detergents can importantly influence the secondary structure of Aβ and its aggregation and fibrillation in vitro. Thus, studies using NMR, circular dichroism, Fourier transform infrared spectroscopy, molecular dynamics simulations and fluorescence spectroscopy have demonstrated Aβ interactions with vesicles containing anionic phospholipids such as phosphatidylglycerol (Terzi et al. 1997; McLaurin et al. 1998; Nagarajan et al. 2008), as well as with other lipid assemblies such as bilayers, micelles or ganglioside clusters (Matsuzaki and Horikiri 1999; Mandal and Pettegrew 2004; Xu et al. 2005; Matsuzaki 2007; Miyashita et al. 2009; Matsuzaki et al. 2010), and showed that these interactions can alter the secondary and/or the quaternary structure of the peptide.

Aβ binding to lipid can involve electrostatic interactions between the polar and charged residues of the peptide and the phospholipid head groups, van-der-Waals interactions between the hydrophobic C-terminal tail of the peptide and the apolar lipid moieties, or combinations thereof. To address the nature of these interactions, model systems using zwitterionic or anionic phospholipid liposomes have been used to determine the effects of lipid surface charge on the secondary structure and aggregation properties of Aβ peptides. Anionic liposomes have been proposed to provide templates for the in vitro analysis of Aβ aggregation upon interaction with anionic lipids that can become exposed to the outer membrane leaflet in injured cells.

Anionic phospholipids such as phosphatidylglycerol or phosphatidylinositol have been shown to ultimately shift the conformation in Aβ peptides from the soluble largely unstructured monomer towards β-sheet-rich aggregates (McLaurin et al. 1998; Chi et al. 2008). In contrast, zwitterionic phospholipids, such as phosphatidylcholine (PC) or sphingomyelin, were reported in some studies to show very weak or no interaction with

Aβ peptides and induce little if any ordered secondary structure (Terzi et al. 1997; Chauhan et al. 2000; Matsuzaki 2007; Chi et al. 2008; Nagarajan et al. 2008). In contrast, other studies showed significant effects of PCs on the secondary structure and aggregation of Aβ; this includes our own work described in Sect. 3.4.3 below (Dahse et al. 2010). Such a diversity in the results reported in different studies illustrates extreme sensitivity of the Aβ solution conformation not only to the presence of specific lipids but also to many other factors, such as the solvent composition, ionic strength, pH, and lipid to peptide ratio, to name a few (Matsuzaki 2007; Garvey et al. 2011).

Fluorescence studies by Kremer et al. also highlight the contribution of hydrophobic interactions to Aβ-membrane interactions. They describe the increased hydrophobicity of Aβ as it forms aggregates which expose surface hydrophobic patches that interact with the core of membrane bilayers decreasing their fluidity (Kremer et al. 2000, 2001).

In sum, the consensus in the field is that Aβ binding to the negatively charged phospholipid surfaces, which could become exposed upon cell injury, consistently favors amyloid formation. Some studies proposed that Aβ adsorption to such lipid surfaces is driven mainly by electrostatic interactions at the lipid head group level where the β-sheet starts to form (Maltseva et al. 2005; Chi et al. 2008). Other studies suggested that Aβ aggregation and amyloid formation on the negatively charged lipid surfaces can be mediated via the transient α-helical conformation (Terzi et al. 1997; Fezoui et al. 2000) described below.

3.4.2 Lipid-Induced Helical Structure in Aβ: Relevance to Amyloid Formation

Binding of soluble Aβ to anionic lipid surfaces can induce transient α-helical structure in the peptide (Terzi et al. 1997; Davis and Berkowitz 2009a, b). This helical structure was proposed to be induced by the alignment of the positively charged groups in Aβ and the negatively charged

membrane template (Terzi et al. 1997). Similarly, micelles of the anionic detergent sodium dodecyl sulfate, but not individual detergent molecules, induced an α-helical conformation in Aβ (Wahlström et al. 2008). An α-helical structural intermediate can be induced upon lipid surface binding not only in Aβ but also in many other intrinsically disordered peptides (described by Uversky in Chap. 2 of this volume).

Importantly, depending on its stability, the α-helical conformation in proteins and peptides such as Aβ can either augment or mitigate amyloid formation. Generally, if the α-helical structure is only marginally stable, it can facilitate intermolecular interactions among the peptides and thereby promote their aggregation and conversion into β-structured aggregates (Fezoui et al. 2000; Morgado and Faendrich 2011). However, if the α-helical structure is relatively more stable, it can block β-aggregation and amyloid formation by the peptide. For example, PEGylated phospholipid nanomicelles were found to stabilize the α-helical conformation in Aβ, block its aggregation into β-sheet rich amyloid, and thereby block the peptide neurotoxicity in vitro (Pai et al. 2006). This finding suggested a potential therapeutic strategy to design lipid-based blockers of Aβ aggregation (Pai et al. 2006).

3.4.3 Interactions of Aβ with Lipid Molecules and Surfaces: A Test Case of DHPC

Interactions of amyloidogenic proteins and peptides such as Aβ with lipid membranes and with individual lipid molecules are distinctly different. Elucidating these differences may help establish the molecular basis for the complex effects of lipids in amyloidogenesis. To this end, short-chain phospholipids such as 1,2-dihexanoyl-sn-glycero-3-phosphocholine (DHPC) provide useful model systems. In aqueous solution below their critical micelle concentration (CMC), these lipids are found mainly as soluble monomers, whereas above CMC they are predominantly present as micelles with surface properties mimicking those of the lipid membranes.

In our study, the effects of DHPC on the structure and self-association properties of Aβ(1–40) were analyzed as a function of DHPC concentration by using a wide range of biophysical techniques (Dahse et al. 2010). The results showed that DHPC promoted Aβ aggregation at concentrations well below and above its CMC. The implication is that Aβ interacts not only with lipid surfaces but also with individual lipid molecules, and these interactions promote amyloid formation. The effect is illustrated in Fig. 3.4 showing the kinetics of Aβ aggregation in the absence (A) or the presence (B) of 1.9 mM DHPC, which is well below the CMC (~13 mM). The time course of Aβ amyloid formation was monitored by thioflavin T fluorescence. In the presence of sub-micellar DHPC, the lag phase corresponding to amyloid nucleation was shortened at least by a factor of 2, and the growth phase was also accelerated by at least a factor of 2. Consequently, interactions of Aβ with individual molecules of DHPC significantly accelerated both amyloid nucleation and the fiber growth.

Circular dichroism (CD) studies indicated that the presence of sub-micellar lipids had no effect on the secondary structure in the soluble peptide, which remained largely disordered prior to fibrillation (Fig. 3.5a, blue line). However, increasing DHPC concentration induced characteristic changes in far-UV CD spectra of Aβ (Fig. 3.5a) indicating random coil to β-sheet conversion with a midpoint *circa* 11 mM of DHPC, which approaches its CMC (Fig. 3.5b). This suggests that peptide binding to DHPC micelles induces β-sheet formation by Aβ. Direct peptide interactions with DHPC micelles were confirmed in surface tension and proton NMR studies (Dahse et al. 2010). Fluorescence studies using diagnostic dye 1-anilino-8-naphthalene sulfonate suggested that the Aβ interactions with individual lipid molecules and with micelles have a hydrophobic component. In sum, monomeric soluble Aβ can form hydrophobic interactions both with individual DHPC molecules and with lipid micelles. These interactions differentially affect the secondary structure and aggregation properties of Aβ and accelerate both nucleation and growth of Aβ fibrils. These results imply that subtle peptide-lipid interactions involving zwitterionic phospholipids can greatly influence the overall aggregation properties of Aβ, as well as alter specific steps in the complex pathway of its fibrillogenesis (Dahse et al. 2010).

3.4.4 Interactions of Aβ Fibrils with Lipid Surfaces

Although this chapter is focused mainly on the interactions between soluble Aβ and lipids, insoluble Aβ fibrils can also interact with lipid surfaces, and these interactions may occur in vivo and potentially contribute to AD pathogenesis. For example, incubation of mature Aβ amyloid fibrils with a liposome suspension of zwitterionic dioleoyl phosphatidylcholine reverted peptide aggregation causing the fibrils to disassemble into highly neurotoxic protofibrilar species (Martins et al. 2008). Furthermore, atomistic molecular dynamic simulations suggested that electrostatic interactions between charged peptide residues and phospholipid head groups importantly modulate the insertion of the C-terminal tail of the Aβ fibrils or oligomers into lipid bilayers, leading to formation of amyloid channels with polymorphic structures containing β-strands as well as α-helices in direct contact with the lipid (Tofoleanu and Buchete 2012). Other examples of fibril-lipid interactions, such as fibril unwinding on the lipid membrane, are described by Gorbenko and colleagues in Chap. 6 of this volume.

3.5 Effects of Aβ on Membrane Integrity and Cell Toxicity

3.5.1 Aβ Interactions with Lipid Membranes: The Role of Gangliosides

Apart from their key role in the amyloidogenic processing of APP, lipid rafts can also provide platforms for Aβ aggregation in AD. Gangliosides, which preferentially cluster in lipid rafts (Fig. 3.6), are particularly prominent in this process.

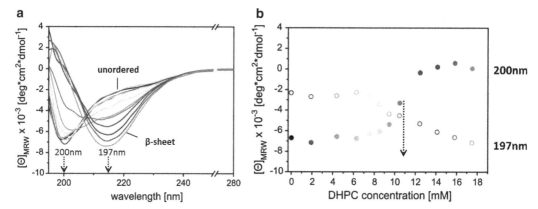

Fig. 3.5 Unordered to β-sheet transition in Aβ induced by binding to DHPC micelles. (**a**) Far-UV circular dichroism spectra of Aβ(1–40) (50 μM peptide in 50 mM Na phosphate buffer at pH 7.4) in the presence of 0–18 mM DHPC. Lipid concentrations are represented by the color scheme in panel (**b**). In the absence of DHPC (*blue*) the peptide is substantially unfolded, with a characteristic CD minimum near 200 nm. Increasing DHPC concentrations to about its CMC (*red* and *orange*) induces β-sheet formation in Aβ, as evident from the characteristic CD minimum near 197 nm. The absence of an isochromatic point where the spectra intersect indicates the complex non-two-state character of the random coil to β-sheet transition. (**b**) Mean residue ellipticity [θ] at 200 nm (*filled circles*) and 217 nm (*open circles*) as a function of lipid concentration. A progressive loss of the negative CD signal at 200 nm and simultaneous increase in the negative CD at 217 upon increasing DHPC concentration indicates random coil to β-sheet transition. The transition is centered at approximately 11 mM (indicated by an arrow), just below the CMC (~13 mM), suggesting a link between the β-sheet formation and peptide binding to the micelle (Adopted with permission from Dahse et al. 2010)

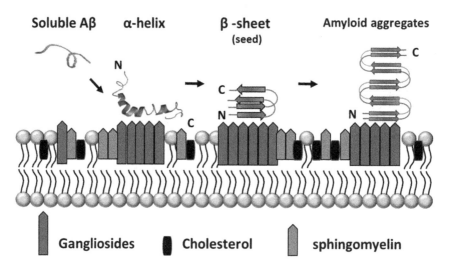

Fig. 3.6 Aβ aggregation through interaction with neuronal membrane gangliosides. According to the model proposed by (Matsuzaki 2011), soluble unstructured Aβ binds to ganglioside clusters in membrane rafts. At low peptide to ganglioside ratios, bound Aβ adopts an α-helix-rich structure. Upon increasing the concentration of the ganglioside-bound peptide, Aβ undergoes a conformational transition to β-sheet structure that acts as a seed to promote further aggregation and formation of amyloid fibrils

Gangliosides are membrane glycolipids that are important for neuroplasticity and are thought to be the primary modulators of Aβ aggregation and cytotoxicity (Choo-Smith et al. 1997; Choo-Smith and Surewicz 1997; Hayashi et al. 2004; Kakio et al. 2002; Hong et al. 2014). Gangliosides are located in the outer leaflet of the plasma membrane as well as in the luminal leaflet of several cellular organelles. The most abundant species in the brain are monosialotetrahexosyl ganglioside (GM1), along with GD1a, GD1b and GT1b gangliosides (Posse de Chaves and Sipione 2010).

Analysis of clinical samples from the brains of the AD patients identified Aβ species bound to GM1 (Yanagisawa et al. 1995) and found that this binding strongly potentiated Aβ fibril assembly (Choo-Smith et al. 1997). In vitro studies showed that Aβ binding to GM1-containing PC vesicles induces α-helical peptide structure that undergoes an α-helix to β-sheet transition upon increasing the peptide concentration (McLaurin and Chakrabartty 1996; Choo-Smith and Surewicz 1997; Matsuzaki and Horikiri 1999). The resulting GM1-Aβ complex was proposed to act as an endogenous seed for the formation of toxic amyloids that perturb cell membranes (Hayashi et al. 2004; Okada et al. 2008; Matsuzaki et al. 2010) (Fig. 3.6). The importance of Aβ binding to GM1 in cell membranes is underscored by the finding that soluble Aβ oligomers are rapidly captured from the brain interstitial fluid onto synaptic membranes where they interact with GM1 and perhaps other lipids, leading to progressive perturbations in the membrane structure and function (Hong et al. 2014).

Under physiological conditions, soluble Aβ displays high affinity for membrane rafts that contain gangliosides, such as GM1 (Straub and Thirumalai 2014; Bucciantini et al. 2014). In contrast, Aβ normally does not show a strong interaction with non-rafted membranes, even if in the presence of GM1 (Kakio et al. 2001, 2002). Apparently, the specificity of the Aβ-GM1 complex formation in lipid rafts is determined by the local concentrations of cholesterol and GM1 (Kakio et al. 2001, 2002). Interestingly, the specificity for GM1 binding can be replaced by other

gangliosides in different Aβ mutants; for example, amyloid formation by Dutch (E22Q) and Italian (E22K) Aβ mutants can be potentiated by GM3 rather than GM1 (Yamamoto et al. 2005). These observations suggest that Aβ binding to gangliosides involves specific interactions.

NMR analysis of Aβ interactions with lyso-GM1 micelles revealed that the ganglioside regions that are perturbed upon Aβ binding are located at the sugar-lipid interface (Yagi-Utsumi et al. 2010). Aβ was implicated to bind to the inner part of the ganglioside cluster via the hydrophobic interactions involving the C-terminal tail of the peptide and its two α-helices that span residues 14–24 and 31–36 and form upon micelle binding (Utsumi et al. 2009; Yagi-Utsumi et al. 2010). This binding, which was proposed to mimic key aspects of Aβ-ganglioside interactions, suggests how the ganglioside clusters can serve as platforms for coupled folding and binding of Aβ peptide, ultimately leading to its aggregation on the cell membrane.

3.5.2 Aβ Aggregation on Cell Membranes: Mutual Effects

A close link between Aβ toxicity and aggregation on cellular membranes is well established, and the mutually disruptive effects of such interactions have been reported in many studies (McLaurin and Chakrabartty 1996; McLaurin et al. 2000; Relini et al. 2009). While cell membranes can provide a surface template for Aβ misfolding and aggregation, Aβ aggregates can disrupt the structural integrity of the membrane and alter its permeability (McLaurin and Chakrabartty 1996; Murphy 2007; Relini et al. 2009). In fact, some ex vivo studies of the plasma membranes isolated from the brains of AD patients found Aβ(1–42) deposits on the cell surface as well as in the hydrophobic core region of the bilayer (Yamaguchi et al. 2000; Oshima et al. 2001). Other studies reported that aggregated Aβ localized to the polar head group region in synaptic plasma membranes, while soluble Aβ partitioned into the hydrophobic acyl chain region (Mason et al. 1996, 1999). Such partitioning inside the membrane is expected to interfere with its structural integrity.

Cell membranes can act as catalysts of fibril formation in which the local chemical environment can lower the free-energy barrier to peptide aggregation (Chi et al. 2008). As described above, binding of soluble Aβ to anionic membranes was proposed to be primarily driven by electrostatic interactions between the negatively charged lipid head groups and the positively charged side chains of the amphipathic Aβ peptide (Terzi et al. 1997; McLaurin et al. 2000; Kakio et al. 2002; Aisenbrey et al. 2008), followed by hydrophobic contacts at the acyl chain level (Aisenbrey et al. 2008; Butterfield and Lashuel 2010). Furthermore, peptide adsorption to the two-dimensional membrane surface may impose structural constraints bringing different residues together. These effects, together with the reduction in the local dielectric constant at the membrane surface, can promote hydrogen bonding and secondary structure formation in intrinsically disordered proteins, such as Aβ (see Chap. 2 by Uversky in this volume). Moreover, "molecular crowding", which results from the local increase in peptide concentrations upon adsorption to a 2D surface, as well as the local decrease in pH at the membrane surface, are also expected to promote intermolecular β-sheet formation and Aβ aggregation (Bokvist and Groebner 2007; Bystroem et al. 2008). Together, these effects can induce a specific misfolding pathway in Aβ, generating toxic aggregates (Aisenbrey et al. 2008; Bystroem et al. 2008).

The extent to which Aβ interacts with lipid membranes depends on its aggregation state (Murphy 2007). Oligomeric Aβ shows strong irreversible binding to liposomes, as opposed to monomeric Aβ which demonstrates a rapid reversible adsorption (Good and Murphy 1995; Kremer and Murphy 2003). This behaviour may arise from the presence of hydrophobic surfaces that promote Aβ-membrane interactions (Kremer et al. 2000). The lack of such accessible surfaces in Aβ monomers or fibrils helps explain their lower toxicity as compared to Aβ oligomers. The initial conformation of Aβ can also influence the kinetics of aggregation, as well as the

membrane disruption. In sum, Aβ adsorption to the membrane surface can promote peptide aggregation in a process that disrupts the membrane integrity and ultimately leads to fibril formation. In contrast, adsorption of pre-formed fibrils does not significantly perturb the structural integrity of the membrane (Yip and McLaurin 2001; Bokvist et al. 2004). The latter is consistent with relatively low fibril toxicity as compared to pre-fibrillar aggregates.

3.5.3 Aβ Toxicity and Perturbation of Cell Membranes

Aβ aggregation on the cell membrane is believed to exert neurotoxicity by altering membrane structure and permeability. This hypothesis is supported by early studies reporting that Aβ-induced membrane disruption leads to decreased acyl chain fluidity (McLaurin and Chakrabartty 1996; Mattson 1997; Yip and McLaurin 2001). However, many studies addressing the effects Aβ on cell membranes were performed in vitro and hence, may not reflect the physiological situation in vivo. The accumulation of Aβ on the membranes of living cells was firstly confirmed in studies showing that labeled Aβ accumulated in a concentration-dependent manner on the surface of PC12 rat cells, and induced cytotoxicity (Wakabayashi et al. 2005). Such membrane-binding ability, as well as the toxicity, was greatly increased for Aβ aggregated on the membrane rafts containing GM1, sphingomyelin and cholesterol (Okada et al. 2008). Later studies revealed that incubation of living cells (non-neuronal HEK293 and neuroblastoma SH-SY5Y transfected with human APP695) with exogenous soluble Aβ(1–40) decreased membrane fluidity (Peters et al. 2009). This effect was much more pronounced for Aβ oligomers as compared to monomers or fibrils, suggesting that pre-fibrillar oligomers can interfere with the acyl chain packing. Furthermore, production of endogenous Aβ was also reported to reduce membrane fluidity (Peters et al. 2009). Consistent with these findings, pre-fibrillar Aβ oligomers were found to affect the dielectric properties of the membranes, apparently by partitioning in their hydrophobic interior (Valincius et al. 2008).

Other experimental studies consistently showed that Aβ disturbs integrity, fluidity and permeability of cellular membranes, although it is not clear in which manner. Many studies using different Aβ peptides and tissue extracts from AD brains reported that Aβ induces decrease in membrane fluidity (Muller et al. 1998; Eckert et al. 2000, 2010; Yip and McLaurin 2001; Peters et al. 2009). However, other studies reported an increase in the annular and bulk lipid fluidity of synaptosomal plasma membranes (Avdulov et al. 1997; Mason et al. 1999; Chochina et al. 2001). In hippocampal membranes obtained from AD patients, changes in fluidity and membrane disruption correlated with cholesterol content (Eckert et al. 2000). Such differences among different studies could arise from variations in the experimental conditions in the cell lines used, in the physical properties of the membranes, and in the aggregation state of Aβ, among other factors (Eckert et al. 2010; Stefani 2010).

In particular, Aβ oligomers were reported to have strong affinity for synaptic membranes (Lacor et al. 2007) and thus, can induce synaptic dysfunction, which is proposed to be the primary cause of cognitive impairment in AD (Haass and Selkoe 2007; Palop and Mucke 2010). Although the likely cause of synaptic toxicity and cell death is disruption and altered permeability of cell membranes, the exact mechanisms remain unclear. Multiple lines of evidence suggest that cell membrane interaction with Aβ species can alter intracellular Ca^{2+} levels (Stefani and Dobson 2003; Dobson 2003; Cecchi et al. 2005), leading to depolarization and disruption of the membrane (Arispe et al. 1993a, b; Quist et al. 2005) and its increased conductance (Kayed et al. 2004; Demuro et al. 2005; Deshpande et al. 2006). These effects perturb ion homeostasis and neuronal signal transduction, which can lead to cell death (Lashuel et al. 2002; Soto 2003). For zwitterionic bilayers, changes in conductance and membrane integrity have been attributed to a non-specific bilayer perturbation by small Aβ oligomers without pore formation (de Planque et al. 2007). However, many other studies have

attributed a channel-like activity to Aβ (Pollard et al. 1993; Arispe et al. 1993a, b), leading to a hypothesis that Aβ-mediated membrane disruption and toxicity occur mainly through pore formation.

Membrane pore formation by Aβ is thought to be driven mainly by pre-fibrillar aggregates, such as oligomers or protofibrils, which is consistent with high cellular toxicity of these aggregates. Electron and atomic force microscopic observations of the annular pore-like shape in certain Aβ aggregates suggests that these aggregates can act as non-regulated membrane pores (Lashuel et al. 2002; Quist et al. 2005; Kayed et al. 2009; Lal et al. 2007). Consistent with this idea is the proposed molecular model of toxic Aβ oligomers containing a central hole (Stroud et al. 2012). Other studies, using microscopy, electrophysiology recording, biophysical and biochemical analysis and molecular dynamics simulations have demonstrated the formation of "amyloid channels" upon incorporation of Aβ oligomers or disordered Aβ into membranes (Lin et al. 2001; Quist et al. 2005; Jang et al. 2010, 2014; Prangkio et al. 2012). Such channel-like structures are permeable to Ca^{2+} and eventually disrupt Ca^{2+} trafficking and homeostasis (Lin et al. 2001; Kawahara et al. 2011). The resulting abnormal elevation of intracellular Ca^{2+} levels can lead to the activation of apoptotic pathways and neuritic degeneration (Kawahara et al. 2011). Other possible consequences of altered neuronal Ca^{2+} homeostasis include membrane disruption and changes in the dendritic spine number, morphology and synaptic plasticity, eventually leading to synaptic impairment (Kato-Negishi et al. 2003). Ultimately, cytosolic and mitochondrial Ca^{2+} overload can occur, contributing to the oxidative damage and apoptosis (Querfurth and LaFerla 2010).

Recent studies revealed that membrane cholesterol plays an essential role in Aβ-mediated pore formation. Cholesterol-binding peptide fragment, Aβ(22–35), was reported to induce large Ca^{2+} influx to neuroblastoma cells (Di Scala et al. 2014b), an effect that was completely abrogated upon the depletion of membrane cholesterol (Fantini et al. 2014). The authors used molecular dynamic simulations to propose a mechanism for the pore formation, in which a hydrogen-bonded network was formed by Asn27 and Lys28 of Aβ(22–35) (Di Scala et al. 2014b). According to this model, the interactions between the peptide molecules are favored by the tilted conformation Aβ acquires upon cholesterol binding. The resulting peptide-cholesterol complexes form an annular channel in which Glu22 and Asp23 side chains form the central pore, while cholesterol faces the membrane interior. The electronegative mouth of the pore attracts calcium ions, which is consistent with Ca^{2+} permeability.

Similarly, cholesterol was proposed to influence membrane insertion and pore formation by full-length Aβ peptides. Biophysical and molecular dynamics studies reported that cholesterol mediates peptide insertion into phospholipid membranes (Yu and Zheng 2012) and channel formation (Qiu et al. 2009). *In silico* models of oligomeric Aβ(1–40) and Aβ(1–42) have suggested the formation of an acidic central pore involving Glu3, Asp7 and Glu11 (Diaz et al. 2006). A tetrameric channel was proposed as a model for oligomeric Aβ(1–42) bound to cholesterol in the membrane (Di Scala et al. 2014a).

While accumulative evidence suggests that cholesterol is a critical modulator of membrane pore formation by Aβ oligomers, such pore-containing oligomers can also form in the absence of cholesterol. Several structural models of Aβ oligomers have been proposed that are independent of cholesterol. For example, on the basis of the X-ray diffraction and electron microscopic studies, a molecular model for toxic Aβ oligomers has been proposed that contains a central hole (Stroud et al. 2012). Other *in silico* studies have also proposed channel-like structures for truncated Aβ peptides which form a β-strand – turn – β-strand motif that is suited for pore formation (Jang et al. 2010). Formation of transmembrane pores via the β-barrel-like Aβ hexamers that can perfuse the membrane has also been proposed (Shafrir et al. 2010). The molecular mechanism of membrane insertion of such β-barrels, the role of cholesterol in their formation, and the relevance of these processes to neurotoxicity in vivo remains to be established.

Apart from the lipid surface binding and pore formation, other deleterious effects of Aβ at the cell surface may include peptide binding to various membrane proteins, which interferes with receptor recognition and signaling as well as other important metabolic processes (reviewed in Verdier et al. 2004; Bamberger et al. 2003). Other potential damaging effects include oxidative stress, inflammation, vascular damage, disruption of protein function and apoptosis (Verdier et al. 2004; Querfurth and LaFerla 2010). Further studies will be needed to establish the relative roles of these effects in neuronal damage and AD pathogenesis in vivo.

3.6   Concluding Remarks

Alzheimer's disease is a complex pathology with multiple contributing factors, among which lipids and lipid homeostasis play a central role. It is well established that lipids, particularly at the level of the cell membrane, and their interactions with Aβ and APP, are among the crucial factors in Aβ biogenesis, aggregation and clearance, which ultimately determines the AD pathology. Efforts to elucidate the underlying mechanisms are underway and have provided a plethora of scenarios for the lipid-Aβ interactions which have been observed in vitro and in silico. However, relevance of such interactions to in vivo conditions is often unclear, and many important questions remain unanswered. One major thrust of the on-going studies is to clarify the mechanisms of amyloidogenic versus non-amyloidogenic processing of APP and the modulatory role of cholesterol and other membrane lipids in this processing. Another is a better understanding of the role of specific membrane lipids, such as anionic lipids and gangliosides, in the pathologic aggregation of Aβ on the cell membrane, as well as the impact of such aggregation on the physical and functional properties of the membrane. Yet another crucial aspect is to better characterize the clearance mechanisms of Aβ and the role of apolipoproteins, lipoproteins and lipoprotein receptors in these complex processes. A continuous effort in elucidating these and other aspects of

AD pathology is necessary to identify potential targets for disease-modifying interventions. In addition, specific lipid signatures or changes in the lipid environment as disease-predisposing factors can possibly be revealed, providing new potential methods for improved diagnostics and ultimate prevention of AD.

Acknowledgements The authors are very grateful to Prof. Gursky for her invaluable advice, writing assistance, critical review and proof reading of the chapter. This work was supported, in part, by Marie Curie International Outgoing Fellowship (IOF) 628077 "Structural and Biochemical Basis of Protein Amyloid Evolution" from the European Union to I. M.; M. G. acknowledges support from the BMBF Forschungsinitiative "BioEnergie 2021 - Forschung für die Nutzung von Biomasse" (0315487A-C) and the Cluster of Excellence "Tailor-made Fuels from Biomass" (EXC 236).

References

Acharya P, Segall ML, Zaiou M, Morrow J, Weisgraber KH, Phillips MC, Lund-Katz S, Snow J (2002) Comparison of the stabilities and unfolding pathways of human apolipoprotein E isoforms by differential scanning calorimetry and circular dichroism. Biochim Biophys Acta 1584(1):9–19

Aisenbrey C, Borowik T, Bystrom R, Bokvist M, Lindstrom F, Misiak H, Sani MA, Grobner G (2008) How is protein aggregation in amyloidogenic diseases modulated by biological membranes? Eur Biophys J 37(3):247–255

Alberdi E, Sáchez-Gómez MV, Cavaliere F, Pérez-Samartín A, Zugaza JL, Trullas R, Domercq M, Matute C (2010) Amyloid β oligomers induce Ca2+ dysregulation and neuronal death through activation of ionotropic glutamate receptors. Cell Calcium 47(3):264–272

Aleshkov S, Abraham CR, Zannis VI (1997) Interaction of nascent ApoE2, ApoE3, and ApoE4 isoforms expressed in mammalian cells with amyloid peptide Aβ(1–40). Relevance to Alzheimer's disease. Biochemistry 36(34):10571–10580

Alzheimer A (1911) über eigenartige Krankheitsfälle des späteren Alters. Z Gesamte Neurol Psychiatr 4(1):356–385

Arispe N, Pollard HB, Rojas E (1993a) Giant multilevel cation channels formed by Alzheimer disease amyloid β-protein [AβP-(1–40)] in bilayer membranes. Proc Natl Acad Sci U S A 90(22):10573–10577

Arispe N, Rojas E, Pollard HB (1993b) Alzheimer disease amyloid β protein forms calcium channels in bilayer membranes: blockade by tromethamine and aluminum. Proc Natl Acad Sci U S A 90(2):567–571

Atwood CS, Perry G, Smith MA (2003) Cerebral hemorrhage and amyloid-β. Science 299(5609):1014

Avdulov NA, Chochina SV, Igbavboa U, O'Hare EO, Schroeder F, Cleary JP, Wood WG (1997) Amyloid β-peptides increase annular and bulk fluidity and induce lipid peroxidation in brain synaptic plasma membranes. J Neurochem 68(5):2086–2091

Bailey JA, Maloney B, Ge Y-W, Lahiri DK (2011) Functional activity of the novel Alzheimer's amyloid β-peptide interacting domain (AβID) in the APP and BACE1 promoter sequences and implications in activating apoptotic genes and in amyloidogenesis. Gene 488(1–2):13–22

Bales KR, Liu F, Wu S, Lin S, Koger D, DeLong C, Hansen JC, Sullivan PM, Paul SM (2009) Human APOE isoform-dependent effects on brain β-amyloid levels in PDAPP transgenic mice. J Neurosci 29(21):6771–6779

Bamberger ME, Harris ME, McDonald DR, Husemann J, Landreth GE (2003) A cell surface receptor complex for fibrillar b-amyloid mediates microglial activation. J Neurosci 23(7):2665–2674

Barrett PJ, Song Y, Van Horn WD, Hustedt EJ, Schafer JM, Hadziselimovic A, Beel AJ, Sanders CR (2012) The amyloid precursor protein has a flexible transmembrane domain and binds cholesterol. Science 336(6085):1168–1171

Baruch-Suchodolsky R, Fischer B (2009) Aβ40, either soluble or aggregated, is a remarkably potent antioxidant in cell-free oxidative systems. Biochemistry 48(20):4354–4370

Bateman RJ, Munsell LY, Morris JC, Swarm R, Yarasheski KE, Holtzman DM (2006) Human amyloid-β synthesis and clearance rates as measured in cerebrospinal fluid in vivo. Nat Med 12(7):856–861

Beel AJ, Mobley CK, Kim HJ, Tian F, Hadziselimovic A, Jap B, Prestegard JH, Sanders CR (2008) Structural studies of the transmembrane C-terminal domain of the amyloid precursor protein (APP): does APP function as a cholesterol sensor? Biochemistry 47(36):9428–9446

Bell RD, Sagare AP, Friedman AE, Bedi GS, Holtzman DM, Deane R, Zlokovic BV (2007) Transport pathways for clearance of human Alzheimer's amyloid β-peptide and apolipoproteins E and J in the mouse central nervous system. J Cereb Blood Flow Metab 27:909–918

Benseny-Cases N, Cocera M, Cladera J (2007) Conversion of non-fibrillar β-sheet oligomers into amyloid fibrils in Alzheimer's disease amyloid peptide aggregation. Biochem Biophys Res Commun 361:916–921

Bloom GS (2014) Amyloid-β and tau: the trigger and bullet in Alzheimer disease pathogenesis. J Am Med Assoc Neurol 71(4):505–508

Bodovitz S, Klein WL (1996) Cholesterol modulates a-secretase cleavage of amyloid precursor protein. J Biol Chem 271(8):4436–4440

Bokvist M, Groebner G (2007) Misfolding of amyloidogenic proteins at membrane surfaces: the impact of macromolecular crowding. J Am Chem Soc 129(48):14848–14849

Bokvist M, Lindstroem F, Watts A, Groebner G (2004) Two types of Alzheimer's β-amyloid (1–40) peptide membrane interactions: aggregation preventing transmembrane anchoring versus accelerated surface fibril formation. J Mol Biol 335(4):1039–1049

Brouwers N, Sleegers K, Van Broeckhoven C (2008) Molecular genetics of Alzheimer's disease: an update. Ann Med 40(8):562–583

Brown DA, London E (2000) Structure and function of sphingolipid- and cholesterol-rich membrane rafts. J Biol Chem 275(23):17221–17224

Bucciantini M, Rigacci S, Stefani M (2014) Amyloid aggregation: role of biological membranes and the aggregated membrane system. J Phys Chem Lett 5(3):517–527

Butterfield SM, Lashuel HA (2010) Amyloidogenic protein–membrane interactions: mechanistic insight from model systems. Angew Chem Int Ed 49(33):5628–5654

Bystroem R, Aisenbrey C, Borowik T, Bokvist M, Lindstroem F, Sani M-A, Olofsson A, Groebner G (2008) Disordered proteins: biological membranes as two-dimensional aggregation matrices. Cell Biochem Biophys 52(3):175–189

Castellano JM, Kim J, Stewart FR, Jiang H, DeMattos RB, Patterson BW, Fagan AM, Morris JC, Mawuenyega KG, Cruchaga C, Goate AM, Bales KR, Paul SM, Bateman RJ, Holtzman DM (2011) Human apoE isoforms differentially regulate brain amyloid-β peptide clearance. Sci Transl Med 3:89ra57

Cecchi C, Baglioni S, Fiorillo C, Pensalfini A, Liguri G, Nosi D, Rigacci S, Bucciantini M, Stefani M (2005) Insights into the molecular basis of the differing susceptibility of varying cell types to the toxicity of amyloid aggregates. J Cell Sci 118(15):3459–3470

Cedazo-Mínguez A, Cowburn RF (2001) Apolipoprotein E: a major piece in the Alzheimer's disease puzzle. J Cell Mol Med 5(3):254–266

Chauhan NB (2003) Membrane dynamics, cholesterol homeostasis, and Alzheimer's disease. J Lipid Res 44(11):2019–2029

Chauhan A, Ray I, Chauhan VP (2000) Interaction of amyloid β-protein with anionic phospholipids: possible involvement of Lys28 and C-terminus aliphatic amino acids. Neurochem Res 25(3):423–429

Chen J, Li Q, Wang J (2011) Topology of human apolipoprotein E3 uniquely regulates its diverse biological functions. Proc Natl Acad Sci U S A 108(36):14813–14818

Chi EY, Ege C, Winans A, Majewski J, Wu G, Kjaer K, Lee KYC (2008) Lipid membrane templates the ordering and induces the fibrillogenesis of Alzheimer's disease amyloid-β peptide. Proteins 72(1):1–24

Chochina SV, Avdulov NA, Igbavboa U, Cleary JP, O'Hare EO, Wood WG (2001) Amyloid b-peptide 1–40 increases neuronal membrane fluidity: role of cholesterol and brain region. J Lipid Res 42(8):1292–1297

Choo-Smith L-P, Surewicz WK (1997) The interaction between Alzheimer amyloid Aβ(1–40) peptide and ganglioside GM1-containing membranes. FEBS Lett 402(2–3):95–98

Choo-Smith L-P, Garzon-Rodriguez W, Glabe CG, Surewicz WK (1997) Acceleration of amyloid fibril formation by specific binding of Aβ-(1–40) peptide to ganglioside-containing membrane vesicles. J Biol Chem 272(37):22987–22990

Cleary JP, Walsh DM, Hofmeister JJ, Shankar GM, Kuskowski MA, Selkoe DJ, Ashe KH (2005) Natural oligomers of the amyloid-b protein specifically disrupt cognitive function. Nat Neurosci 8:79–84

Cohen SIA, Linse S, Luheshi LM, Hellstrand E, White DA, Rajah L, Otzen DE, Vendruscolo M, Dobson CM, Knowles TPJ (2013) Proliferation of amyloid-β42 aggregates occurs through a secondary nucleation mechanism. Proc Natl Acad Sci U S A 110(24):9758–9763

Cole GM, Ard MD (2000) Influence of lipoproteins on microglial degradation of Alzheimer's amyloid β-protein. Microsc Res Tech 50(4):316–324

Coles M, Bicknell W, Watson AA, Fairlie DP, Craik DJ (1998) Solution structure of amyloid β-peptide(1–40) in a water-micelle environment. Is the membrane-spanning domain where we think it is? Biochemistry 37(31):11064–11077

Conejero-Goldberg C, Gomar JJ, Bobes-Bascaran T, Hyde TM, Kleinman JE, Herman MM, Chen S, Davies P, Goldberg TE (2014) APOE2 enhances neuroprotection against Alzheimer's disease through multiple molecular mechanisms. Mol Psychiatry 19(11):1243–1250

Corder EH, Saunders AM, Strittmatter WJ, Schmechel DE, Gaskell PC, Small GW, Roses AD, Haines JL, Pericak-Vance MA (1993) Gene dose of apolipoprotein E type 4 allele and the risk of Alzheimer's disease in late onset families. Science 261:921–923

Cruts M, Van Broeckhoven C (1998) Molecular genetics of Alzheimer's disease. Ann Med 30(6):560–565

Cutler RG, Kelly J, Storie K, Pedersen WA, Tammara A, Hatanpaa K, Troncoso JC, Mattson MP (2004) Involvement of oxidative stress-induced abnormalities in ceramide and cholesterol metabolism in brain aging and Alzheimer's disease. Proc Natl Acad Sci U S A 101(7):2070–2075

Dahse K, Garvey M, Kovermann M, Vogel A, Balbach J, Fandrich M, Fahr A (2010) DHPC strongly affects the structure and oligomerization propensity of Alzheimer's Aβ(1–40) peptide. J Mol Biol 403(4):643–659

Davis CH, Berkowitz ML (2009a) Interaction between amyloid-b (1–42) peptide and phospholipid bilayers: a molecular dynamics study. Biophys J 96(3):785–797

Davis CH, Berkowitz ML (2009b) Structure of the amyloid-b (1b-42) monomer absorbed to model phospholipid bilayers: a molecular dynamics study. J Phys Chem B 113(43):14480–14486

de Planque MRR, Raussens V, Contera SA, Rijkers DTS, Liskamp RMJ, Ruysschaert J-M, Ryan JF, Separovic F,

Watts A (2007) β-sheet structured β-amyloid(1–40) perturbs phosphatidylcholine model membranes. J Mol Biol 368(4):982–997

Deane R, Wu Z, Sagare A, Davis J, Du Yan S, Hamm K, Xu F, Parisi M, LaRue B, Hu HW, Spijkers P, Guo H, Song X, Lenting PJ, Van Nostrand WE, Zlokovic BV (2004) LRP/amyloid β-peptide interaction mediates differential brain efflux of Aβ isoforms. Neuron 43(3):333–344

Deane R, Sagare A, Hamm K, Parisi M, Lane S, Finn MB, Holtzman DM, Zlokovic BV (2008) ApoE isoform-specific disruption of amyloid β peptide clearance from mouse brain. J Clin Invest 118:4002–4013

Demuro A, Mina E, Kayed R, Milton SC, Parker I, Glabe CG (2005) Calcium dysregulation and membrane disruption as a ubiquitous neurotoxic mechanism of soluble amyloid oligomers. J Biol Chem 280(17):17294–17300

Deshpande A, Mina E, Glabe C, Busciglio J (2006) Different conformations of amyloid β induce neurotoxicity by distinct mechanisms in human cortical neurons. J Neurosci 26(22):6011–6018

Di Scala C, Chahinian H, Yahi N, Garmy N, Fantini J (2014a) Interaction of Alzheimer's β-amyloid peptides with cholesterol: mechanistic insights into amyloid pore formation. Biochemistry 53(28):4489–4502

Di Scala C, Troadec J-D, Lelièvre C, Garmy N, Fantini J, Chahinian H (2014b) Mechanism of cholesterol-assisted oligomeric channel formation by a short Alzheimer β-amyloid peptide. J Neurochem 128(1):186–195

Diaz JC, Linnehan J, Pollard H, Arispe N (2006) Histidines 13 and 14 in the Aβ sequence are targets for inhibition of Alzheimer's disease Aβ ion channel and cytotoxicity. Biol Res 39:447–460

Dobson CM (2003) Protein folding and misfolding. Nature 426(6968):884–890

Doody RS, Raman R, Farlow M, Iwatsubo T, Vellas B, Joffe S, Kieburtz K, He F, Sun X, Thomas RG, Aisen PS, Siemers E, Sethuraman G, Mohs R (2013) A phase 3 trial of semagacestat for treatment of Alzheimer's disease. N Engl J Med 369(4):341–350

Dovidchenko NV, Finkelstein AV, Galzitskaya OV (2014) How to determine the size of folding nuclei of protofibrils from the concentration dependence of the rate and lag-time of aggregation. I. Modeling the amyloid protofibril formation. J Phys Chem B 118(5):1189–1197

Eckert GP, Cairns NJ, Maras A, Gattaz WF, Müller WE (2000) Cholesterol modulates the membrane- disordering effects of β-amyloid peptides in the hippocampus: specific changes in Alzheimer's disease. Dement Geriatr Cogn Disord 11(4):181–186

Eckert GP, Wood WG, Muller WE (2010) Lipid membranes and β-amyloid: a harmful connection. Curr Protein Pept Sci 11(5):319–325

Eriksen JL, Sagi SA, Smith TE, Weggen S, Das P, McLendon DC, Ozols VV, Jessing KW, Zavitz KH, Koo EH, Golde TE (2003) NSAIDs and enantiomers of flurbiprofen target g-secretase and lower Aβ42 in vivo. J Clin Invest 112(3):440–449

Esler WP, Stimson ER, Jennings JM, Vinters HV, Ghilardi JR, Lee JP, Mantyh PW, Maggio JE (2000) Alzheimer's disease amyloid propagation by a template-dependent dock-lock mechanism. Biochemistry 39(21):6288–6295

Fantini J, Garmy N, Mahfoud R, Yahi N (2002) Lipid rafts: structure, function and role in HIV, Alzheimer's and prion diseases. Expert Rev Mol Med 4(27):1–22

Fantini J, Di Scala C, Yahi N, Troadec J-D, Sadelli K, Chahinian H, Garmy N (2014) Bexarotene blocks calcium-permeable ion channels formed by neurotoxic Alzheimer's β-amyloid peptides. ACS Chem Neurosci 5(3):216–224

Fassbender K, Simons M, Bergmann C, Stroick M, Lütjohann D, Keller P, Runz H, Kühl S, Bertsch T, von Bergmann K, Hennerici M, Beyreuther K, Hartmann T (2001) Simvastatin strongly reduces levels of Alzheimer's disease β-amyloid peptides Aβ42 and Aβ40 in vitro and in vivo. Proc Natl Acad Sci U S A 98(10):5856–5861

Fezoui Y, Teplow DB (2002) Kinetic studies of amyloid β-protein fibril assembly: differential effects of α-helix stabilization. J Biol Chem 277(40):36948–36954

Fezoui Y, Hartley DM, Walsh DM, Selkoe DJ, Osterhout JJ, Teplow DB (2000) A de novo designed helix-turn-helix peptide forms non-toxic amyloid fibrils. Nat Struct Biol 7:1095–1099

Foley P (2010) Lipids in Alzheimer's disease: a century-old story. Biochim Biophys Acta 1801(8):750–753

Frears ER, Stephens DJ, Walters CE, Davies H, Austen BM (1999) The role of cholesterol in the biosynthesis of β-amyloid. Neuroreport 10(8):1699–1705

Gandy S, Simon AJ, Steele JW, Lublin AL, Lah JJ, Walker LC, Levey AI, Krafft GA, Levy E, Checler F, Glabe C, Bilker WB, Abel T, Schmeidler J, Ehrlich ME (2010) Days to criterion as an indicator of toxicity associated with human Alzheimer amyloid-β oligomers. Ann Neurol 68(2):220–230

Garvey M, Tepper K, Haupt C, Knuepfer U, Klement K, Meinhardt J, Horn U, Balbach J, Faendrich M (2011) Phosphate and HEPES buffers potently affect the fibrillation and oligomerization mechanism of Alzheimer's Aβ peptide. Biochem Biophys Res Commun 409(3):385–388

Ghiso J, Frangione B (2002) Amyloidosis and Alzheimer's disease. Adv Drug Deliv Rev 54(12):1539–1551

Gilbert BJ (2013) The role of amyloid β in the pathogenesis of Alzheimer's disease. J Clin Pathol 66(5):362–366

Golde TE, Eckman CB (2001) Cholesterol modulation as an emerging strategy for the treatment of Alzheimer's disease. Drug Discov Today 6(20):1049–1055

Golde TE, Petrucelli L, Lewis J (2010) Targeting Ab and tau in Alzheimer's disease, an early interim report. Exp Neurol 223(2):252–266

Good TA, Murphy RM (1995) Aggregation state-dependent binding of β-amyloid peptide to protein and lipid components of rat cortical homogenates. Biochem Biophys Res Commun 207(1):209–215

Grootendorst J, Bour A, Vogel E, Kelche C, Sullivan PM, Dodart J-C, Bales K, Mathis C (2005) Human apoE targeted replacement mouse lines: h-apoE4 and h-apoE3 mice differ on spatial memory performance and avoidance behavior. Behav Brain Res 159(1):1–14

Haass C, Selkoe DJ (2007) Soluble protein oligomers in neurodegeneration: lessons from the Alzheimer's amyloid [β]-peptide. Nat Rev Mol Cell Biol 8(2):101–112

Haass C, Koo EH, Mellon A, Hung AY, Selkoe DJ (1992) Targeting of cell-surface b-amyloid precursor protein to lysosomes: alternative processing into amyloid-bearing fragments. Nature 357(6378):500–503

Hardy JA, Higgins GA (1992) Alzheimer's disease: the amyloid cascade hypothesis. Science 256:184–185

Hardy J, Selkoe DJ (2002) The amyloid hypothesis of Alzheimer's disease: progress and problems on the road to therapeutics. Science 297:353–356

Harper JD, Lansbury PT (1997) Models of amyloid seeding in Alzheimer's disease and scrapie: mechanistic truths and physiological consequences of the time-dependent solubility of amyloid proteins. Annu Rev Biochem 66:385–407

Hartmann T, Kuchenbecker J, Grimm MOW (2007) Alzheimer's disease: the lipid connection. J Neurochem 103:159–170

Hatters DM, Peters-Libeu CA, Weisgraber KH (2006) Apolipoprotein E structure: insights into function. Trends Biochem Sci 31(8):445–454

Hauser PS, Ryan RO (2013) Impact of apolipoprotein E on Alzheimer's disease. Curr Alzheim Res 10(8):809–817

Hayashi H, Kimura N, Yamaguchi H, Hasegawa K, Yokoseki T, Shibata M, Yamamoto N, Michikawa M, Yoshikawa Y, Terao K, Matsuzaki K, Lemere CA, Selkoe DJ, Naiki H, Yanagisawa K (2004) A seed for Alzheimer amyloid in the brain. J Neurosci 24(20):4894–4902

Holtzman DM (2001) Role of apoe/Aβ interactions in the pathogenesis of Alzheimer's disease and cerebral amyloid angiopathy. J Mol Neurosci 17:147–155

Holtzman DM, Bales KR, Tenkova T, Fagan AM, Parsadanian M, Sartorius LJ, Mackey B, Olney J, McKeel D, Wozniak D, Paul SM (2000) Apolipoprotein E isoform-dependent amyloid deposition and neuritic degeneration in a mouse model of Alzheimer's disease. Proc Natl Acad Sci U S A 97:2892–2897

Hong S, Ostaszewski Beth L, Yang T, O' Malley TT, Jin M, Yanagisawa K, Li S, Bartels T, Selkoe DJ (2014) Soluble Aβ oligomers are rapidly sequestered from brain ISF in vivo and bind GM1 ganglioside on cellular membranes. Neuron 82(2):308–319

Hooper N, Trew A, Parkin E, Turner A (2002) The role of proteolysis in Alzheimer's disease. In: Langner J, Siegfried, A (ed) Cellular peptidases in immune functions and diseases 2, vol 477. Advances in experimental medicine and biology. Springer, New York, pp 379–390

Hou L, Shao H, Zhang Y, Li H, Menon NK, Neuhaus EB, Brewer JM, Ray DG, Vitek MP, Iwashita T, Makula RA, Przybyla AB, Zagorski MG (2004) Solution NMR studies of the Aβ(1–40) and Aβ(1–42) peptides establish that the Met35 oxidation state affects the mechanism of amyloid formation. J Am Chem Soc 126:1992–2005

Igbavboa U, Sun GY, Weisman GA, He Y, Wood WG (2009) Amyloid β-protein stimulates trafficking of cholesterol and caveolin-1 from the plasma membrane to the Golgi complex in mouse primary astrocytes. Neuroscience 162(2):328–338

Jang H, Arce FT, Ramachandran S, Capone R, Azimova R, Kagan BL, Nussinov R, Lal R (2010) Truncated β-amyloid peptide channels provide an alternative mechanism for Alzheimer's disease and down syndrome. Proc Natl Acad Sci U S A 107(14):6538–6543

Jang H, Teran Arce F, Ramachandran S, Kagan BL, Lal R, Nussinov R (2014) Disordered amyloidogenic peptides may insert into the membrane and assemble into common cyclic structural motifs. Chem Soc Rev 43(19):6750–6764

Kakio A, S-i N, Yanagisawa K, Kozutsumi Y, Matsuzaki K (2001) Cholesterol-dependent formation of GM1 ganglioside-bound amyloid β-protein, an endogenous seed for Alzheimer amyloid. J Biol Chem 276(27):24985–24990

Kakio A, S-i N, Yanagisawa K, Kozutsumi Y, Matsuzaki K (2002) Interactions of Amyloid β-protein with various gangliosides in raft-like membranes: importance of GM1 ganglioside-bound form as an endogenous seed for Alzheimer amyloid. Biochemistry 41(23):7385–7390

Karran E, Mercken M, Strooper BD (2011) The amyloid cascade hypothesis for Alzheimer's disease: an appraisal for the development of therapeutics. Nat Rev Drug Discov 10(9):698–712

Kato-Negishi M, Muramoto K, Kawahara M, Hosoda R, Kuroda Y, Ichikawa M (2003) Bicuculline induces synapse formation on primary cultured accessory olfactory bulb neurons. Eur J Neurosci 18(6):1343–1352

Kawahara M, Ohtsuka I, Yokoyama S, Kato-Negishi M, Sadakane Y (2011) Membrane incorporation, channel formation, and disruption of calcium homeostasis by Alzheimer's β-amyloid protein. Int J Alzheimers Dis 2011:17

Kayed R, Head E, Thompson JL, McIntire TM, Milton SC, Cotman CW, Glabe CG (2003) Common structure of soluble amyloid oligomers implies common mechanism of pathogenesis. Science 300:486–489

Kayed R, Sokolov Y, Edmonds B, McIntire TM, Milton SC, Hall JE, Glabe CG (2004) Permeabilization of lipid bilayers is a common conformation-dependent activity of soluble amyloid oligomers in protein misfolding diseases. J Biol Chem 279(45):46363–46366

Kayed R, Pensalfini A, Margol L, Sokolov Y, Sarsoza F, Head E, Hall J, Glabe C (2009) Annular protofibrils are a structurally and functionally distinct type of amyloid oligomer. J Biol Chem 284(7):4230–4237

Kepp KP (2012) Bioinorganic chemistry of Alzheimer's disease. Chem Rev 112(10):5193–5239

Kim S, Jeon T-J, Oberai A, Yang D, Schmidt JJ, Bowie JU (2005) Transmembrane glycine zippers: physiological and pathological roles in membrane proteins. Proc Natl Acad Sci U S A 102(40):14278–14283

Kim J, Basak JM, Holtzman DM (2009) The role of apolipoprotein E in Alzheimer's disease. Neuron 63(3):287–303

Kivipelto M, Helkala E-L, Laakso MP, Hänninen T, Hallikainen M, Alhainen K, Soininen H, Tuomilehto J, Nissinen A (2001) Midlife vascular risk factors and Alzheimer's disease in later life: longitudinal, population based study. BMJ 322(7300):1447–1451

Klement K, Wieligmann K, Meinhardt J, Hortschansky P, Richter W, Faendrich M (2007) Effect of different salt Ions on the propensity of aggregation and on the structure of Alzheimer's Aβ(1–40) amyloid fibrils. J Mol Biol 373(5):1321–1333

Koistinaho M, Lin S, Wu X, Esterman M, Koger D, Hanson J, Higgs R, Liu F, Malkani S, Bales KR, Paul SM (2004) Apolipoprotein E promotes astrocyte colocalization and degradation of deposited amyloid-b peptides. Nat Med 10(7):719–726

Korwek K, Trotter J, LaDu M, Sullivan P, Weeber E (2009) ApoE isoform-dependent changes in hippocampal synaptic function. Mol Neurodegener 4(1):21

Kremer JJ, Murphy RM (2003) Kinetics of adsorption of β-amyloid peptide Aβ(1–40) to lipid bilayers. J Biochem Biophys Meth 57(2):159–169

Kremer JJ, Pallitto MM, Sklansky DJ, Murphy RM (2000) Correlation of β-amyloid aggregate size and hydrophobicity with decreased bilayer fluidity of model membranes. Biochemistry 39(33):10309–10318

Kremer JJ, Sklansky DJ, Murphy RM (2001) Profile of changes in lipid bilayer structure caused by β-amyloid peptide. Biochemistry 40(29):8563–8571

Krishnaswamy S, Verdile G, Groth D, Kanyenda L, Martins RN (2009) The structure and function of Alzheimer's g secretase enzyme complex. Crit Rev Clin Lab Sci 46(5–6):282–301

Kukar TL, Ladd TB, Bann MA, Fraering PC, Narlawar R, Maharvi GM, Healy B, Chapman R, Welzel AT, Price RW, Moore B, Rangachari V, Cusack B, Eriksen J, Jansen-West K, Verbeeck C, Yager D, Eckman C, Ye W, Sagi S, Cottrell BA, Torpey J, Rosenberry TL, Fauq A, Wolfe MS, Schmidt B, Walsh DM, Koo EH, Golde TE (2008) Substrate-targeting γ-secretase modulators. Nature 453(7197):925–929

Kukar TL, Ladd TB, Robertson P, Pintchovski SA, Moore B, Bann MA, Ren Z, Jansen-West K, Malphrus K, Eggert S, Maruyama H, Cottrell BA, Das P, Basi GS, Koo EH, Golde TE (2011) Lysine 624 of the amyloid precursor protein (APP) is a critical determinant of amyloid β peptide length: support for a sequential model of γ-secretase intramembrane proteolysis and regulation by the amyloid β precursor protein (APP) juxtamembrane region. J Biol Chem 286(46):39804–39812

Kuo Y-M, Emmerling MR, Bisgaier CL, Essenburg AD, Lampert HC, Drumm D, Roher AE (1998) Elevated low-density lipoprotein in Alzheimer's disease correlates with brain Aβ 1–42 levels. Biochem Biophys Res Commun 252(3):711–715

Lacor PN, Buniel MC, Furlow PW, Clemente AS, Velasco PT, Wood M, Viola KL, Klein WL (2007) Aβ oligomer-induced aberrations in synapse composition, shape, and density provide a molecular basis for loss of connectivity in Alzheimer's disease. J Neurosci 27:796–807

LaDu MJ, Falduto MT, Manelli AM, Reardon CA, Getz GS, Frail DE (1994) Isoform-specific binding of apolipoprotein E to β-amyloid. J Biol Chem 269(38):23403–23406

Lal R, Lin H, Quist AP (2007) Amyloid beta ion channel: 3D structure and relevance to amyloid channel paradigm. Biochim Biophys Acta 1768(8):1966–1975

Lannfelt L, Moller C, Basun H, Osswald G, Sehlin D, Satlin A, Logovinsky V, Gellerfors P (2014) Perspectives on future Alzheimer therapies: amyloid-β protofibrils – a new target for immunotherapy with BAN2401 in Alzheimer's disease. Alzheimer Res Ther 6(2):16

Lashuel HA, Hartley D, Petre BM, Walz T, Lansbury PT (2002) Neurodegenerative disease: amyloid pores from pathogenic mutations. Nature 418(6895):291–291

Lee S-J, Liyanage U, Bickel PE, Xia W, Lansbury PT, Kosik KS (1998) A detergent-insoluble membrane compartment contains Aβ in vivo. Nat Med 4(6):730–734

Li Y, Lu W, Marzolo MP, Bu G (2001) Differential functions of members of the low density lipoprotein receptor family suggested by their distinct endocytosis rates. J Biol Chem 276(21):18000–18006

Lin HAI, Bhatia R, Lal R (2001) Amyloid β protein forms ion channels: implications for Alzheimer's disease pathophysiology. FASEB J 15(13):2433–2444

Lu J-X, Qiang W, Yau W-M, Schwieters CD, Meredith SC, Tycko R (2013) Molecular structure of β-amyloid fibrils in Alzheimer's disease brain tissue. Cell 154(6):1257–1268. doi:10.1016/j.cell.2013.1008.1035

Luhrs T, Ritter C, Adrian M, Riek-Loher D, Bohrmann B, Dobeli H, Schubert D, Riek R (2005) 3D structure of Alzheimer's amyloid-b(1–42) fibrils. Proc Natl Acad Sci U S A 102(48):17342–17347

Mahley RW (1988) Apolipoprotein E: cholesterol transport protein with expanding role in cell biology. Science 240(4852):622–630

Mahley RW, Weisgraber KH, Huang Y (2006) Apolipoprotein E4: a causative factor and therapeutic target in neuropathology, including Alzheimer's disease. Proc Natl Acad Sci U S A 103(15): 5644–5651

Maltseva E, Kerth A, Blume A, Möhwald H, Brezesinski G (2005) Adsorption of amyloid β (1–40) peptide at phospholipid monolayers. ChemBioChem 6(10):1817–1824

Mandal PK, Pettegrew JW (2004) Alzheimer's disease: NMR studies of asialo (GM1) and trisialo (GT1b) ganglioside interactions with Aβ(1–40) peptide in a membrane mimic environment. Neurochem Res 29(2):447–453

Martins IC, Kuperstein I, Wilkinson H, Maes E, Vanbrabant M, Jonckheere W, Van Gelder P, Hartmann D, D'Hooge R, De Strooper B, Schymkowitz J, Rousseau F (2008) Lipids revert inert Aβ amyloid fibrils to neurotoxic protofibrils that affect learning in mice. EMBO J 27(1):224–233

Mason RP, Estermyer JD, Kelly JF, Mason PE (1996) Alzheimer's disease amyloid β peptide 25–35 is localized in the membrane hydrocarbon core: x-ray diffraction analysis. Biochem Biophys Res Commun 222(1):78–82

Mason RP, Jacob RF, Walter MF, Mason PE, Avdulov NA, Chochina SV, Igbavboa U, Wood WG (1999) Distribution and fluidizing action of soluble and aggregated amyloid β-peptide in rat synaptic plasma membranes. J Biol Chem 274(26):18801–18807

Matsuzaki K (2007) Physicochemical interactions of amyloid β-peptide with lipid bilayers. Biochim Biophys Acta Biomembr 1768(8):1935–1942

Matsuzaki K (2011) Formation of toxic amyloid fibrils by amyloid β-protein on ganglioside clusters. Int J Alzheimers Dis 956104:1–7

Matsuzaki K, Horikiri C (1999) Interactions of amyloid β-Peptide (1–40) with ganglioside-containing membranes. Biochemistry 38(13):4137–4142

Matsuzaki K, Kato K, Yanagisawa K (2010) Aβ polymerization through interaction with membrane gangliosides. Biochim Biophys Acta Mol Cell Biol L 1801(8):868–877

Mattson MP (1997) Cellular actions of β-amyloid precursor protein and its soluble and fibrillogenic derivatives. Physiol Rev 77(4):1081–1132

Mawuenyega KG, Sigurdson W, Ovod V, Munsell L, Kasten T, Morris JC, Yarasheski KE, Bateman RJ (2010) Decreased clearance of CNS β-amyloid in Alzheimer's disease. Science 330(6012):1774

McGuinness B, Passmore P (2010) Can statins prevent or help treat Alzheimer's disease? J Alzheimers Dis 20(3):925–933

McLaurin J, Chakrabartty A (1996) Membrane disruption by Alzheimer β-amyloid peptides mediated through specific binding to either phospholipids or gangliosides. Implications for neurotoxicity. J Biol Chem 271(43):26482–26489

McLaurin J, Franklin T, Chakrabartty A, Fraser PE (1998) Phosphatidylinositol and inositol involvement in Alzheimer amyloid-β fibril growth and arrest. J Mol Biol 278(1):183–194

McLaurin J, Yang D, Yip CM, Fraser PE (2000) Review: modulating factors in amyloid-β fibril formation. J Struct Biol 130(2–3):259–270

Meinhardt J, Sachse C, Hortschansky P, Grigorieff N, Fändrich M (2009) Ab(1–40) fibril polymorphism implies diverse interaction patterns in amyloid fibrils. J Mol Biol 386(3):869–877

Meisl G, Yang X, Hellstrand E, Frohm B, Kirkegaard JB, Cohen SIA, Dobson CM, Linse S, Knowles TPJ (2014) Differences in nucleation behavior underlie the contrasting aggregation kinetics of the Aβ40 and Aβ42 peptides. Proc Natl Acad Sci U S A 111(26):9384–9389

Miyashita N, Straub JE, Thirumalai D (2009) Structures of β-amyloid peptide 1–40, 1–42, and 1–55, the 672–726 fragment of APP-in a membrane environment with implications for interactions with γ-Secretase. J Am Chem Soc 131(49):17843–17852

Morgado I, Faendrich M (2011) Assembly of Alzheimer's Aβ peptide into nanostructured amyloid fibrils. Curr Opin Colloid Interface Sci 16(6):508–514

Mulder SD, Nielsen HM, Blankenstein MA, Eikelenboom P, Veerhuis R (2014) Apolipoproteins E and J interfere with amyloid-β uptake by primary human astrocytes and microglia in vitro. Glia 62(4):493–503

Muller WE, Eckert GP, Scheuer K, Cairns NJ, Maras A, Gattaz WF (1998) Effects of β-amyloid peptides on the fluidity of membranes from frontal and parietal lobes of human brain. High potencies of Aβ 1–42 and Aβ 1–43. Amyloid 5(1):10–15

Murphy RM (2007) Kinetics of amyloid formation and membrane interaction with amyloidogenic proteins. Biochim Biophys Acta 1768(8):1923–1934

Nagarajan S, Ramalingam K, Neelakanta Reddy P, Cereghetti DM, Padma Malar EJ, Rajadas J (2008) Lipid-induced conformational transition of the amyloid core fragment Aβ(28–35) and its A30G and A30I mutants. FEBS J 275(10):2415–2427

Okada T, Ikeda K, Wakabayashi M, Ogawa M, Matsuzaki K (2008) Formation of toxic Aβ(1–40) fibrils on GM1 ganglioside-containing membranes mimicking lipid rafts: polymorphisms in Aβ(1–40) fibrils. J Mol Biol 382(4):1066–1074

Oshima N, Morishima-Kawashima M, Yamaguchi H, Yoshimura M, Sugihara S, Khan K, Games D, Schenk D, Ihara Y (2001) Accumulation of amyloid β-protein in the low-density membrane domain accurately reflects the extent of β-amyloid deposition in the brain. Am J Pathol 158(6):2209–2218

Pai AS, Rubinstein I, Oniuksel H (2006) PEGylated phospholipid nanomicelles interact with β-amyloid (1–42) and mitigate its β-sheet formation, aggregation and neurotoxicity in vitro. Peptides 27(11):2858–2866

Palop JJ, Mucke L (2010) Amyloid-β-induced neuronal dysfunction in Alzheimer's disease: from synapses toward neural networks. Nat Neurosci 13(7):812–818

Pauwels K, Williams TL, Morris KL, Jonckheere W, Vandersteen A, Kelly G, Schymkowitz J, Rousseau F, Pastore A, Serpell LC, Broersen K (2012) Structural basis for increased toxicity of pathological Aβ42: aβ40 ratios in Alzheimer disease. J Biol Chem 287(8):5650–5660

Pellarin R, Caflisch A (2006) Interpreting the aggregation kinetics of amyloid peptides. J Mol Biol 360(4):882–892

Peters I, Igbavboa U, Schutt T, Haidari S, Hartig U, Rosello X, Bottner S, Copanaki E, Deller T, Kogel D,

Wood WG, Muller WE, Eckert GP (2009) The interaction of β-amyloid protein with cellular membranes stimulates its own production. Biochim Biophys Acta 1788(5):964–972

Petkova AT, Ishii Y, Balbach JJ, Antzutkin ON, Leapman RD, Delaglio F, Tycko R (2002) A structural model for Alzheimer's b-amyloid fibrils based on experimental constraints from solid state NMR. Proc Natl Acad Sci U S A 99(26):16742–16747

Petkova AT, Leapman RD, Guo Z, Yau W-M, Mattson MP, Tycko R (2005) Self-Propagating, molecular-level polymorphism in Alzheimer's β-amyloid fibrils. Science 307(5707):262–265

Phillips MC (2013) New insights into the determination of HDL structure by apolipoproteins: thematic review series: high density lipoprotein structure, function, and metabolism. J Lipid Res 54(8):2034–2048

Phillips MC (2014) Apolipoprotein E isoforms and lipoprotein metabolism. IUBMB Life 66(9):616–623

Pitas RE, Boyles JK, Lee SH, Hui D, Weisgraber KH (1987) Lipoproteins and their receptors in the central nervous system. Characterization of the lipoproteins in cerebrospinal fluid and identification of apolipoprotein B, E(LDL) receptors in the brain. J Biol Chem 262(29):14352–14360

Pollard HB, Rojas E, Arispe N (1993) A new hypothesis for the mechanism of amyloid toxicity, based on the calcium channel activity of amyloid β protein (AβP) in phospholipid bilayer membranes. Ann N Y Acad Sci 695(1):165–168

Posse de Chaves E, Sipione S (2010) Sphingolipids and gangliosides of the nervous system in membrane function and dysfunction. FEBS Lett 584(9):1748–1759

Prangkio P, Yusko EC, Sept D, Yang J, Mayer M (2012) Multivariate analyses of amyloid-β oligomer populations indicate a connection between pore formation and cytotoxicity. PLoS ONE 7(10):e47261

Proitsi P, Lupton M, Velayudhan L, Newhouse S, Fogh I, Tsolaki M, Daniilidou M, Pritchard M, Kloszewska I, Soininen H, Mecocci P, Vellas B, Williams J, Stewart R, Sham P, Lovestone S, Powell J, Alzheimer's Disease Neuroimaging I, GC (2014) Genetic predisposition to increased blood cholesterol and triglyceride lipid levels and risk of Alzheimer disease: a Mendelian randomization analysis. PLoS Med 11(9):e1001713

Puglielli L, Konopka G, Pack-Chung E, Ingano LAM, Berezovska O, Hyman BT, Chang TY, Tanzi RE, Kovacs DM (2001) Acyl-coenzyme A: cholesterol acyltransferase modulates the generation of the amyloid [beta]-peptide. Nat Cell Biol 3(10):905–912

Qiu L, Lewis A, Como J, Vaughn MW, Huang J, Somerharju P, Virtanen J, Cheng KH (2009) Cholesterol modulates the interaction of β-amyloid peptide with lipid bilayers. Biophys J 96(10):4299–4307

Querfurth HW, LaFerla FM (2010) Alzheimer's disease. N Engl J Med 362(4):329–344

Quist A, Doudevski I, Lin H, Azimova R, Ng D, Frangione B, Kagan B, Ghiso J, Lal R (2005) Amyloid ion channels: a common structural link for protein-misfolding disease. Proc Natl Acad Sci U S A 102(30):10427–10432

Rebeck GW, Kindy M, LaDu MJ (2002) Apolipoprotein E and Alzheimer's disease: the protective effects of ApoE2 and E3. J Alzheimers Dis 4(3):145–154

Reed MN, Hofmeister JJ, Jungbauer L, Welzel AT, Yu C, Sherman MA, Lesne S, LaDu MJ, Walsh DM, Ashe KH, Cleary JP (2011) Cognitive effects of cell-derived and synthetically derived Aβ oligomers. Neurobiol Aging 32(10):1784–1794

Refolo LM, Pappolla MA, Malester B, LaFrancois J, Bryant-Thomas T, Wang R, Tint GS, Sambamurti K, Duff K (2000) Hypercholesterolemia accelerates the Alzheimer's amyloid pathology in a transgenic mouse model. Neurobiol Dis 7(4):321–331

Reitz C, Tang MX, Luchsinger J, Mayeux R (2004) Relation of plasma lipids to Alzheimer disease and vascular dementia. Arch Neurol 61(5):705–714

Relini A, Cavalleri O, Rolandi R, Gliozzi A (2009) The two-fold aspect of the interplay of amyloidogenic proteins with lipid membranes. Chem Phys Lipids 158(1):1–9

Richardson K, Schoen M, French B, Umscheid CA, Mitchell MD, Arnold SE, Heidenreich PA, Rader DJ, deGoma EM (2013) Statins and cognitive function: a systematic review. Ann Intern Med 159(10):688–697

Riddell DR, Christie G, Hussain I, Dingwall C (2001) Compartmentalization of β-secretase (Asp2) into low-buoyant density, noncaveolar lipid rafts. Curr Biol 11(16):1288–1293

Rockwood K, Kirkland S, Hogan DB, MacKnight C, Merry H, Verreault R, Wolfson C, McDowell I (2002) Use of lipid-lowering agents, indication bias, and the risk of dementia in community-dwelling elderly people. Arch Neurol 59(2):223–227

Roher AE, Palmer KC, Yurewicz EC, Ball MJ, Greenberg BD (1993) Morphological and biochemical analyses of amyloid plaque core proteins purified from Alzheimer disease brain tissue. J Neurochem 61(5):1916–1926

Roses MDAD (1996) Apolipoprotein E alleles as risk factors in Alzheimer's disease. Annu Rev Med 47(1):387–400

Sachse C, Fandrich M, Grigorieff N (2008) Paired β-sheet structure of an A(1–40) amyloid fibril revealed by electron microscopy. Proc Natl Acad Sci U S A 105(21):7462–7466

Saher G, Brugger B, Lappe-Siefke C, Mobius W, R-i T, Wehr MC, Wieland F, Ishibashi S, Nave K-A (2005) High cholesterol level is essential for myelin membrane growth. Nat Neurosci 8(4):468–475

Saunders AM, Strittmatter WJ, Schmechel D, George-Hyslop PH, Pericak-Vance MA, Joo SH, Rosi BL, Gusella JF, Crapper-MacLachlan DR, Alberts MJ et al (1993) Association of apolipoprotein E allele epsilon 4 with late-onset familial and sporadic Alzheimer's disease. Neurology 43(8):1467–1472

Schmechel DE, Saunders AM, Strittmatter WJ, Crain BJ, Hulette CM, Joo SH, Pericak-Vance MA, Goldgaber D, Roses AD (1993) Increased amyloid β-peptide deposition in cerebral cortex as a consequence of apolipoprotein E genotype in late-onset alzheimer disease. Proc Natl Acad Sci U S A 90(20):9649–9653

Seubert P, Vigo-Pelfrey C, Esch F, Lee M, Dovey H, Davis D, Sinha S, Schiossmacher M, Whaley J, Swindlehurst C, McCormack R, Wolfert R, Selkoe D, Lieberburg I, Schenk D (1992) Isolation and quantification of soluble Alzheimer's β-peptide from biological fluids. Nature 359(6393):325–327

Shafrir Y, Durell S, Arispe N, Guy HR (2010) Models of membrane-bound Alzheimer's Aβ peptide assemblies. Protein: Struct Funct Bioinf 78(16):3473–3487

Shibata M, Yamada S, Kumar SR, Calero M, Bading J, Frangione B, Holtzman DM, Miller CA, Strickland DK, Ghiso J, Zlokovic BV (2000) Clearance of Alzheimer's amyloid-β(1–40) peptide from brain by LDL receptor–related protein-1 at the blood-brain barrier. J Clin Invest 106(12):1489–1499

Shie FS, Jin LW, Cook DG, Leverenz JB, LeBoeuf RC (2002) Diet-induced hypercholesterolemia enhances brain Aβ accumulation in transgenic mice. Neuroreport 13(4):455–459

Simons K, Ikonen E (1997) Functional rafts in cell membranes. Nature 387(6633):569–572

Simons M, Keller P, De Strooper B, Beyreuther K, Dotti CG, Simons K (1998) Cholesterol depletion inhibits the generation of β-amyloid in hippocampal neurons. Proc Natl Acad Sci U S A 95(11):6460–6464

Simons M, Schwarzler F, Lutjohann D, von Bergmann K, Beyreuther K, Dichgans J, Wormstall H, Hartmann T, Schulz JB (2002) Treatment with simvastatin in normocholesterolemic patients with Alzheimer's disease: a 26-week randomized, placebo-controlled, double-blind trial. Ann Neurol 52(3):346–350

Soscia SJ, Kirby JE, Washicosky KJ, Tucker SM, Ingelsson M, Hyman B, Burton MA, Goldstein LE, Duong S, Tanzi RE, Moir RD (2010) The Alzheimer's disease-associated amyloid β-protein is an antimicrobial peptide. PLoS ONE 3(5):e9505

Soto C (2003) Unfolding the role of protein misfolding in neurodegenerative diseases. Nat Rev Neurosci 4(1):49–60

Sparks DL, Sabbagh MN, Connor DJ, Lopez J, Launer LJ, Browne P, Wasser D, Johnson-Traver S, Lochhead J, Ziolwolski C (2005) Atorvastatin for the treatment of mild to moderate Alzheimer disease: preliminary results. Arch Neurol 62(5):753–757

Stefani M (2010) Biochemical and biophysical features of both oligomer/fibril and cell membrane in amyloid cytotoxicity. FEBS J 277(22):4602–4613

Stefani M, Dobson C (2003) Protein aggregation and aggregate toxicity: new insights into protein folding, misfolding diseases and biological evolution. J Mol Med 81(11):678–699

Straub JE, Thirumalai D (2014) Membrane-protein interactions are key to understanding amyloid formation. J Phys Chem Lett 5(3):633–635

Strittmatter WJ, Saunders AM, Schmechel D, Pericak-Vance M, Enghild J, Salvesen GS, Roses AD (1993) Apolipoprotein E: high-avidity binding to β-amyloid and increased frequency of type 4 allele in late-onset familial Alzheimer disease. Proc Natl Acad Sci U S A 90(5):1977–1981

Stroud JC, Liu C, Teng PK, Eisenberg D (2012) Toxic fibrillar oligomers of amyloid-b have cross-b structure. Proc Natl Acad Sci U S A 109(20):7717–7722

Sunde M, Serpell LC, Bartlam M, Fraser PE, Pepys MB, Blake CCF (1997) Common core structure of amyloid fibrils by synchrotron X-ray diffraction. J Mol Biol 273(3):729–739

Suri S, Heise V, Trachtenberg AJ, Mackay CE (2013) The forgotten APOE allele: a review of the evidence and suggested mechanisms for the protective effect of APOE e2. Neurosci Biobehav Rev 37(10, Part 2):2878–2886

Takami M, Nagashima Y, Sano Y, Ishihara S, Morishima-Kawashima M, Funamoto S, Ihara Y (2009) g-secretase: successive tripeptide and tetrapeptide release from the transmembrane domain of β-carboxyl terminal fragment. J Neurosci 29(41):13042–13052

Tamamizu-Kato S, Cohen JK, Drake CB, Kosaraju MG, Drury J, Narayanaswami V (2008) Interaction with amyloid β peptide compromises the lipid binding function of apolipoprotein E. Biochemistry 47(18):5225–5234

Teplow DB, Lazo ND, Bitan G, Bernstein S, Wyttenbach T, Bowers MT, Baumketner A, Shea J-E, Urbanc B, Cruz L, Borreguero J, Stanley HE (2006) Elucidating amyloid β-protein folding and assembly: a multidisciplinary approach. Acc Chem Res 39(9):635–645

Terzi E, Holzemann G, Seelig J (1995) Self-association of β-amyloid peptide (1–40) in solution and binding to lipid membranes. J Mol Biol 252(5):633–642

Terzi E, Holzemann G, Seelig J (1997) Interaction of Alzheimer β-amyloid peptide(1–40) with lipid membranes. Biochemistry 36(48):14845–14852

Tiraboschi P, Hansen LA, Masliah E, Alford M, Thal LJ, Corey-Bloom J (2004) Impact of APOE genotype on neuropathologic and neurochemical markers of Alzheimer disease. Neurology 62(11):1977–1983

Tofoleanu F, Buchete N-V (2012) Alzheimer Aβ peptide interactions with lipid membranes: fibrils, oligomers and polymorphic amyloid channels. Prion 6(4):339–345

Tokuda T, Calero M, Matsubara E, Vidal R, Kumar A, Permanne B, Zlokovic B, Smith JD, Ladu MJ, Rostagno A, Frangione B, Ghiso J (2000) Lipidation of apolipoprotein E influences its isoform-specific interaction with Alzheimer's amyloid β peptides. Biochem J 348(2):359–365

Tomaselli S, Esposito V, Vangone P, van Nuland NAJ, Bonvin AMJJ, Guerrini R, Tancredi T, Temussi PA, Picone D (2006) The α-to-β conformational transition of Alzheimer's Aβ-(1–42) peptide in aqueous media is reversible: a step by step conformational analysis suggests the location of β conformation seeding. ChemBioChem 7(2):257–267

Utsumi M, Yamaguchi Y, Sasakawa H, Yamamoto N, Yanagisawa K, Kato K (2009) Up-and-down topological mode of amyloid b-peptide lying on hydrophilic/hydrophobic interface of ganglioside clusters. Glycoconj J 26(8):999–1006

Valincius G, Heinrich F, Budvytyte R, Vanderah DJ, McGillivray DJ, Sokolov Y, Hall JE, Loesche M (2008) Soluble amyloid β-oligomers affect dielectric membrane properties by bilayer insertion and domain formation: implications for cell toxicity. Biophys J 95(10):4845–4861

Verdier Y, Zarándi M, Penke B (2004) Amyloid β-peptide interactions with neuronal and glial cell plasma membrane: binding sites and implications for Alzheimer's disease. J Pept Sci 10(5):229–248

Verghese PB, Castellano JM, Holtzman DM (2011) Apolipoprotein E in Alzheimer's disease and other neurological disorders. Lancet Neurol 10(3):241–252

Verghese PB, Castellano JM, Garai K, Wang Y, Jiang H, Shah A, Bu G, Frieden C, Holtzman DM (2013) ApoE influences amyloid-β (Aβ) clearance despite minimal apoE/Aβ association in physiological conditions. Proc Natl Acad Sci U S A 110:E1807–E1816

Vetrivel KS, Thinakaran G (2010) Membrane rafts in Alzheimer's disease β-amyloid production. Biochim Biophys Acta 1801(8):860–867

Vivekanandan S, Brender JR, Lee SY, Ramamoorthy A (2011) A partially folded structure of amyloid-β(1–40) in an aqueous environment. Biochem Biophys Res Commun 411(2):312–316

Wahlström A, Hugonin L, Perálvarez-Marín A, Jarvet J, Gräslund A (2008) Secondary structure conversions of Alzheimer's Aβ(1–40) peptide induced by membrane-mimicking detergents. FEBS J 275(20):5117–5128

Wakabayashi M, Okada T, Kozutsumi Y, Matsuzaki K (2005) GM1 ganglioside-mediated accumulation of amyloid b-protein on cell membranes. Biochem Biophys Res Commun 328(4):1019–1023

Walsh DM, Selkoe DJ (2004) Oligomers on the brain: the emerging role of soluble protein aggregates in neuro-degeneration. Protein Pept Lett 11:213–228

Walsh DM, Klyubin I, Fadeeva JV, Cullen WK, Anwyl R, Wolfe MS, Rowan MJ, Selkoe DJ (2002) Naturally secreted oligomers of amyloid β protein potently inhibit hippocampal long-term potentiation in vivo. Nature 416(6880):535–539

Weisgraber KH (1994) Apolipoprotein E: structure-function relationships. Adv Protein Chem 45:249–302

Wetzel R (2006) Kinetics and thermodynamics of amyloid fibril assembly. Acc Chem Res 39(9):671–679

Wildsmith K, Holley M, Savage J, Skerrett R, Landreth G (2013) Evidence for impaired amyloid β clearance in Alzheimer's disease. Alzheimer Res Ther 5(4):33

William Rebeck G, Reiter JS, Strickland DK, Hyman BT (1993) Apolipoprotein E in sporadic Alzheimer's disease: allelic variation and receptor interactions. Neuron 11(4):575–580

Wisniewski T, Castano EM, Golabek A, Vogel T, Frangione B (1994) Acceleration of Alzheimer's fibril formation by apolipoprotein E in vitro. Am J Pathol 145(5):1030–1035

Wittnam JL, Portelius E, Zetterberg H, Gustavsson MK, Schilling S, Koch B, Demuth H-U, Blennow K, Wirths O, Bayer TA (2012) Pyroglutamate amyloid β (Aβ) aggravates behavioral deficits in transgenic amyloid mouse model for Alzheimer disease. J Biol Chem 287(11):8154–8162

Wolozin B, Kellman W, Ruosseau P, Celesia GG, Siegel G (2000) Decreased prevalence of Alzheimer disease associated with 3-hydroxy-3-methyglutaryl coenzyme A reductase inhibitors. Arch Neurol 57(10):1439–1443

Wood WG, Schroeder F, Igbavboa U, Avdulov NA, Chochina SV (2002) Brain membrane cholesterol domains, aging and amyloid β-peptides. Neurobiol Aging 23(5):685–694

Xu Y, Shen J, Luo X, Zhu W, Chen K, Ma J, Jiang H (2005) Conformational transition of amyloid β-peptide. Proc Natl Acad Sci U S A 102:5403–5407

Xue W-F, Homans SW, Radford SE (2008) Systematic analysis of nucleation-dependent polymerization reveals new insights into the mechanism of amyloid self-assembly. Proc Natl Acad Sci U S A 105(26):8926–8931

Yagi-Utsumi M, Kameda T, Yamaguchi Y, Kato K (2010) NMR characterization of the interactions between lyso-GM1 aqueous micelles and amyloid β. FEBS Lett 584(4):831–836

Yamaguchi H, Maat-Schieman ML, van Duinen SG, Prins FA, Neeskens P, Natte R, Roos RA (2000) Amyloid β protein (Aβ) starts to deposit as plasma membrane-bound form in diffuse plaques of brains from hereditary cerebral hemorrhage with amyloidosis-Dutch type, Alzheimer disease and nondemented aged subjects. J Neuropathol Exp Neurol 59(8):723–732

Yamamoto N, Hirabayashi Y, Amari M, Yamaguchi H, Romanov G, Van Nostrand WE, Yanagisawa K (2005) Assembly of hereditary amyloid β-protein variants in the presence of favorable gangliosides. FEBS Lett 579(10):2185–2190

Yamauchi K, Tozuka M, Hidaka H, Nakabayashi T, Sugano M, Katsuyama T (2002) Isoform-specific effect of apolipoprotein E on endocytosis of Aβ-amyloid in cultures of neuroblastoma cells. Ann Clin Lab Sci 32(1):65–74

Yan P, Bero AW, Cirrito JR, Xiao Q, Hu X, Wang Y, Gonzales E, Holtzman DM, Lee J-M (2009) Characterizing the appearance and growth of amyloid plaques in APP/PS1 mice. J Neurosci 29(34):10706–10714

Yanagisawa K, Odaka A, Suzuki N, Ihara Y (1995) GM1 ganglioside-bound amyloid β-protein (Aβ): a possible form of preamyloid in Alzheimer's disease. Nat Med 1(10):1062–1066

Yao Z-X, Papadopoulos V (2002) Function of β-amyloid in cholesterol transport: a lead to neurotoxicity. FASEB J 16:677–679

Yip CM, McLaurin J (2001) Amyloid-β peptide assembly: a critical step in fibrillogenesis and membrane disruption. Biophys J 80(3):1359–1371

Yu X, Zheng J (2012) Cholesterol promotes the interaction of Alzheimer β-amyloid monomer with lipid bilayer. J Mol Biol 421(4–5):561–571

Zannis VI, Kardassis D, Zanni EE (1993) Genetic mutations affecting human lipoproteins, their receptors, and their enzymes. Adv Hum Genet 21:145–319

Zha Q, Ruan Y, Hartmann T, Beyreuther K, Zhang D (2004) GM1 ganglioside regulates the proteolysis of amyloid precursor protein. Mol Psychiatry 9(10):946–952

Zhang S, Iwata K, Lachenmann MJ, Peng JW, Li S, Stimson ER, Lu Y, Felix AM, Maggio JE, Lee JP (2000) The Alzheimer's peptide Aβ adopts a collapsed structure in water. J Struct Biol 130:130–141

Zhao LN, Long H, Mu Y, Chew LY (2012) The toxicity of amyloid β oligomers. Int J Mol Sci 13(6):7303–7327

Zhong N, Weisgraber KH (2009) Understanding the association of apolipoprotein E4 with Alzheimer disease: clues from its structure. J Biol Chem 284(10):6027–6031

Zlokovic BV (2008) The blood-brain barrier in health and chronic neurodegenerative disorders. Neuron 57(2):178–201

Role of Cholesterol and Phospholipids in Amylin Misfolding, Aggregation and Etiology of Islet Amyloidosis

4

Sanghamitra Singh, Saurabh Trikha, Diti Chatterjee Bhowmick, Anjali A. Sarkar, and Aleksandar M. Jeremic

Abstract

Amyloidosis is a biological event in which proteins undergo structural transitions from soluble monomers and oligomers to insoluble fibrillar aggregates that are often toxic to cells. Exactly how amyloid proteins, such as the pancreatic hormone amylin, aggregate and kill cells is still unclear. Islet amyloid polypeptide, or amylin, is a recently discovered hormone that is stored and co-released with insulin from pancreatic islet β-cells. The pathology of type 2 diabetes mellitus (T2DM) is characterized by an excessive extracellular and intracellular accumulation of toxic amylin species, soluble oligomers and insoluble fibrils, in islets, eventually leading to β-cell loss. Obesity and elevated serum cholesterol levels are additional risk factors implicated in the development of T2DM. Because the homeostatic balance between cholesterol synthesis and uptake is lost in diabetics, and amylin aggregation is a hallmark of T2DM, this chapter focuses on the biophysical and cell biology studies exploring molecular mechanisms by which cholesterol and phospholipids modulate secondary structure, folding and aggregation of human amylin and other amyloid proteins on membranes and in cells. Amylin turnover and toxicity in pancreatic cells and the regulatory role of cholesterol in these processes are also discussed.

Keywords

Atomic force microscopy • Phospholipids • Cholesterol • Membranes • Islet amyloid • Type-2 diabetes mellitus

S. Trikha
Department of Biological Sciences,
The George Washington University,
2023 G Street NW, Washington, DC 20052, USA

Present Affiliation: Endocrinology Division,
Boston Children's Hospital,
Harvard Medical School, Boston, MA 02115, USA
e-mail: Saurabh.Trikha@childrens.harvard.edu

S. Singh • D.C. Bhowmick • A.A. Sarkar
A.M. Jeremic (✉)
Department of Biological Sciences, The George
Washington University, 2023 G Street NW,
Washington, DC 20052, USA
e-mail: jerema@gwu.edu

© Springer International Publishing Switzerland 2015
O. Gursky (ed.), *Lipids in Protein Misfolding*, Advances in Experimental
Medicine and Biology 855, DOI 10.1007/978-3-319-17344-3_4

Abbreviations

AFM   Atomic force microscopy
BCD   Methylbetacyclodextrin
CD   Circular dichroism
CTX   Cholera toxin
DOPC   1,2-dioleoyl-phosphatidylcholine
DOPS   1,2-dioleoylphosphatidylserine
EM   Electron microscopy
HFIP   Hexafluoride isopropanol
hIAPP   Human islet amyloid peptide
Lov   Lovostatin
PM   Plasma membranes
T2DM   Type 2 diabetes mellitus
ThT   Thioflavin-T
Trf   Transferrin

4.1 Amylin Biology and Function

Amylin, also known as islet amyloid polypeptide (IAPP), is a 37 amino acid hormone produced and co-secreted with insulin from pancreatic β-cells (Hoppener and Lips 2006; Clark and Nilsson 2004). IAPP is also expressed in pancreatic islet δ-cells in rat and mouse, in gastrointestinal tract of rat, mouse, cat, and human, as well as in sensory neurons of rat and mouse. In chicken, IAPP is mainly expressed in the brain and intestine and, at much lower levels, in the pancreas (Fan et al. 1994; Miyazato et al. 1991; Mulder et al. 1996). Amylin is expressed in both human and rat placenta primarily during early pregnancy (Piper et al. 2004). Placental amylin mRNA expression is highest in the third trimester of pregnancy in humans and 16 days of gestational age in rats, and gets lower as the gestation progresses (Caminos et al. 2009).

Although the exact hormonal function of amylin is still unclear, it has been proposed that amylin controls food intake and energy homeostasis (Lutz 2006, 2010). Amylin primarily regulates nutrient fluxes by acting as a potent satiation signal that reduces secretion of gastric juices and the glucagon hormone, and also reduces the rate of gastric emptying (Young and Denaro 1998). Peripheral amylin regulates satiation signal by directly binding to area postrema neurons, which are rich in amylin receptors and subsequently convey this signal to other brain areas (Lutz 2006). Interestingly, the central regulatory pathway by which other gastrointestinal peptides like cholecystokinin, glucagon-like peptide 1 and peptide YY 3–36 suppress eating widely overlaps with that of amylin (Lutz 2006; Riediger et al. 2004). In addition, amylin is also involved in adiposity signaling and, similar to leptin, in body weight regulation all through adult life (Lutz 2010). Studies with animal and human subjects showed that combinational application of leptin and amylin increases leptin responsiveness in anti-obesity treatments, which suggests the synergistic function of these hormones (Lutz 2010). Recent studies suggest that amylin-mediated regulation of energy balance is not limited to the control of nutrient flux but also involves the body's energy expenditure (Lutz 2010). However, the exact mechanism and the physiological relevance are still under scrutiny (Lutz 2010).

As the first two trimesters of pregnancy in humans and the equivalent period in rats are considered to be highly anabolic, and placental amylin expression is highest during this time, amylin could potentially play important roles in anabolic control of food intake in both the mother and the fetus during pregnancy (Caminos et al. 2009). Amylin also plays a developmental role by contributing to the development of bone, kidney and pancreas (Wookey et al. 2006). Finally, amylin regulates the early postnatal development of hindbrain in mouse though its positive neurotropic effects (Lutz 2010). In addition to its hormonal role, amylin also imposes important paracrine and autocrine effects in islets by regulating glucagon and insulin release from α- and β-cells, respectively (Trikha and Jeremic 2013; Wagoner et al. 1993).

Amylin and insulin genes share common promoter elements, and the transcription factor PDX1 regulates glucose-stimulated secretion of both these genes (German et al. 1992). It has been reported that in rodent models, glucose stimulation of pancreatic β-cells results in parallel expression patterns of insulin and IAPP, although in experimental diabetic models of rodents this parallel expression pattern is altered (Mulder et al. 1996). Amylin is synthesized in cells as an 89-residue pre-pro-protein (Nakazato et al. 1990; Nishi et al. 1989). The 22-residue signal peptide of immature form is cleaved off in the endoplasmic reticulum (ER). Further processing of pro-IAPP, along with pro-insulin, takes place in the Golgi and the secretory vesicles in a pH-dependent manner using two endoproteases: prohormone convertase 2 (PC2) and prohormone convertase 1/3 (PC1/3) (Westermark et al. 2011). PC2 and PC1/3 cleave pro-IAPP after Lys10 and/or Arg11 (Wang et al. 2001) and after Lys 50 and Arg51, respectively (Marzban et al. 2004). After PC1/3-mediated cleavage, the two C-terminal amino acid residues are then removed by carboxypeptidase E, which results in an exposed glycine residue at the C-terminus of pro-IAPP (Westermark et al. 2011). This glycine is used as a signal for C-terminal amidation; finally, a disulfide bridge is formed between Cys2 and Cys7. Both C-terminal amide and this disulfide bridge are important for full biological activity of IAPP (Westermark et al. 2011). Fully processed IAPP is a 37-residue polypeptide stored in secretory granules of pancreatic islet β-cells along with fully processed insulin. Upon physiological stimulation such as glucose spike in serum, insulin and amylin are co-secreted at a molar ratio of 20:1 (Martin 2006).

The hormone amylin, which is similar to the neuropeptide calcitonin gene-related peptide, should have specific receptors to mediate its physiological function. However, efforts to identify specific amylin receptors were futile for a long time until the identification of a family of single- domain proteins called "receptor activity-modifying proteins", or RAMPs, which do not function as receptors by themselves (McLatchie et al. 1998). Amylin receptor utilizes a novel principle that has so far been detected only among the family of calcitonin receptors. RAMPs bind to the calcitonin receptors, and hetero-dimerization of RAMP with calcitonin receptor yields a unique high-affinity amylin receptor (AM-R) phenotype (Poyner et al. 2002). The three known AM-R isoforms discovered so far have been shown to exhibit distinct pharmacological and functional properties (Morfis et al. 2008). AM-R expression in different organs and

tissues, particularly in the brain and in the pancreas, suggests its regulatory role in glucose homeostasis, hormone and neurotransmitter release and signaling (Martinez et al. 2000; Trikha and Jeremic 2013).

4.2   Amylin Aggregation, Islet Amyloidosis and Type 2 Diabetes Mellitus

Islet amyloid was first reported in 1901 (Opie 1901) as thick proteinaceous deposits in the pancreas of diabetics, and was initially named "islet hyalinization" because of its hyaline-like or glassy appearance. It was later renamed "amyloid", which means "starch-like", because islet amyloids were initially believed to be carbohydrates as they could take up dyes which are typically used to stain starch (Clark and Nilsson 2004). Despite numerous studies, the origin and nature of islet amyloid remained enigmatic for a long time (Westermark et al. 2011). Purification and characterization of amyloid aggregates from the amyloid-rich insulinoma cells and islets of human and feline origin identified amylin as the main component (Cooper et al. 1987; Westermark et al. 1986). It is now known that amylin-derived amyloid aggregates often associate with apolipoprotein E (apoE) and heparan sulfate proteoglycans (Ancsin 2003; Clark and Nilsson 2004; Hoppener et al. 2000).

Type 2 diabetes mellitus, one of the most common metabolic diseases in the world, is characterized by "insulin resistance" in the target organs, mainly muscle and liver, and by the decline in the production and secretion of insulin, loss of β-cell mass and formation of islet amyloid (Clark and Nilsson 2004; Hoppener et al. 2000). The role of islet amyloidosis in the pathogenesis of T2DM is supported by several studies showing the presence of amylin-derived amyloid plaques in over 90 % of diabetics (Clark and Nilsson 2004; Hoppener et al. 2000). While amylin has been detected in monkeys and cats, species known to develop T2DM, it is absent in rodents and mice, species which do not develop T2DM (Clark and Nilsson 2004; Hoppener et al. 2000).

This is strong yet indirect evidence correlating T2DM and islet amyloidosis. Whether islet amyloidosis is a cause or a consequence of the disease is still unclear.

In vitro studies revealed that human but not rat amylin undergoes rapid aggregation in physiological buffers and that insulin, but not pro-insulin, inhibits IAPP aggregation by forming hetero-molecular complexes (Clark and Nilsson 2004; Westermark et al. 1999; Kayed et al. 1999). Therefore, faulty insulin processing in diabetics could partially explain amylin aggregation in T2DM. Defective processing of pro-IAPP into IAPP is another candidate for amylin aggregation in T2DM, as N-terminal intact pro-IAPP has been identified in islet β-cells of diabetics (Clark and Nilsson 2004). In fact, pro-peptides have strong self-association properties and are capable of forming amyloid aggregates (Krampert et al. 2000). However, compared to fully processed IAPP, proIAPP has less amyloidogenicity and less toxicity (Jha et al. 2009; Krampert et al. 2000) This suggests that pro region of pro-peptide may play a protective role in amyloidogenic and toxic potentials of fully processed IAPP (Krampert et al. 2000). Increased accumulation of amyloid aggregates inside and outside the cells accounts for downstream pathological events such as calcium overload, cell membrane disruption, ER stress, mitochondrial dysfunction, defects in autophagy, oxidative stress and activation of JNK and caspase-3 death signaling pathways (Abedini and Schmidt 2013; Cao et al. 2013a; Costes et al. 2014; Huang et al. 2011; Konarkowska et al. 2006; Rivera et al. 2014; Zhang et al. 2003). Since the ability of IAPP to penetrate through lipid membranes depends on the lipid-to-peptide ratio, the toxicity of IAPP is thought to be enhanced due to an increase in its local concentration (Cao et al. 2013b; Clark and Nilsson 2004).

4.3   Amylin Misfolding, Aggregation, and Toxicity: A Dangerous Trio

The primary sequences of mature (fully processed) rat (rIAPP) and human amylin (hIAPP) are depicted in Fig. 4.1a. Although human and rat amylin share high sequence homology, the presence or absence of just a few key amino acids in the amyloidogenic region of the peptide (residues 18–29, Fig. 4.1a) may drastically alter protein's aggregation and cytotoxic properties. Computational and mutational studies confirmed that 18–29 aa segment of mature hIAPP is highly amyloidogenic (Chiu et al. 2013; Moriarty and Raleigh 1999; Westermark et al. 1990). For instance, the presence of His at position 18 in human amylin is required for amylin-plasma membrane interactions, aggregation and toxicity (Abedini and Raleigh 2005; Brender et al. 2008a; Tu and Raleigh 2013). The presence of three Pro residues in positions 25, 28 and 29 renders rat amylin soluble (non-amyloidogenic) and non-toxic (Fig. 4.1b, c) (Westermark et al. 2011). Likewise, substitutions of Asn22, Gly24, and residues 26–28 with Pro markedly reduced aggregation of 20–29 hIAPP fragment (Moriarty and Raleigh 1999). Thus, an absence of His and the presence of Pro in the sensitive residue segment 18–29 of rat as compared to human amylin is believed to prevent its aggregation and toxicity in rodents.

In addition to His and Pro, other polar amino acids from the amyloidogenic region (Fig. 4.1a), such as Ser20, may also play a regulatory role in human amylin aggregation and islet amyloid formation. In fact, Ser20 to Gly mutation in mature human amylin was observed in a small subset of Chinese and Japanese populations who are at an increased risk of developing T2DM. Interestingly, in vitro studies revealed that Ser20Gly substitution accelerated amylin aggregation in solution (Cao et al. 2012), which may help explain increased incidence of diabetes in these two ethnic groups. Notably, amylin has a characteristic intramolecular disulfide bond between Cys2 and Cys7, which does not initially contribute to the aggregation (nucleation) process, although its absence reduces fibril formation (Khemtemourian et al. 2008; Koo and Miranker 2005). The rate of amylin fibrillization parallels the onset and the extent of membrane damage in vitro (Engel et al. 2008). These findings support the *fibril hypothesis* of amylin's toxicity in pancreatic islets. However, recent studies point to an important role of pre-fibrillar, soluble oligomeric species in

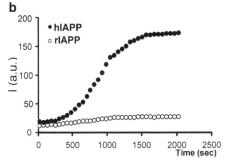

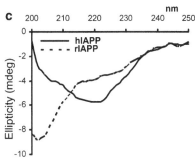

Fig. 4.1 Aggregation of human amylin and changes of its secondary structure coincides in time. (**a**) Primary structures of mature human (hIAPP) and rat (rIAPP) amylin are depicted. Species-specific amino-acids within the amyloidoigenic region (*underlined*) of the polypeptide chain are bolded for clarity. (**b**) Kinetics and extent of aggregation of human and rat amylin in PBS as a function of time. Thioflavin-T fluorescent assay reveals fibrilogenesis of 20 μM human amylin in solution (*closed circles*) and lack of aggregation of non-amyloidogenic rat amylin (20 μM; *open circles*). (**c**) Far-UV CD spectra of human amylin (*solid line*) and rat amylin (*dashed line*) taken after 20 min. in PBS solution in the presence of 2 % HFIP. Note the absorption minimum at ~220 nm for human but not rat amylin, typical for peptides and proteins adopting β-sheet conformation

human amylin-induced membrane damage and β-cell death (Cao et al. 2013a; Haataja et al. 2008; Janson et al. 1999; Konarkowska et al. 2006; Ritzel et al. 2007; Trikha and Jeremic 2011; Zhang et al. 2014). This process, commonly referred as toxic *oligomer hypothesis*, together with the *fibril hypothesis*, will be addressed in this chapter.

4.4 Conformation Changes and Aggregation of Amylin: A Causal Link

Because the dynamics and the extent of amylin oligomerization and aggregation were shown to be important parameters of amylin's toxicity (Cao et al. 2013a; Engel et al. 2008; Ritzel et al. 2007), implementation of biophysical methods such as fluorescence and circular dichroism (CD) spectroscopy and high-resolution microscopy capable of tracking these changes in real time have become a norm in recent years. Such biophysical studies are essential to understand amylin aggregation at the molecular level and to determine how certain cellular factors such as pH, ionic content and temperature may contribute to the formation of amyloid plaques in the pancreas and other organs.

Thioflavin (ThT) fluorescence assay (Fig. 4.1b) is a commonly used method to monitor the extent and the kinetics of aggregation of various amyloid peptides and proteins in vitro in cell-free environment (Munishkina and Fink 2007). In the absence of amyloid, the diagnostic dye ThT is weakly fluorescent in solution. However, during amyloid formation, the ThT molecules intercalate into the growing amyloid fibers, rendering the probe more fluorescent (for details see Gorbenko et al., Chap. 6, this volume). Thus, increase in ThT fluorescence over time reflects the fibrillization process that is amenable for experimental manipulations. Lag (nucleation) phase followed by sigmoidal (fibril growth) phase are two common traits shared by amyloid proteins undergoing aggregation (Fig. 4.1b). The ThT assay was previously used by many investigators to understand how changes in pH, temperature or presence or absence of certain metals affect the rate and the extent of amylin aggregation. For example, increasing the salt concentrations in the incubation medium to screen out electrostatic interactions in solution decreased both the rate and the extent of human amylin aggregation (Cho et al. 2008). Thus, amylin aggregation inversely correlates to the solvent ionic strength, which suggests that intra- and inter-molecular non-covalent interactions among certain residues play a major role in self-association and polymerization of human amylin in solution. Aromatic and hydrophobic interactions were proposed to play a major role in amylin polymerization in solution (Gazit 2002; Tu and Raleigh 2013). These two non-covalent interactions also play an important role in self-assembly (oligomerization) of peptides into channel-like structures in the membrane, the efficacy of which inversely correlates to ionic strength of the solution (Zhao et al. 2008). In this study formation of protein pores was inhibited when ionic strength of solution increased, whereas both hydrophobic and aromatic interactions were retarded with the increase of salt concentration (Zhao et al. 2008). Thus, it is highly conceivable that amylin oligomerization, the first step in amylin aggregation, is retarded in solution with increased ionic strength due to the inhibitory effect of salts on aromatic and hydrophobic interactions, two major driving forces in amylin polymerization (Gazit 2002; Tu and Raleigh 2013). This eventually would diminish aggregation of human amylin, as showed recently (Cho et al 2008).

Together with ThT assay, the structural studies revealed a causal link between conformational changes in amylin and its propensity to aggregate (Fig. 4.1c) (Brender et al. 2008b; Cho et al. 2008, 2009; Wiltzius et al. 2008). Similar to many other small proteins and peptides (described by Uversky in Chap. 2 of this volume), amylin is natively unfolded in solution. However, amylin can polymerize in a cross-β-sheet conformation upon aggregation in amyloid fibers. CD analysis revealed that aggregation of human amylin is accompanied by secondary structural changes, from random coil in the monomeric from to the β-sheet-enriched fibrillar form characterized by a single minimum at ~220 nm (Fig. 4.1c). In con-

trast, rat amylin retains its random coil conformation in solution, characterized by a minimum at 202 nm (Fig. 4.1c), which prevents its aggregation (Fig. 4.1b). The likely reason for this difference is that rat amylin contains three structure-breaking prolines, Pro25, Pro28 and Pro29, in the residue segment that probably initiates amyloid formation of human amylin (Fig. 4.1a); these three prolines are expected and observed to prevent β-aggregation (Fig. 4.1b, c) (Moriarty and Raleigh 1999). Inhibition of human amylin transition towards β-sheet conformation by certain inhibitors (divalent metals, insulin or cholesterol) also prevents its aggregation (Cho et al. 2008, 2009; Salamekh et al. 2011; Susa et al. 2014). Collectively, these biophysical studies reveal that aggregation of amylin, like other amyloid proteins, is strongly conformation-dependent and that transition to β-sheet is a requirement for the formation of fibrils.

Although the aforementioned bulk spectroscopy studies provided important information on the dynamics and conformational changes associated with protein misfolding and aggregation, they could neither provide information on the nature and architecture of pre-aggregated species, nor explain how they assemble into fibrils. Without this information, the process of amylin aggregation and amyloid formation in tissues cannot be fully understood. Therefore, visualization of amylin aggregation became imperative. Given the small size of aggregated species, and in order to visualize peptide/protein transition from monomers to oligomers to large aggregates, a new real-time imaging tool capable of imaging at nm-resolution was needed. The development of atomic force microscope (AFM), a 3D lens imaging instrument, allowed investigators to examine, for the first time, the process of amyloid formation with unprecedented clarity and specificity. Formation and growth (extension) of a single fibril has been monitored using this technology (Fig. 4.2) (Cho et al. 2008; Goldsbury et al. 1999; Green et al. 2004). The unique capability of AFM to directly monitor changes in the conformation or aggregation state of macromolecules, and to study dynamic aspects of molecular interactions in their physiological buffer environment has

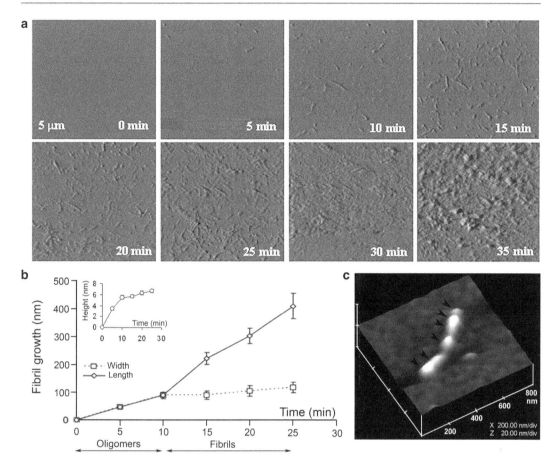

Fig. 4.2 Dynamics of amylin aggregation on solid surface. (**a**) Structural intermediates, oligomers and fibrils, are resolved during amylin aggregation on mica by time-lapse AFM (tapping mode amplitude images). Note a time-dependent transition of human amylin from small round oligomers (0–10 min) into fibrils during early-mid stage of amylin aggregation (10–25 min). Late–stage of amylin aggregation (25–35 min) is characterized by accumulation of massive peptide deposits on the mica surface. All micrographs are 5×5 μm. (**b**) Fibril growth curves reveal two phases of amylin aggregation, an early oligomeric phase (0–10 min) characterized by oligomers formation, followed by oligomers incorporation into growing fibrils (10–25 min, second phase or fibril maturation). Note the significant increase in oligomer heights (*inset*) and widths during the first phase of amylin aggregation, and an abrupt increase in fibrils length following formation of full-size oligomers. Data represents mean particle size at each time point (mean ± SEM), obtained from three independent time-lapse AFM experiments. (**c**) 3-D AFM image of a single full-grown fibril on mica showing arrangement of several amylin oligomers and their bi-directional extension into a fibril (depicted by *arrowheads*). Micrograph is 800×800 nm scale

allowed examination of amylin aggregates at ~5 nm lateral and ≤1 nm vertical resolution (Figs. 4.2, 4.4, and 4.5) (Cho et al. 2008, 2009; Green et al. 2004). This novel imaging technology has provided new insights into the molecular mechanism of amyloid assembly.

In our studies, time-lapse AFM operating either in contact or tapping mode was used to investigate the organization of amylin aggregates on solid surface such as mica (Fig. 4.2) and on planar lipid membranes (Figs. 4.4 and 4.5), two surfaces bearing distinct physicochemical properties. With the scanner speed set at 1 Hz and image acquisition time of ~5 min/image, and using high-resolution scanning parameters (512×512 lines per image), the dynamics, polymorphism and the extent of amylin fibrillization can be obtained ("the fibril growth" in Fig. 4.2b).

Time-lapse amplitude AFM micrographs revealed structural transitions of amylin on mica, from small spherical oligomers to extended fibrils, over a 30 min time period (Fig. 4.2a).

After acquiring micrographs, the size of individual fibrils and oligomers (i.e. radius, length and height) that were deposited on mica (Fig. 4.2b) or on planar membranes (Fig. 4.4) could be determined using a section analysis tool (Veeco, Santa Barbara, CA). Cross-sectional analysis revealed that amylin fibrils varied by length and consistently measured 90–110 nm in width and 5–6 nm in height (Fig. 4.2a, b) (Cho et al. 2008). In addition to amplitude AFM images (Fig. 4.2a), the height-AFM micrographs revealed changes in fibril height during amylin aggregation (Figs. 4.2c and 4.4). Changes in particle height are more visible in the height imaging mode as compared to amplitude images, which are better suited for imaging the fine morphological details of amylin aggregates. Some fibrils were relatively short (less than 200 nm), whereas others extended over 500 nm in length (Fig. 4.2a, c). To construct fibril growth curves, the average size of oligomers and fibrils was determined and plotted for each time point (Figs. 4.2b and 4.4c). In the presence of 1–2 % hexafluoride isopropanol, which accelerates amylin aggregation, massive amyloid-like amylin deposits generally developed after 30 min of incubation (Fig. 4.2a, 30–35 min), thus precluding the monitoring of fibril growth for an extended period of time. However, AFM resolved amylin structural intermediates prior to amyloid accumulation. Formation of fibrils occurred in two distinct phases. The initial phase involved deposition of small spherical oligomers with diameter (width) of 47 ± 7 nm and height of 3.4 ± 0.3 nm (Fig. 4.2a, b, 0–5 min). During the next 5 min, oligomers almost doubled in size (diameter 89 ± 13 nm, height 5.5 ± 0.4 nm; Fig. 4.2a, b), followed by the oligomer's bi-directional extension into a fibril (Fig. 4.2c) at an average fibrillization rate of 21 nm/min (Fig. 4.2a, b, 10–25 min). Growth curves revealed two distinct phases in amylin aggregation: the first phase, or oligomer growth, was characterized by the large change in the oligomer's height and width within the first

10 min of aggregation, reaching ~90 % of their maximal value (Fig. 4.2b, inset) but accounting for only ~20 % of maximum fibril length during that period; and second phase or fibril growth, when fibrils rapidly elongated by doubling their extension rate from 9 nm/min (0–10 min interval, Fig. 4.2b) to 21 nm/min (10–25 min interval, Fig. 4.2b). Taken together, these results suggest that fibrils are formed on mica by longitudinal extension of full-grown oligomers. Hence, amylin fibrillization depends on formation of "building block" oligomers, or nuclei, measuring approximately ~6 nm in height and ~90 nm in diameter. Once formed, these nuclei align and elongate into a fibril (Fig. 4.2a–c), a scenario originally proposed by Aebi and co-workers (Green et al. 2004).

4.5 Amylin Folding and Aggregation in Solution Are Strongly Modulated by Anionic Lipids and Cholesterol

The above-mentioned spectroscopy and microscopy studies revealed species and molecular mechanism of amylin aggregation in solution and on solid surface. However, amylin aggregates were also found in close proximity to islet β-cells, with some fibrils integrated into the β-cell plasma membranes (PM) (MacArthur et al. 1999). This finding suggests that amylin-membrane interactions may be important for both amylin aggregation on the cell surface and for the integrity and function of the β-cell PM. The regulatory role and involvement of the membrane's main constituents, phospholipids and cholesterol, in amylin aggregation were explored during the last decade, prompted by findings that amylin toxicity stems, at least in part, from its ability to disrupt fluidity and organization of cellular membranes (Brender et al. 2008a, b; Khemtemourian et al. 2008). Thus, understanding the process of amylin aggregation on membranes has a direct implication for the etiology of islet amyloidosis and T2DM.

Given that human amylin is a positively charged (cationic) peptide, one can expect that lipids bearing strong negative charge, such as phosphatidylserine (PS) and other anionic lipids, interact with human amylin to modulate its aggregation in solution and/or on membranes. This idea, first tested by Miranker and colleagues (Knight and Miranker 2004), was later verified by several independent biochemical studies showing accelerated amylin aggregation and conformational changes in solution following addition of anionic lipids such as PS (Cho et al. 2008, 2009; Jayasinghe and Langen 2005, 2007; Knight et al. 2006). Supporting this notion, the presence of negatively charged liposomes composed of phosphatidylcholine (PC) and PS (PC:PS, 2.8:1.2 mol:mol) in the incubation solution potentiated amylin aggregation by increasing both the extent (Fig. 4.3a) and the rate of amylin aggregation (Cho et al. 2008). Not only do the PS-enriched liposomes increase the rate of amylin aggregation, but they also shorten the lag phase, suggesting that electrostatic interactions between the peptide and the lipid accelerate nucleation, which is a rate-limiting step in aggregation (Fig. 4.3a).

In contrast to anionic liposomes, zwitterionic (neutral) PC liposomes did not significantly affect the rate or the extent of amylin aggregation in solution, further implicating electrostatic interactions as a culprit in amylin aggregation (Cho et al. 2008). Interestingly, inclusion of cholesterol, another essential component of cellular membranes, into anionic liposomes (PC:PS:Chol, 2.3:1:0.8 mol:mol:mol) attenuated the stimulatory effect of PS on amylin aggregation in solution by ~30 % (PC:PS versus PC:PS:Chol, Fig. 4.3a), a decrease comparable to the inhibitory effect of cholesterol on amylin deposition across planar membranes (PC:PS versus PC:PS:Chol, Fig. 4.4a). By applying the calculated rate constants to the Arrhenius equation it was inferred that the presence of negatively charged PC:PS vesicles decreases the activation energy (E_a) of aggregation by $\Delta E_a = -3.7$ kJ/mol as compared to amylin alone, which in turn increases the rate of amylin aggregation by more than four times (Cho et al. 2008). In contrast, inclusion of cholesterol in PC:PS vesicles reversed their stimulatory effect on

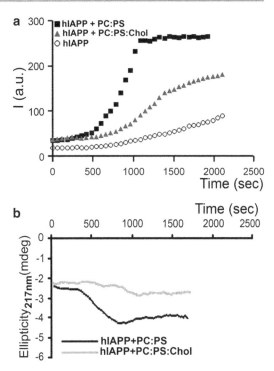

Fig. 4.3 Membrane cholesterol and anionic phospholipids oppositely regulate amylin aggregation and misfolding in solution. (**a**) Thioflavin-T fluorescent assay reveals slow aggregation of 10 μM human amylin in solution (*circles*). Presence of 100 μM anionic liposomes (PC:PS, 2.8:1.2 mol:mol, *squares*) in incubating solution (PBS, 1 % HFIP) accelerates amylin aggregation, the effect of which was reversed by inclusion of cholesterol in the lipid vesicles (PC:PS:Chol, 2.3:1:0.8 mol:mol:mol, *triangles*). (**b**) Dynamics of amylin secondary structural transitions in solution are regulated by membranes. CD spectra of human amylin (10 μM) incubated with 100 μM PC:PS liposomes (2.8:1.2 mol:mol, *black trace*) or PC:PS:Chol liposomes (2.3:1:0.8 mol:mol:mol, *gray trace*) were continuously acquired at 220 nm to monitor appearance of β-sheet conformation. Note that the onset of amylin aggregation and the transition from random coil to β-sheet coincide (hA+PC:PS, Fig. 4.3a, b). Inclusion of cholesterol prolonged amylin's transition to β-sheets evoked by anionic liposomes (Fig. 4.3b), ultimately reducing kinetics and the extent of amylin aggregation (Fig. 4.3a)

amylin aggregation by increasing the activation energy by $\Delta E_a = 845$ J/mol. The ΔE_a values were calculated for amylin aggregation at room temperature (25 °C) at which the experiment was performed (Cho et al. 2008). Remarkably, despite marked difference in their aggregation rates, all three reactions exhibited first-order kinetics (Cho et al. 2008), indicating that anionic lipids and cholesterol do not change the mechanism of amylin aggregation but act as catalyst and inhibitor, respectively. This finding also suggests that phospholipids and cholesterol in membranes oppositely modulate amylin aggregation by changing the activation energy of the amylin transition from random coil to β-sheets. In support of this conclusion, inclusion of cholesterol reduced the stimulatory effect of PS-containing liposomes on the conformational transition of amylin from random coil to β-sheets (Fig. 4.3b) (Cho et al. 2008).

4.6 Cholesterol Regulates Amylin Aggregation on Planar Lipid Membranes

We used AFM to investigate the supramolecular organization and dynamics of amylin aggregates on model membranes (Cho et al. 2009) that resemble the cell PM in composition and fluidity. Amylin aggregation on neutral and negatively-charged planar membranes that contained or lacked cholesterol was investigated by time-lapse AFM followed by single-particle analysis as described above. AFM revealed transition of small 25–35 nm diameter spherical oligomers (formed during the first 5 min) into larger 90–130 nm supramolecular complexes on anionic PC:PS (2.8:1.2 mol:mol) membranes (Fig. 4.4a, 10 min, arrowheads). Amylin oligomers formed during the first 5 min (Fig. 4.4a, left panel) and were morphologically similar to the oligomers initially assembled on mica (Fig. 4.2a). In marked contrast to amylin fibrillization on mica, amylin oligomers did not align and elongate into fibrils on anionic membranes but rather assembled into channel- or pore-like structures (Fig. 4.4a, 10 min, arrowheads). Interestingly, AFM revealed that amylin oligomers preferentially deposited on planar PC:PS membranes (Fig. 4.4a, 5 min) and much less frequently (<3 % of all particles) on mica surfaces (Fig. 4.2a, 5 min). This result demonstrates that amylin interacts with anionic membranes earlier during the oligomeric stage of aggregation, which most likely serve as a catalyst for amylin oligomerization. 3D image analysis

revealed a characteristic fourfold rotational symmetry of self-assembled supramolecular complexes of amylin featuring a central pore (Fig. 4.4b, PC:PS, 10 min; inset). The majority of self-assembled amylin complexes on planar PC:PS membranes exhibited tetrameric and, at times, pentameric globular organization (Fig. 4.4b, right panels). Tetrameric amylin complexes accounted for ~95 % of all supramolecular amylin structures assembled on the membranes. Less than 5 % of amylin complexes were pentamers (7 out of 163 particles examined, n = 3).

A twofold symmetrical organization of amylin and other amyloid proteins, incorporated first into liposomes and subsequently into planar membranes, has been previously demonstrated (Quist et al. 2005). In our study, which replicates amylin interactions with the pancreatic β-cells, we showed that amylin oligomers in solution can also directly assemble into symmetrical channel-like structures on pre-formed planar membranes. This may be relevant for the pathology of diabetes, as amylin and other amyloid proteins interact with cellular membranes and are cytotoxic when assembled into oligomers (Haataja et al. 2008; Ritzel et al. 2007; Trikha and Jeremic 2011). Interestingly, the sizes of amylin globular particles assembled on planar membranes (Fig. 4.4b) were in the same range (20–40 nm) as the soluble intermediate-sized cytotoxic amylin particles reported earlier (Janson et al. 1999). As expected from the peptide's amphiphilic nature, cytotoxic amylin oligomers readily form ion-permeable channels in bilayers and in cell membranes (Mirzabekov et al. 1996; Quist et al. 2005; Trikha and Jeremic 2011; Zhao et al. 2014). Besides forming channel-like structures, amylin also accumulates on membranes as unstructured amorphous aggregates (Fig. 4.4a, left panel; arrows), resembling in size (300–500 nm) and morphology the amyloid deposits often associated with T2DM (Jaikaran and Clark 2001). Incorporation of cholesterol into anionic membranes (PC:PS:Chol, 2.3:1:0.8 mol:mol:mol) redirected the surface distribution of amylin aggregates (Fig. 4.4a, right panel). While oligomers were observed again during the first 5 min of aggregation on cholesterol-containing mem-

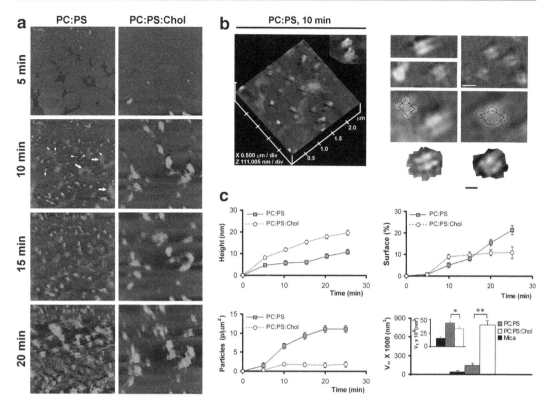

Fig. 4.4 Dynamics and organization of amylin aggregates on planar membranes. (**a**) Amylin (20 μM) at time zero was injected into the imaging chamber and the peptide membrane assembly was monitored in real time by time-lapse AFM. All micrographs are 5×5 μm, and are taken at the same time intervals of 5 min. (**b**) Section analysis of amylin aggregates on anionic membranes. Channel-like topology of two amylin supramolecular complexes featuring a central pore is shown (**b**, PC:PS, 2.8:1.2 mol:mol, 10 min, *inset*). High-resolution 2D AFM micrographs of several amylin supramolecular complexes on PC:PS membranes are shown. Tetrameric and pentameric subunits are outlined (**b**, *right panel*). Bar is 50 nm. (**c**) Quantitative analysis of cholesterol-regulated amylin assembly on anionic membranes. Presence of cholesterol in planar membranes (PC:PS:Chol,2.3:1:0.8 mol:mol:mol, *circles*) stimulates a significant increase in the size (*height*) of amylin aggregates over time when compared with cholesterol-depleted membranes (PC:PS, 2.8:1.2 mol:mol, *filed squares*, **c**, *height plot*). Cholesterol abrogates amylin deposition and overall membrane surface coverage with amylin (**c**, *surface plot*). Cholesterol also inhibits seeding of amylin aggregates on PC:PS membranes (**c**, *particle plot*). The mean (V_m) and the total particle volume (V_t) of amylin aggregates on PC:PS membranes are significantly different in the presence of cholesterol after 20 min (*$p < 0.05$ and **$p < 0.01$, $n = 3$ Student's *t*-test) (**c**, *volume plot*)

branes, amylin further aggregated and concentrated in discrete areas measuring 300–500 nm in diameter (Fig. 4.4a, 10–20 min; PC:PS:Chol). Presence or absence of channel-like structures on these membranes could not be verified due to a strong clustering effect of cholesterol from the very early stages of amylin aggregation.

To further understand how phospholipids and cholesterol modulate amylin surface deposition and to learn more about the regulatory mechanisms driving aggregation, we measured amylin accumulation on planar membranes using single-particle analysis as explained above (see Fig. 4.2b). The regulatory effect of cholesterol was quite obvious from the early stages of amylin aggregation (10 min and thereafter; Fig. 4.4a–c). Over time there was a significant increase in the height of amylin aggregates due to the large clustering effect of cholesterol (Fig. 4.4c, height analysis). This was accompanied by an overall decrease in amylin deposition across the planar membranes (Fig. 4.4c, surface analysis). As amylin aggregated and accumulated in some membrane areas, other regions of the membrane were virtually devoid of the protein aggregates (Fig. 4.4a, right panel). Consequently, amylin's capacity to form an extensive network of amyloid aggregates on the membrane was diminished in membranes that contained cholesterol (Fig. 4.4a, c, surface analysis).

Similar to most amyloid proteins, amylin aggregation is nucleation-dependent (Khemtemourian et al. 2008; Padrick and Miranker 2002). Consistent with the "nucleation" hypothesis, amylin seeding was diminished in the presence of cholesterol: a sevenfold decrease in the number of amylin particles was observed on planar anionic membranes that contained cholesterol as compared to those that lacked cholesterol (Fig. 4.4c, particle analysis). This phenomenon may also account for the observed inhibitory effect of membrane cholesterol on initial (nucleation) phase of amylin polymerization in solution evoked by anionic liposomes (Fig. 4.3a). As the number of amylin particles diminished in the presence of cholesterol (Fig. 4.4a), their size and average volume increased over time (Fig. 4.4a, c, volume analysis). The mean volume V_m of amylin particles was larger on lipid membranes than on mica. A further, much larger increase in particle size was detected on membranes containing cholesterol (PC:PS:Chol. Fig. 4.4c, volume analysis). However, the total volume V_t of amylin aggregates on the cholesterol-containing membranes significantly decreased when compared to the cholesterol-free membranes (Fig. 4.4c volume analysis, inset) due to a large decrease in the amylin seeding capacity (Fig. 4.4c, particle analysis). Thus, in the presence of cholesterol, amylin aggregated and accumulated on planar membranes as submicron-sized protein clusters, which served as templates for the ongoing amylin binding and aggregation.

Comparisons of AFM micrographs and amylin growth curves on surfaces bearing different physicochemical properties (Fig. 4.5, upper panel) revealed that amylin monomers polymerize via two distinct mechanisms: on stiff and polar mica, amylin formed fibrils by longitudinal bi-directional extension of full-grown spherical oligomers, or nuclei, measuring ~6 nm in height and ~90 nm in diameter (Fig. 4.2), and on soft negatively-charged PC:PS planar membranes amylin formed pore-like supramolecular structures that self-assembled from ~25 to 35 nm diameter globular subunits or oligomers (Figs. 4.4 and 4.5, lower panel). AFM revealed another important feature of amylin aggregates on planar membranes. Amorphous deposits and channel-like structures that were formed during the early (5–10 min) stages of amylin aggregation did not transition into new structures, although the total amount and size of amorphous aggregates increased over time (Figs. 4.4 and 4.5). These findings signify the important contribution and long-lasting effect of lipids and cholesterol in the regulation of amylin aggregation, summarized in Fig. 4.5 (lower panel).

To test if the regulatory effect of cholesterol is charge-specific, amylin aggregation was also studied on neutral PC membranes that lack or include cholesterol (Fig. 4.5, upper panel). As in anionic PC:PS-membranes (Figs. 4.4c and 4.5), the presence of cholesterol in neutral PC membranes (PC:Chol, 3.2:0.8 mol:mol) stimulated an increase in the height of amylin aggregates by almost twofold, from 6.2 ± 0.9 to 11.9 ± 1.7 nm (mean ± SEM). Similar to anionic membranes (Fig. 4.4, PC:PS:Chol), the presence of cholesterol in neutral membranes resulted in heterogeneous amylin distribution and clustering over the membrane surface with a mean particle (cluster) radius of 196 ± 19 nm based on section analysis reported previously (Fig. 4.5, PC:Chol upper panel) (Cho et al. 2009). Thus, the amylin clusters on neutral membranes (PC:Chol) were 20–30 % smaller compared to those formed on cholesterol-containing anionic (PC:PS:Chol) membranes (Fig. 4.5, upper panel). Collectively, our results demonstrate the intrinsic ability of cholesterol to regulate amylin aggregation and deposition on membranes, irrespective of the chemical composition or charge of the membrane (Cho et al. 2009).

Amylin's self-assembly into non-fibrillogenic, channel-like structures on neutral and anionic membranes (Figs. 4.4 and 4.5) suggests that physical properties, rather than specific chemical

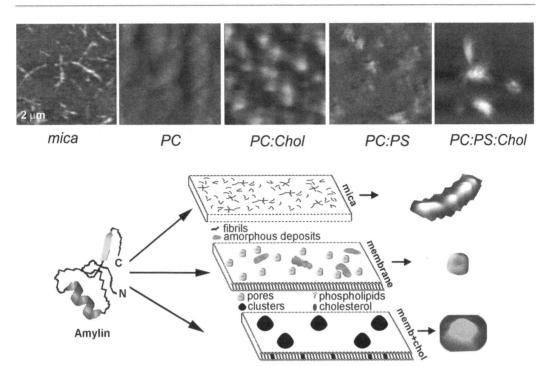

Fig. 4.5 AFM analysis of membrane-directed amylin self-assembly. High-resolution 2D AFM micrographs (*top panel*) reveal distinct patterns of amylin aggregation and deposition on different surfaces. Note the clustering of amylin aggregates on cholesterol-containing membranes, PC:Chol (3.2:0.8 mol:mol) and PC:PS:Chol (2.3:1:0.8 mol:mol:mol) and their homogenous distribution/aggregation on cholesterol-free membranes, PC and PC:PS (2.8:1.2 mol:mol). In contrast to mica, no fibrils were detected on either membranes. Micrographs are 2×2 µm. *Bottom panel*: Proposed pathway of amylin polymerization and accumulation on different surfaces. The form and amount of amylin deposits correlate with the physicochemical properties of the supporting surface. On stiff polar mica surface, amylin monomers (*left*) asso-ciate into spherical oligomers that align and elongate overtime to produce mature fibrils that randomly distribute across the surface (mica). On soft planar membranes, amylin self-assembles into globular highly symmetrical supramolecular structures featuring a central pore. Unstructured amorphous amylin aggregates are also formed on this surface (membrane). Incorporation of cholesterol into planar membranes redirects amylin surface deposition by stimulating formation of larger, but fewer amylin clusters (memb+chol). Consequently, the membrane surface area free of amylin deposits increases significantly, which diminishes amylin accumulation. Three major polymorphic forms, a fibril, a pore and a single cluster formed during amylin polymerization on different surfaces, are presented top to bottom (*bottom panel*)

properties of the membrane, determine its susceptibility to distinct amyloid forms. For example, β-amyloid oligomers, but not fibrils, were found to be enriched in neuronal membrane lipid rafts (Schneider et al. 2006), which are discrete membrane segments with phospholipids in a liquid-ordered phase. Cholesterol is a known regulator of membrane fluidity. It is found in lipid rafts and many endomembranes, where it establishes a sorting platform for a scaffold of various protein-lipid complexes important for cell signaling, endocytosis and other essential physiological processes (Simons and Toomre 2000).

It is therefore quite possible that, by modulating membrane fluidity and/or membrane curvature (Chen and Rand 1997; Smith et al. 2009), cholesterol also regulates amylin-lipid interactions and amylin aggregation observed in our studies (Cho et al. 2008, 2009; Trikha and Jeremic 2011).

Another possibility is that amylin directly interacts with cholesterol in the membrane. Amylin, as a monomer, has a strong tendency to insert into phospholipid monolayers (Engel et al. 2006). This event may set a stage for direct peptide-cholesterol interactions in the membrane core. To test if cholesterol can directly interact with amylin, we evaluated their interaction by ThT aggregation assay and CD spectroscopy (Cho et al. 2008, 2009). The presence of soluble cholesterol (1:1 and 1:2 mol/mol peptide/sterol ratios) impeded amylin aggregation in solution in two major ways: it prolonged the lag (nucleation) phase and it decreased the fibrillization rate by 16-fold, from $k = 8.32 \times 10^{-3}$ s^{-1} (amylin) to $k = 5.2 \times 10^{-4}$ s^{-1} (amylin plus cholesterol) (Cho et al. 2008, 2009). These results are consistent with inhibition of both fibril nucleation and elongation. These reconstituted studies have demonstrated that cholesterol may directly affect amylin conformational changes in solution and possibly in the membranes. However, whether human amylin directly interacts with cholesterol in plasma and/or endomembranes still remains to be confirmed.

4.7 Plasma Membrane Cholesterol Restricts Aggregation of Human Amylin on Cellular Membranes and Amylin's Toxicity

Experiments performed with synthetic liposomes and planar membranes provided important although indirect evidence for the role of membranes in aggregation of amylin and other amyloid proteins. To confirm that native membranes modulate amylin's turnover and toxicity in situ, we resorted to cellular studies (Trikha and Jeremic 2011, 2013). In our experimental setup, monomeric human amylin was added to pancreatic rat and human islet cells in which we systematically varied plasma membrane cholesterol levels using cholesterol biosynthesis inhibitor lovastatin (Lov) and/or cholesterol-depleting agent, beta-cyclodextrin (BCD). The extent of human amylin aggregation (Fig. 4.6) and toxicity (Fig. 4.7) in cholesterol-containing and cholesterol-depleted cells was assessed over 24 h by confocal microscopy (Trikha and Jeremic 2011). Oligomers were detected with the oligomer-specific A_{11} antibody (Fig. 4.6a) that does not react with either monomers or fibrils

(Kayed et al. 2003). To detect human amylin monomer distribution on the PM and inside the cells (Fig. 4.6b), we used a human-specific amylin antibody that does not cross-react with the rat isoform or large oligomers/aggregates (Trikha and Jeremic 2011). In addition to human amylin, we used the lipid raft marker, cholera toxin (CTX), and the clathrin endocytotic marker, transferrin (not shown), to determine the specificity of amylin monomer and oligomer binding to the cell PM (Fig. 4.6).

Amylin and CTX were sequentially (Fig. 4.6) or concurrently (Trikha and Jeremic 2011) incubated with cultured pancreatic insulinoma RIN-m5F cells for the indicated periods of time, fixed and processed for immunochemical analysis. In experiments in which amylin and CTX were concurrently incubated with cells, immuno-confocal microscopy revealed a punctuated staining pattern of CTX and amylin oligomers on the cell PM, exhibiting high spectral overlap (yellow) and a high co-localization coefficient ($R = 0.74 \pm 0.09$) in discrete membrane regions (Trikha and Jeremic 2011). Changes in the cell morphology characterized by the appearances of elongated protrusions were frequently observed in cultures treated with human amylin (Trikha and Jeremic 2011). Amylin oligomer binding and uptake were particularly noticeable in protrusions. However, amylin oligomers also co-patched with the lipid raft marker CTX in other PM regions (Trikha and Jeremic 2011, 2013). This suggests that amylin oligomers accumulate at specific microdomains, possibly lipid rafts, on the cell PM prior to their uptake.

To exclude a possible modulatory effect of CTX on amylin binding, sequential incubations of rat insulinoma cells with amylin and CTX were performed. Following the incubation with amylin, cells were further incubated with CTX for 30 min at 4 °C. Amylin oligomers and CTX once more co-localized on the cell PM (Fig. 4.6a, 30 min). Prolonging the incubation period from 30 min to 24 h allowed amylin oligomers in the microdomains to internalize, as demonstrated by a ~50 % drop in amylin oligomer/CTX colocalization values (Fig. 4.6a) (Trikha and Jeremic 2011). Upon depletion of PM cholesterol with BCD/Lov,

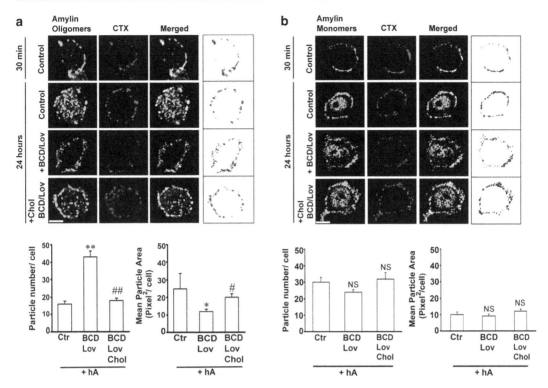

Fig. 4.6 Binding and clustering of amylin oligomers into microdomains on the cell PM requires cholesterol. (**a**) Confocal microscopy analysis of amylin oligomer and cholera toxin (CTX) distribution on the cell PM. Characteristic binding profiles of amylin oligomers on the cell PM for each treatment (within *boxes*, *right panel*) are rendered in *gray tones* for easier particle comparisons, which are presented side by side with the original fluorescence images (*left panels*). Note the time-dependent increase in the number of internalized amylin oligomers (control, 30 min vs. 24 h, *left panel*), which prevents accumulation of amylin oligomers on the cell PM (control, 30 min vs. 24 h, *right box*). Single particle analysis demonstrates a threefold increase in the number of amylin oligomer clusters, or puncta, on the PM (particle no.), and their dispersion across the PM in cholesterol-depleted cells (mean particle area). Significance established at *p < 0.05, **P < 0.01 control vs. BCD/Lov, and #p < 0.05, ##p < 0.01 BCD/Lov vs. BCD/Lov/Chol. (**b**) Amylin monomer internalization is not blocked by cholesterol depletion. Confocal microscopy demonstrates internalization of amylin monomers both in controls (hA) and cholesterol-depleted cells (hA + BCD/Lov). No significant (NS) change in PM-binding pattern of amylin monomers (*right boxes*) was noticed upon cholesterol depletion. PM cholesterol does not modulate amylin deposition on the PM. The number of amylin puncta on PM (particle no.) and their area (mean particle area) did not change significantly upon depletion of PM cholesterol by BCD/Lov. Bars are 5 μm

a significant decrease in colocalization of amylin oligomers with cholera toxin on the cell PM was observed indicating amylin/CTX particle de-clustering and their dispersions across the cell surface (Fig. 4.6a). In line with this finding, image and single-particle analysis revealed that the mean particle area of PM-bound amylin oligomers (Fig. 4.6a, right boxed panel) in cholesterol-depleted cells (BCD/Lov) decreased significantly as compared to control cells (Fig. 4.6a, particle analysis). Conversely, the number of amylin oligomer clusters, or puncta, on the PM of cholesterol-depleted cells increased by threefold relative to control cells increased by threefold relative to control cells (Fig. 4.6a, particle analysis). Consequently, cell surface coverage by amylin oligomers increased by ~twofold in cells with reduced PM cholesterol content as compared to control cells (Fig. 4.6a, right boxed panel). This inhibitory effect of BCD/Lov on clustering of amylin oligomers on the PM was also observed at earlier time points, 30 min and 3 h (data not shown). These results demonstrate that the seeding (nucleation) capacity of amylin oligomers and their ability to form a dense network of amyloid aggregates on the PM were augmented in cells with impaired cholesterol homeostasis.

Clustering of amylin oligomers on the PM was fully restored following replenishment of PM cholesterol (Fig. 4.6a), indicating that amylin oligomer deposition on the PM is modulated by cholesterol and is reversible. These findings are in full compliance with the inhibitory and clustering effect of cholesterol on amylin aggregation on synthetic membranes (Figs. 4.4 and 4.5) (Cho et al. 2008, 2009).

In contrast to oligomers (Fig. 4.6a), fewer amylin monomers co-localized with the lipid raft marker CTX on the PM during the first 30 min of incubation with cells (Fig. 4.6b). However, analogous to the binding of oligomers (control 30 min, Fig. 4.6a right boxed panel), amylin monomers exhibited a discrete, punctuated staining pattern on the cell PM during that period (control 30 min, Fig. 4.6b right boxed panel). Interestingly, the degree of colocalization between amylin monomers and CTX also decreased significantly upon cholesterol depletion with BCD/Lov (Fig. 4.6b), albeit to much lesser extent than for amylin oligomers (Fig. 4.6a). In contrast to oligomers, the number and the mean particle area of amylin monomer puncta at the cell PM remained almost the same in control and BCD/Lov-treated cells (Fig. 4.6b image and graphs) (Trikha and Jeremic 2011), indicating that clustering of amylin monomers into microdomains on the cell PM is not affected by PM cholesterol. Specificity of amylin and CTX binding is further demonstrated in experiments in which transferrin, a marker of clathrin-dependent endocytosis, was used (Trikha and Jeremic 2011, 2013). Colocalization analysis revealed that neither amylin monomers nor oligomers colocalize with transferrin on the cell PM (Trikha and Jeremic 2011). Similarly, CTX and Trf were found to bind to distinct regions on the cell PM. Collectively, our studies depicted in Fig. 4.6 (Trikha and Jeremic 2011) imply that PM cholesterol reversibly and specifically determines binding and distribution of amylin oligomers, but not monomers, on the cell PM.

We also investigated whether, and to what extent, variations in PM cholesterol levels affect amylin toxicity in rat and human pancreatic islet cells. Cells were treated or not with BCD and/or Lov and then exposed to human amylin. Following 24 h incubation, cells were analyzed by confocal microscopy for the presence or absence of apoptotic markers, cleaved caspase-3 and PS externalization (Fig. 4.7). Amylin was toxic to cultured human islet cells, albeit with notably higher potency ($LD_{50} = 2.5$ μM) as compared to its toxic effect in rat insulinoma cells ($LD_{50} = 15$ μM). Human amylin (hA, 2 μM) evoked apoptosis in 43 ± 5 % of cultured human islet cells. The combination of BCD (5 mM, 10 min) and lovostatin (0.5 μM, 24 h), reduces PM cholesterol to 67 ± 4 % relative to controls (Trikha and Jeremic 2011), which in turn increases amylin toxicity by a 34 ± 5 % as compared to cells with normal membrane cholesterol content (Fig. 4.7, image and histogram). Addition of soluble cholesterol together with BCD/Lov replenishes cholesterol levels (94 ± 5 % in BCD/Lov/Chol relative to controls), which reverses the stimulatory effect of BCD/Lov on amylin toxicity. As demonstrated in rat insulinoma cells (Trikha and Jeremic 2011), Lov alone had a small insignificant stimulatory effect on amylin toxicity in human islet cells (<10 %, data not shown). Finally, incubation of human islet cells with soluble cholesterol (50 μg/ml) increases PM cholesterol levels by 14 ± 6 % relative to control cells (Trikha and Jeremic 2011) and further attenuates amylin toxicity (Fig. 4.7). A strong inverse relationship between PM cholesterol levels and amylin toxicity in human islets is obtained (Fig. 4.7, graph). The inhibitory effect of PM cholesterol on amylin toxicity in human islets was confirmed with the LDH-release assay (Trikha and Jeremic 2011). Similarly, Western blot analysis demonstrates extracellular accumulation of low-molecular-weight amylin monomers and dimers and intermediate-sized oligomers upon PM-cholesterol depletion with BCD/Lov, and their efficient clearance by islet cells upon PM cholesterol reloading with soluble cholesterol (BCD/Lov/Chol). In line with these findings, confocal microscopy shows that cholesterol supplementation stimulates, while PM cholesterol depletion decreases amylin oligomer internalization in human islet cells (Trikha and Jeremic 2011).

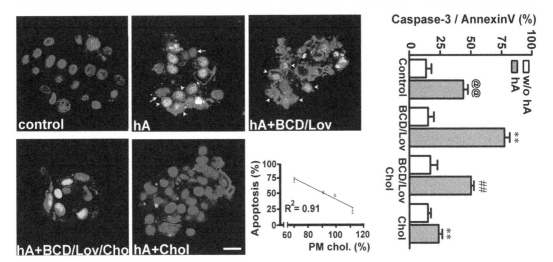

Fig. 4.7 PM cholesterol prevents toxicity of soluble amylin oligomers in cultured human islet cells. Confocal microscopy analysis of phosphatidylserine (PS) externalization and caspase-3 proteolytic activation by amylin in cells with normal, depleted and enriched cholesterol levels (micrographs, *left panel*). *Arrows* depict non-apoptotic nuclei (*blue*) in viable cells, whereas arrowheads depict fluorogenic caspase-3 substrate found in the nuclei of apoptotic cells, giving these nuclei a *green/blue* appearance. The majority of PS-positive cells (*red*) show shrinkage and nuclear condensation (*arrowheads*) indicative of apoptosis (hA and hA+BCD/Lov). Bar is 10 μm. Linear regression analysis shows an inverse relationship between amylin-induced cell death and PM cholesterol levels. The extent of cell death evoked by amylin was plotted as a function of variable PM cholesterol levels in controls and treatments (apoptosis vs. cholesterol, graph). Quantitative analysis of amylin-induced apoptosis in human islets (caspase-3/annexin, graph) reveals a significant increase in amylin toxicity in cholesterol-depleted cells as compared to controls (@@p < 0.01 hA vs. control, **p < 0.01 hA vs. treatments, *right panel*). Replenishment of PM cholesterol levels significantly decreased amylin toxicity (##p < 0.01 hA+BCD/lov vs. hA+BCD/Lov/Chol)

These results demonstrate a major role of PM cholesterol in amylin turnover and toxicity in human islets.

4.8 Role of Cholesterol and Phospholipids in Aggregation and Toxicity of Other Amyloid Proteins

Amyloid formation, or amyloidogenesis, an intrinsic property of all polypeptides (Chiti and Dobson 2006) is a process by which soluble proteins aggregate into insoluble, structurally conserved unbranched fibers that are characterized by resistance to proteinase K digestion, dye binding specificity, and ordered β-sheet-rich structure (Sipe et al. 2010). Amyloids can be broadly categorized into detrimental and functional. Detrimental amyloids, which cause protein misfolding in amyloid diseases, include huntingtin implicated in Huntington's disease, α-synuclein implicated in Parkinson's disease, prion protein implicated in Creutzfield-Jacob's disease, superoxide dismutase implicated in amyotrophic lateral sclerosis, and amyloid-β (Aβ) peptide implicated in Alzheimer's disease, transthyretin implicated in transthyretin familial amyloidosis, Tau implicated in frontotemporal lobar degeneration etc. Some of these proteins, including Aβ, α-synuclein, serum amyloid A (implicated in inflammation-linked amyloidosis) and other apolipoproteins (implicated in systemic amyloidosis and atherosclerosis) are described in other chapters in this volume. Functional amyloids are an integral part of the normal physiology of the cell and include curli, chaplin, URE2p and PmeL17. Curli found in *E. coli* plays a role in biofilm formation and mediates infection. Chaplin found in Streptomyces plays a role in protection against water surface tension. URE2p found in *S. Cerevisiae* plays a role in nitrogen catabolism, and Pmel17 found in humans play a role in melanin synthesis (Granzotto et al. 2011; Lau et al. 2007; Rymer and Good 2000).

High local concentrations of proteins accumulating on biological surfaces, such as the plasma or organelle membranes, and the physicochemical properties of membrane lipids that interact with these proteins may accelerate aggregation and fibril formation (Burke et al. 2013). Interestingly, while there have been reports of cholesterol and phospholipids stimulating amyloid aggregation and enhancing amyloid-mediated toxicity, there have also been contrasting reports describing that cholesterol and phospholipids reduce aggregation and toxicity of a variety of amyloid proteins. Moreover, distinct mechanisms are reported for the effects of cholesterol and phospholipids on amyloid protein aggregation and toxicity. It is widely reported that high levels of cholesterol aggravate Alzheimer's disease (Cossec et al. 2010) by promoting the generation of Aβ peptide (Barrett et al. 2012) and stimulating its aggregation. Barrett et al. have shown that the carboxyl terminal transmembrane domain of amyloid precursor protein binds cholesterol and is cleaved by γ-secretase to generate Aβ peptides. This promotes the possible use of cholesterol lowering drugs, like statins, in the control of Alzheimer's pathology (McGuinness and Passmore 2010). Another elegant study that supports the enhancing role of cholesterol in the aggregation of amyloid precursor proteins (APP) (Hayashi et al. 2000) provides evidence that APP is not present in caveolae or caveoli-like domains of the cell membrane but in cholesterol rich microdomains which may be linked to the maturation of APP in a cell type specific manner. Exogenously applied amyloid beta peptides and tau protein have also been reported to accumulate in lipid rafts, in turn affecting their mobility (Williamson et al. 2008).

There are multiple reports that anionic phospholipids like phosphatidylserine provide 'docking sites' where amyloid aggregates can be nucleated, resulting in toxicity through disruption of the membrane (Zhao et al. 2004, 2005; Lee et al. 2002). For example, native-like aggregates of the toxic prion protein Ure2p dock preferentially on PS, speeding up aggregation and dampening toxicity (Pieri et al. 2009). The conversion of another normal cellular predominantly α-helical prion protein isoform, PRP^C, to the protease-resistant, β-sheet-rich PRP^{SC} isoform is enhanced by the presence of cholesterol-rich phospholipid membranes at ambient conditions (Rymer and Good 2000). In addition to cholesterol and phospholipids, fatty acids are abundantly found in membranes. For example, docosahexaenoic acid (DHA) is abundant in the vicinity of α-synuclein in neuronal membranes of patients with Parkinson's disease, and has been reported to enhance α-synuclein aggregation (De Franceschi et al. 2011).

At the physiological concentration (5 μM) at which cholesterol is found in the cerebrospinal fluid, it is reported to slow down, but not block, aggregation of β-amyloid complexes, in the presence or absence of metal ions. On the other hand, cholesterol compensates for membrane damage induced by Aβ peptides, preventing their toxic effects (Granzotto et al. 2011). Lipid rafts formed by cholesterol-rich microdomains in membranes are an attachment target for Aβ peptides. Antagonists of platelet-derived growth factor and inhibitors of phospholipase A2 inhibit the release of cholesterol from cholesterol esters, thereby reducing lipid rafts that traffic Aβ peptides, which in turn lead to the increase of cytosolic degradation of Aβ (Simmons et al. 2014). The apoptotic accumulation and redistribution of Aβ peptides on the neuronal cell membrane and in lipid rafts is reported to depend on a non-receptor tyrosine kinase that increases phosphorylation of membrane associated proteins (Williamson et al. 2008). Apolipoprotein E4, an established risk factor for Alzheimer's disease and cerebral amyloid angiopathy, affects clearance of Aβ peptides. ApoE4 preferentially forms plaques in the cerebral vessels rather than cerebral parenchyma by altering the ratio of Aβ(1–40) and Aβ (1–42) (Fryer et al. 2005).

In summary, during the last three decades investigators working in the amyloid field have made major progress by illuminating important regulatory mechanisms and cellular factors implicated in turnover and function of amyloid proteins such as human amylin. Studies reveal exactly how membrane-associated lipids and cholesterol modulate aggregation of human amylin and other amyloid proteins and how islet amyloid is formed and processed in cells, interacts with membranes and affects cellular functions. The knowledge gained in these studies offers exciting opportunity for development of novel approaches and drugs that may eradicate or slow down progression of amyloid-associated diseases.

Acknowledgment This work was supported by the NIH grant RO1DK091845 and the ICR Basic Science Islet Distribution Program (to A.J.).

References

Abedini A, Raleigh DP (2005) The role of His-18 in amyloid formation by human islet amyloid polypeptide. Biochemistry 44(49):16284–16291

Abedini A, Schmidt AM (2013) Mechanisms of islet amyloidosis toxicity in type 2 diabetes. FEBS Lett 587(8):1119–1127

Ancsin JB (2003) Amyloidogenesis: historical and modern observations point to heparan sulfate proteoglycans as a major culprit. Amyloid 10(2):67–79

Barrett PJ, Song Y, Van Horn WD, Hustedt EJ, Schafer JM, Hadziselimovic A et al (2012) The amyloid precursor protein has a flexible transmembrane domain and binds cholesterol. Science 336(6085):1168–1171

Brender JR, Hartman K, Reid KR, Kennedy RT, Ramamoorthy A (2008a) A single mutation in the nonamyloidogenic region of islet amyloid polypeptide greatly reduces toxicity. Biochemistry 47(48):12680–12688

Brender JR, Lee EL, Cavitt MA, Gafni A, Steel DG, Ramamoorthy A (2008b) Amyloid fiber formation and membrane disruption are separate processes localized in two distinct regions of IAPP, the type-2-diabetes-related peptide. J Am Chem Soc 130(20):6424–6429

Burke KA, Yates EA, Legleiter J (2013) Biophysical insights into how surfaces, including lipid membranes, modulate protein aggregation related to neurodegeneration. Front Neurol 4:17

Caminos JE, Bravo SB, Garces MF, Gonzalez CR, Cepeda LA, Gonzalez AC et al (2009) Vaspin and amylin are expressed in human and rat placenta and regulated by nutritional status. Histol Histopathol 24(8):979–990

Cao P, Tu LH, Abedini A, Levsh O, Akter R, Patsalo V et al (2012) Sensitivity of amyloid formation by human islet amyloid polypeptide to mutations at residue 20. J Mol Biol 421(2–3):282–295

Cao P, Abedini A, Wang H, Tu LH, Zhang X, Schmidt AM et al (2013a) Islet amyloid polypeptide toxicity and membrane interactions. Proc Natl Acad Sci U S A 110(48):19279–19284

Cao P, Marek P, Noor H, Patsalo V, Tu LH, Wang H et al (2013b) Islet amyloid: from fundamental biophysics to mechanisms of cytotoxicity. FEBS Lett 587(8):1106–1118

Chen Z, Rand RP (1997) The influence of cholesterol on phospholipid membrane curvature and bending elasticity. Biophys J 73(1):267–276

Chiti F, Dobson CM (2006) Protein misfolding, functional amyloid, and human disease. Annu Rev Biochem 75:333–366

Chiu CC, Singh S, de Pablo JJ (2013) Effect of proline mutations on the monomer conformations of amylin. Biophys J 105(5):1227–1235

Cho WJ, Jena BP, Jeremic AM (2008) Nano-scale imaging and dynamics of amylin-membrane interactions and its implication in type II diabetes mellitus. Methods Cell Biol 90:267–286

Cho WJ, Trikha S, Jeremic AM (2009) Cholesterol regulates assembly of human islet amyloid polypeptide on model membranes. J Mol Biol 393(3):765–775

Clark A, Nilsson MR (2004) Islet amyloid: a complication of islet dysfunction or an aetiological factor in type 2 diabetes? Diabetologia 47(2):157–169

Cooper GJ, Willis AC, Clark A, Turner RC, Sim RB, Reid KB (1987) Purification and characterization of a peptide from amyloid-rich pancreases of type 2 diabetic patients. Proc Natl Acad Sci U S A 84(23):8628–8632

Cossec JC, Marquer C, Panchal M, Lazar AN, Duyckaerts C, Potier MC (2010) Cholesterol changes in Alzheimer's disease: methods of analysis and impact on the formation of enlarged endosomes. Biochim Biophys Acta 1801(8):839–845

Costes S, Gurlo T, Rivera JF, Butler PC (2014) UCHL1 deficiency exacerbates human islet amyloid polypeptide toxicity in beta-cells: evidence of interplay between the ubiquitin/proteasome system and autophagy. Autophagy 10(6):1004–1014

De Franceschi G, Frare E, Pivato M, Relini A, Penco A, Greggio E et al (2011) Structural and morphological characterization of aggregated species of alpha-synuclein induced by docosahexaenoic acid. J Biol Chem 286(25):22262–22274

Engel MF, Yigittop H, Elgersma RC, Rijkers DT, Liskamp RM, de Kruijff B et al (2006) Islet amyloid polypeptide inserts into phospholipid monolayers as monomer. J Mol Biol 356(3):783–789

Engel MF, Khemtemourian L, Kleijer CC, Meeldijk HJ, Jacobs J, Verkleij AJ et al (2008) Membrane damage by human islet amyloid polypeptide through fibril growth at the membrane. Proc Natl Acad Sci U S A 105(16):6033–6038

Fan L, Westermark G, Chan SJ, Steiner DF (1994) Altered gene structure and tissue expression of islet amyloid polypeptide in the chicken. Mol Endocrinol 8(6):713–721

Fryer JD, Simmons K, Parsadanian M, Bales KR, Paul SM, Sullivan PM et al (2005) Human apolipoprotein

E4 alters the amyloid-beta 40:42 ratio and promotes the formation of cerebral amyloid angiopathy in an amyloid precursor protein transgenic model. J Neurosci 25(11):2803–2810

Gazit E (2002) A possible role for pi-stacking in the self-assembly of amyloid fibrils. FASEB J 16(1):77–83

German MS, Moss LG, Wang J, Rutter WJ (1992) The insulin and islet amyloid polypeptide genes contain similar cell-specific promoter elements that bind identical beta-cell nuclear complexes. Mol Cell Biol 12(4):1777–1788

Goldsbury C, Kistler J, Aebi U, Arvinte T, Cooper GJ (1999) Watching amyloid fibrils grow by time-lapse atomic force microscopy. J Mol Biol 285(1):33–39

Granzotto A, Suwalsky M, Zatta P (2011) Physiological cholesterol concentration is a neuroprotective factor against beta-amyloid and beta-amyloid-metal complexes toxicity. J Inorg Biochem 105(8):1066–1072

Green JD, Goldsbury C, Kistler J, Cooper GJ, Aebi U (2004) Human amylin oligomer growth and fibril elongation define two distinct phases in amyloid formation. J Biol Chem 279(13):12206–12212

Haataja L, Gurlo T, Huang CJ, Butler PC (2008) Islet amyloid in type 2 diabetes, and the toxic oligomer hypothesis. Endocr Rev 29(3):303–316

Hayashi H, Mizuno T, Michikawa M, Haass C, Yanagisawa K (2000) Amyloid precursor protein in unique cholesterol-rich microdomains different from caveolae-like domains. Biochim Biophys Acta 1483(1):81–90

Hoppener JW, Lips CJ (2006) Role of islet amyloid in type 2 diabetes mellitus. Int J Biochem Cell Biol 38(5–6):726–736

Hoppener JW, Ahren B, Lips CJ (2000) Islet amyloid and type 2 diabetes mellitus. N Engl J Med 343(6):411–419

Huang CJ, Gurlo T, Haataja L, Costes S, Daval M, Ryazantsev S et al (2011) Calcium-activated calpain-2 is a mediator of beta cell dysfunction and apoptosis in type 2 diabetes. J Biol Chem 285(1):339–348

Jaikaran ET, Clark A (2001) Islet amyloid and type 2 diabetes: from molecular misfolding to islet pathophysiology. Biochim Biophys Acta 1537(3):179–203

Janson J, Ashley RH, Harrison D, McIntyre S, Butler PC (1999) The mechanism of islet amyloid polypeptide toxicity is membrane disruption by intermediate-sized toxic amyloid particles. Diabetes 48(3):491–498

Jayasinghe SA, Langen R (2005) Lipid membranes modulate the structure of islet amyloid polypeptide. Biochemistry 44(36):12113–12119

Jayasinghe SA, Langen R (2007) Membrane interaction of islet amyloid polypeptide. Biochim Biophys Acta 1768(8):2002–2009

Jha S, Sellin D, Seidel R, Winter R (2009) Amyloidogenic propensities and conformational properties of ProIAPP and IAPP in the presence of lipid bilayer membranes. J Mol Biol 389(5):907–920

Kayed R, Bernhagen J, Greenfield N, Sweimeh K, Brunner H, Voelter W et al (1999) Conformational transitions of islet amyloid polypeptide (IAPP) in amyloid formation in vitro. J Mol Biol 287(4):781–796

Kayed R, Head E, Thompson JL, McIntire TM, Milton SC, Cotman CW et al (2003) Common structure of soluble amyloid oligomers implies common mechanism of pathogenesis. Science 300(5618):486–489

Khemtemourian L, Killian JA, Hoppener JW, Engel MF (2008) Recent insights in islet amyloid polypeptide-induced membrane disruption and its role in beta-cell death in type 2 diabetes mellitus. Exp Diabetes Res 2008:421287

Knight JD, Miranker AD (2004) Phospholipid catalysis of diabetic amyloid assembly. J Mol Biol 341(5):1175–1187

Knight JD, Hebda JA, Miranker AD (2006) Conserved and cooperative assembly of membrane-bound alpha-helical states of islet amyloid polypeptide. Biochemistry 45(31):9496–9508

Konarkowska B, Aitken JF, Kistler J, Zhang S, Cooper GJ (2006) The aggregation potential of human amylin determines its cytotoxicity towards islet beta-cells. FEBS J 273(15):3614–3624

Koo BW, Miranker AD (2005) Contribution of the intrinsic disulfide to the assembly mechanism of islet amyloid. Protein Sci 14(1):231–239

Krampert M, Bernhagen J, Schmucker J, Horn A, Schmauder A, Brunner H et al (2000) Amyloidogenicity of recombinant human pro-islet amyloid polypeptide (ProIAPP). Chem Biol 7(11):855–871

Lau TL, Gehman JD, Wade JD, Perez K, Masters CL, Barnham KJ et al (2007) Membrane interactions and the effect of metal ions of the amyloidogenic fragment Abeta(25–35) in comparison to Abeta(1–42). Biochim Biophys Acta 1768(10):2400–2408

Lee G, Pollard HB, Arispe N (2002) Annexin 5 and apolipoprotein E2 protect against Alzheimer's amyloid-beta-peptide cytotoxicity by competitive inhibition at a common phosphatidylserine interaction site. Peptides 23(7):1249–1263

Lutz TA (2006) Amylinergic control of food intake. Physiol Behav 89(4):465–471

Lutz TA (2010) The role of amylin in the control of energy homeostasis. Am J Physiol Regul Integr Comp Physiol 298(6):R1475–R1484

MacArthur DL, de Koning EJ, Verbeek JS, Morris JF, Clark A (1999) Amyloid fibril formation is progressive and correlates with beta-cell secretion in transgenic mouse isolated islets. Diabetologia 42(10):1219–1227

Martin C (2006) The physiology of amylin and insulin: maintaining the balance between glucose secretion and glucose uptake. Diabetes Educ 32(Suppl 3):101S–104S

Martinez A, Kapas S, Miller MJ, Ward Y, Cuttitta F (2000) Coexpression of receptors for adrenomedullin, calcitonin gene-related peptide, and amylin in pancreatic beta-cells. Endocrinology 141(1):406–411

Marzban L, Trigo-Gonzalez G, Zhu X, Rhodes CJ, Halban PA, Steiner DF et al (2004) Role of beta-cell prohormone convertase (PC)1/3 in processing of pro-islet amyloid polypeptide. Diabetes 53(1):141–148

McGuinness B, Passmore P (2010) Can statins prevent or help treat Alzheimer's disease? J Alzheimers Dis 20(3):925–933

McLatchie LM, Fraser NJ, Main MJ, Wise A, Brown J, Thompson N et al (1998) RAMPs regulate the transport and ligand specificity of the calcitonin-receptor-like receptor. Nature 393(6683):333–339

Mirzabekov TA, Lin MC, Kagan BL (1996) Pore formation by the cytotoxic islet amyloid peptide amylin. J Biol Chem 271(4):1988–1992

Miyazato M, Nakazato M, Shiomi K, Aburaya J, Toshimori H, Kangawa K et al (1991) Identification and characterization of islet amyloid polypeptide in mammalian gastrointestinal tract. Biochem Biophys Res Commun 181(1):293–300

Morfis M, Tilakaratne N, Furness SG, Christopoulos G, Werry TD, Christopoulos A et al (2008) Receptor activity-modifying proteins differentially modulate the G protein-coupling efficiency of amylin receptors. Endocrinology 149(11):5423–5431

Moriarty DF, Raleigh DP (1999) Effects of sequential proline substitutions on amyloid formation by human amylin20-29. Biochemistry 38(6):1811–1818

Mulder H, Ahren B, Sundler F (1996) Islet amyloid polypeptide and insulin gene expression are regulated in parallel by glucose in vivo in rats. Am J Physiol 271(6 Pt 1):E1008–E1014

Munishkina LA, Fink AL (2007) Fluorescence as a method to reveal structures and membrane-interactions of amyloidogenic proteins. Biochim Biophys Acta 1768(8):1862–1885

Nakazato M, Asai J, Miyazato M, Matsukura S, Kangawa K, Matsuo H (1990) Isolation and identification of islet amyloid polypeptide in normal human pancreas. Regul Pept 31(3):179–186

Nishi M, Sanke T, Seino S, Eddy RL, Fan YS, Byers MG et al (1989) Human islet amyloid polypeptide gene: complete nucleotide sequence, chromosomal localization, and evolutionary history. Mol Endocrinol 3(11):1775–1781

Opie EL (1901) On the relation of chronic interstitial pancreatitis to the Islands of Langerhans and to diabetes melutus. J Exp Med 5(4):397–428

Padrick SB, Miranker AD (2002) Islet amyloid: phase partitioning and secondary nucleation are central to the mechanism of fibrillogenesis. Biochemistry 41(14):4694–4703

Pieri L, Bucciantini M, Guasti P, Savistchenko J, Melki R, Stefani M (2009) Synthetic lipid vesicles recruit native-like aggregates and affect the aggregation process of the prion Ure2p: insights on vesicle permeabilization and charge selectivity. Biophys J 96(8):3319–3330

Piper K, Brickwood S, Turnpenny LW, Cameron IT, Ball SG, Wilson DI et al (2004) Beta cell differentiation during early human pancreas development. J Endocrinol 181(1):11–23

Poyner DR, Sexton PM, Marshall I, Smith DM, Quirion R, Born W et al (2002) International Union of Pharmacology. XXXII. The mammalian calcitonin gene-related peptides, adrenomedullin, amylin, and calcitonin receptors. Pharmacol Rev 54(2):233–246

Quist A, Doudevski I, Lin H, Azimova R, Ng D, Frangione B et al (2005) Amyloid ion channels: a common structural link for protein-misfolding disease. Proc Natl Acad Sci U S A 102(30):10427–10432

Riediger T, Zuend D, Becskei C, Lutz TA (2004) The anorectic hormone amylin contributes to feeding-related changes of neuronal activity in key structures of the gut-brain axis. Am J Physiol Regul Integr Comp Physiol 286(1):R114–R122

Ritzel RA, Meier JJ, Lin CY, Veldhuis JD, Butler PC (2007) Human islet amyloid polypeptide oligomers disrupt cell coupling, induce apoptosis, and impair insulin secretion in isolated human islets. Diabetes 56(1):65–71

Rivera JF, Costes S, Gurlo T, Glabe CG, Butler PC (2014) Autophagy defends pancreatic beta cells from human islet amyloid polypeptide-induced toxicity. J Clin Invest 124(8):3489–3500

Rymer DL, Good TA (2000) The role of prion peptide structure and aggregation in toxicity and membrane binding. J Neurochem 75(6):2536–2545

Salamekh S, Brender JR, Hyung SJ, Nanga RP, Vivekanandan S, Ruotolo BT et al (2011) A two-site mechanism for the inhibition of IAPP amyloidogenesis by zinc. J Mol Biol 410(2):294–306

Schneider A, Schulz-Schaeffer W, Hartmann T, Schulz JB, Simons M (2006) Cholesterol depletion reduces aggregation of amyloid-beta peptide in hippocampal neurons. Neurobiol Dis 23(3):573–577

Simmons C, Ingham V, Williams A, Bate C (2014) Platelet-activating factor antagonists enhance intracellular degradation of amyloid-beta42 in neurons via regulation of cholesterol ester hydrolases. Alzheimers Res Ther 6(2):15

Simons K, Toomre D (2000) Lipid rafts and signal transduction. Nat Rev Mol Cell Biol 1(1):31–39

Sipe JD, Benson MD, Buxbaum JN, Ikeda S, Merlini G, Saraiva MJ et al (2010) Amyloid fibril protein nomenclature: 2010 recommendations from the nomenclature committee of the International Society of Amyloidosis. Amyloid 17(3–4):101–104

Smith PE, Brender JR, Ramamoorthy A (2009) Induction of negative curvature as a mechanism of cell toxicity by amyloidogenic peptides: the case of islet amyloid polypeptide. J Am Chem Soc 131(12):4470–4478

Susa AC, Wu C, Bernstein SL, Dupuis NF, Wang H, Raleigh DP et al (2014) Defining the molecular basis of amyloid inhibitors: human islet amyloid polypeptide-insulin interactions. J Am Chem Soc 136(37):12912–12919

Trikha S, Jeremic AM (2011) Clustering and internalization of toxic amylin oligomers in pancreatic cells require plasma membrane cholesterol. J Biol Chem 286(41):36086–36097

Trikha S, Jeremic AM (2013) Distinct internalization pathways of human amylin monomers and its cytotoxic oligomers in pancreatic cells. PLoS One 8(9):e73080

Tu LH, Raleigh DP (2013) Role of aromatic interactions in amyloid formation by islet amyloid polypeptide. Biochemistry 52(2):333–342

Wagoner PK, Chen C, Worley JF, Dukes ID, Oxford GS (1993) Amylin modulates beta-cell glucose sensing via effects on stimulus-secretion coupling. Proc Natl Acad Sci U S A 90(19):9145–9149

Wang J, Xu J, Finnerty J, Furuta M, Steiner DF, Verchere CB (2001) The prohormone convertase enzyme 2 (PC2) is essential for processing pro-islet amyloid polypeptide at the NH2-terminal cleavage site. Diabetes 50(3):534–539

Westermark P, Wernstedt C, Wilander E, Sletten K (1986) A novel peptide in the calcitonin gene related peptide family as an amyloid fibril protein in the endocrine pancreas. Biochem Biophys Res Commun 140(3):827–831

Westermark P, Engstrom U, Johnson KH, Westermark GT, Betsholtz C (1990) Islet amyloid polypeptide: pinpointing amino acid residues linked to amyloid fibril formation. Proc Natl Acad Sci U S A 87(13):5036–5040

Westermark G, Westermark P, Eizirik DL, Hellerstrom C, Fox N, Steiner DF et al (1999) Differences in amyloid deposition in islets of transgenic mice expressing human islet amyloid polypeptide versus human islets implanted into nude mice. Metabolism 48(4):448–454

Westermark P, Andersson A, Westermark GT (2011) Islet amyloid polypeptide, islet amyloid, and diabetes mellitus. Physiol Rev 91(3):795–826

Williamson R, Usardi A, Hanger DP, Anderton BH (2008) Membrane-bound beta-amyloid oligomers are recruited into lipid rafts by a fyn-dependent mechanism. FASEB J 22(5):1552–1559

Wiltzius JJ, Sievers SA, Sawaya MR, Cascio D, Popov D, Riekel C et al (2008) Atomic structure of the cross-beta spine of islet amyloid polypeptide (amylin). Protein Sci 17(9):1467–1474

Wookey PJ, Lutz TA, Andrikopoulos S (2006) Amylin in the periphery II: an updated mini-review. Sci World J 6:1642–1655

Young A, Denaro M (1998) Roles of amylin in diabetes and in regulation of nutrient load. Nutrition 14(6):524–527

Zhang S, Liu J, Dragunow M, Cooper GJ (2003) Fibrillogenic amylin evokes islet beta-cell apoptosis through linked activation of a caspase cascade and JNK1. J Biol Chem 278(52):52810–52819

Zhang S, Liu H, Chuang CL, Li X, Au M, Zhang L et al (2014) The pathogenic mechanism of diabetes varies with the degree of overexpression and oligomerization of human amylin in the pancreatic islet beta cells. FASEB J 28(12):5083–5096

Zhao H, Tuominen EK, Kinnunen PK (2004) Formation of amyloid fibers triggered by phosphatidylserine-containing membranes. Biochemistry 43(32):10302–10307

Zhao H, Jutila A, Nurminen T, Wickstrom SA, Keski-Oja J, Kinnunen PK (2005) Binding of endostatin to phosphatidylserine-containing membranes and formation of amyloid-like fibers. Biochemistry 44(8):2857–2863

Zhao Q, Jayawardhana D, Guan X (2008) Stochastic study of the effect of ionic strength on non covalent interactions in protein pores. Biophys J 94(4):1267–1275

Zhao J, Hu R, Sciacca MF, Brender JR, Chen H, Ramamoorthy A et al (2014) Non-selective ion channel activity of polymorphic human islet amyloid polypeptide (amylin) double channels. Phys Chem Chem Phys 16(6):2368–2377

Intrinsic Stability, Oligomerization, and Amyloidogenicity of HDL-Free Serum Amyloid A

5

Wilfredo Colón, J. Javier Aguilera, and Saipraveen Srinivasan

Abstract

Serum amyloid A (SAA) is an acute-phase reactant protein predominantly bound to high-density lipoprotein in serum and presumed to play various biological and pathological roles. Upon tissue trauma or infection, hepatic expression of SAA increases up to 1,000 times the basal levels. Prolonged increased levels of SAA may lead to amyloid A (AA) amyloidosis, a usually fatal systemic disease in which the amyloid deposits are mostly comprised of the N-terminal 1–76 fragment of SAA. SAA isoforms may differ across species in their ability to cause AA amyloidosis, and the mechanism of pathogenicity remains poorly understood. In vitro studies have shown that SAA is a marginally stable protein that folds into various oligomeric species at 4 °C. However, SAA is largely disordered at 37 °C, reminiscent of intrinsically disordered proteins. Non-pathogenic murine (m)SAA2.2 spontaneously forms amyloid fibrils in vitro at 37 °C whereas pathogenic mSAA1.1 has a long lag (nucleation) phase, and eventually forms fibrils of different morphology than mSAA2.2. Remarkably, human SAA1.1 does not form mature fibrils in vitro. Thus, it appears that the intrinsic amyloidogenicity of SAA is not a key determinant of pathogenicity, and that other factors, including fibrillation kinetics, ligand binding effects, fibril stability, nucleation efficiency, and SAA degradation may play key roles. This chapter will focus on the known structural and biophysical properties of SAA and discuss how these properties may help better understand the molecular mechanism of AA amyloidosis.

W. Colón (✉) • J.J. Aguilera • S. Srinivasan
Department of Chemistry and Chemical Biology, and
Center for Biotechnology and Interdisciplinary
Studies, Rensselaer Polytechnic Institute,
Troy, NY 12180, USA
e-mail: colonw@rpi.edu

© Springer International Publishing Switzerland 2015
O. Gursky (ed.), *Lipids in Protein Misfolding*, Advances in Experimental
Medicine and Biology 855, DOI 10.1007/978-3-319-17344-3_5

Keywords

Inflammation • Amyloidosis • Intrinsically disordered protein • Prion-like infectivity • Acute phase • High-density lipoprotein

Abbreviations

AA	Amyloid A
AEF	Amyloid enhancing factor
AFM	Atomic force microscopy
GAGs	Glycosaminoglycans
HDL	High-density lipoprotein
Hep	Heparin
HS	Heparan sulfate
hSAA	Human serum amyloid A
IDP	Intrinsically disordered proteins
mSAA	Mouse serum amyloid A
SAA	Serum amyloid A
SEC	Size-exclusion chromatography

5.1   Introduction

5.1.1   Serum Amyloid A and AA Amyloidosis

Serum amyloid A (SAA, ~12 kDa) belongs to a highly conserved family of proteins that appears to play very important roles, including cholesterol transport (Kisilevsky and Manley 2012; van der Westhuyzen et al. 2007) and various immunological functions (Eklund et al. 2012). Acute phase SAA is predominantly synthesized by the liver and is secreted into plasma where it binds to high-density lipoprotein (HDL) (Benditt and Eriksen 1977). SAA is also expressed in normal tissue (Urieli-Shoval et al. 1998), as well as in cells in atherosclerotic plaques (Meek et al. 1994), tumor tissue (Malle et al. 2009), and the brains of individuals with Alzheimer's disease (Chung et al. 2000; Liang et al. 1997). Considering the wide expression profile of SAA in normal tissue and in disease, it appears that this protein may not just be important during inflammation, but may also play an active role in many disease processes (for reviews see (Urieli-Shoval et al. 2000; Cunnane 2001).

Although the functions of SAA remain poorly understood, the protein appears to play a seminal role in cholesterol metabolism and transport (Kisilevsky and Subrahmanyan 1992; Liang et al. 1996; Steinmetz et al. 1989; Tam et al. 2002). Specifically, SAA binds to HDL and directs it toward cholesterol-loaded macrophages near damaged tissue to allow cholesterol recycling for tissue repair (Kisilevsky and Manley 2012). The normal level of SAA in serum is 1–3 µg/mL (Gabay and Kushner 1999). However, during acute inflammation such as infection, injury, or trauma, the levels of SAA may increase up to 1,000-fold (Gabay and Kushner 1999; McAdam and Sipe 1976). Due to this dramatic increase in SAA concentration, combined with approximately twofold decline in plasma level of HDL during the acute phase response, it is likely that not all SAA is bound to HDL (Webb et al. 1997). However, due to SAA's intrinsic tendency to aggregate, it seems unlikely that SAA would circulate in the plasma unbound to any ligand.

The prolonged high levels of SAA associated with chronic inflammatory conditions sometimes lead to amyloid A (AA) amyloidosis, which is a secondary amyloidosis characterized by the deposition of SAA-derived amyloid fibrils in organs like liver, spleen, and kidney (Pepys 2006; Sipe 1992). AA is one of the most common systemic amyloid diseases worldwide, but is much less prevalent in western nations, ranging from 0.50 to 0.86 % (Simms et al. 1994). The prevalence is higher for particular chronic diseases, such as rheumatoid arthritis. For example, one study showed that 9 % of Finnish individuals with rheumatoid arthritis died of AA amyloidosis (Mutru et al. 1985). Another study reported that 5 % of individuals with rheumatoid arthritis develop AA amyloidosis (Husby 1998). There is no cure for AA amyloidosis, but an effective treatment involves reducing the high levels of SAA circulating in serum via administration of anti-inflammatory drugs (Hazenberg and van Rijswijk 2000; Immonen et al. 2011). It has been shown that the outcome in AA amyloidosis is favorable when the concentration of SAA is maintained below 0.01 mg/mL, but those with SAA concentration of 0.05 mg/mL had poor prognosis (Gillmore et al. 2001).

AA amyloidosis usually involves fibrillar deposits comprising the N-terminal 76-residue fragment of SAA (Sletten and Husby 1974). However, because full-length SAA and other N-terminal SAA fragments have been found in human amyloid deposits (Husebekk et al. 1985; Rocken and Shakespeare 2002; Westermark et al. 1989), and amyloid deposits in duck comprise full-length SAA (Ericsson et al. 1987), the specific role of SAA proteolysis in amyloid formation and the etiology of AA amyloidosis remains unclear. Nevertheless, the absence of the C-terminus of SAA in most AA deposits suggests that the main determinants of fibril formation are found within the N-terminal 3/4 of SAA. This hypothesis is supported by in vitro studies involving synthetic peptides representing different region of human and mouse SAA (Egashira et al. 2011; Lu et al. 2014; Westermark et al. 1992), site-directed mutations at the N-terminus of human SAA (Patel et al. 1996), proteolytic degradation of the N-terminus (Yamada et al. 1995), as well as by the amino acid sequence analysis (see Chap. 8 by Das and Gursky in this volume).

5.1.2 Pathogenic and Nonpathogenic Isoforms of SAA: Mouse Model of AA Amyloidosis

Humans and mice have two acute-phase SAA isoforms that differ in their pathogenicity, presumably due to their different ability to form amyloid deposits in vivo. Human acute phase SAAs include SAA1 and SAA2 that differ at eight residues (SAA1#SAA2: V52A, A57V, D60N, F68L, F69T, H71R, E84K, K90R). SAA1 is mostly found in amyloid deposits associated with AA amyloidosis (Liepnieks et al. 1995), and is known to exist in three different isotypes that differ from each other at residues 52 and/or 57, resulting in SAA1.1 (V52/A57), SAA1.3 (A52/A57), and SAA1.5 (A52/V57). It has been shown that Japanese individuals homozygous for

SAA1.3 have a greater chance of developing AA amyloidosis, and the disease course is more likely to be aggressive with a poorer outcome (Nakamura et al. 2006). Furthermore, individuals with SAA1.1/1.1 genotype are three to seven times more likely to develop AA amyloidosis (van der Hilst 2011). In mice, most strains have SAA1.1 and SAA2.1 isoforms (formerly known as SAA2 and SAA1, respectively (Sipe 1999)) that differ at nine residues (SAA1.1#SAA2.1: I6V, G7H, G27N, D30N, G31S, A60G, S63A, M76I, A101D). AA amyloidosis can be induced in mice by injection of inflammatory reagents (Ishihara 1973; Skinner et al. 1977). Most studies have been performed in type A mice (Ishihara 1973), which produce equal amounts of SAA2.1 and SAA1.1 isoforms. Only SAA1.1 in mouse is found in amyloid deposits (Hoffman et al. 1984; Meek et al. 1986; Shiroo et al. 1987). In addition, other animal species differ in their SAA sequence and consequently, are either highly susceptible to AA amyloidosis or resistant to the disease (Sipe et al. 1993; Woldemeskel 2012; Zhang et al. 2008). Interestingly, the CE/J strain of mice expressing just one isoform, SAA2.2, was discovered to be resistant to AA amyloidosis (de Beer et al. 1993; Sipe et al. 1993). Thus, even though the sequence of SAA is highly conserved in vertebrates, differences in primary structure appear to account for the divergent pathogenicity of SAA in various species. The mechanism for this structure-pathogenic function relationship remains poorly understood.

This chapter will review the biophysical properties of HDL-free (apo) SAA and discuss the potential implications in AA amyloidosis. Despite the wealth of cellular and clinical studies involving SAA and AA amyloidosis (Obici and Merlini 2012; Westermark et al. 2015), fundamental questions remain. For example, how the biophysical properties of SAA isoforms influence their potential pathogenicity, as well as the disease onset and progression, is poorly understood. Although the biological relevance of in vitro experiments is limited and is often difficult to assess, much of our understanding about the molecular basis of amyloid diseases have benefited from such studies. Similarly, a better understanding of the structural and biophysical properties of SAA isoforms may reveal the basis for their diverse pathogenicity, and provide novel insight leading to a better understanding of the molecular mechanism of AA amyloidosis and developing novel therapeutic strategies.

5.2 AA Amyloidosis: What Is the Pathogenic Mechanism?

5.2.1 Molecular Mechanism of SAA Fibril Formation In Vivo

Amyloid formation occurs by a nucleation-dependent or seeding mechanism, where the rate-limiting step is the formation of a nucleus/seed building block that propagates fibril growth. Here, we are referring to nuclei as any oligomeric to short fibrillar species that can seed fibril formation. The time required for the nuclei to form is known as the lag phase, and it usually takes from hours to days in in vitro experiments, depending on the protein concentration and the solvent conditions. In AA amyloidosis, the disease onset in mice can be accelerated when an inflammatory stimulus is accompanied by an injection of protein extracted from AA amyloid-laden mouse (Axelrad et al. 1982; Kisilevsky and Boudreau 1983). This elusive "amyloid enhancing factor" (AEF) in AA amyloidosis was found to comprise SAA fibrils (Lundmark et al. 2002). However, the molecular entity responsible for AEF activity has not been defined, and may comprise one or more degradation-resistant species, including amyloid nuclei or small fibrils readily generated from the breakup of larger SAA fibrils. Remarkably, AEF was shown to be effective when administered orally, indicating that AA amyloidosis may be transmissible via a prion-like mechanism (Lundmark et al. 2002). A hallmark of prion transmissibility is the high stability of the prion species, which enables it to resist denaturation and proteolytic degradation in the digestive tract, thereby allowing oral transmissibility. Thus, the stability of the SAA-based amyloid nuclei or protofibrils, which are the putative prion-genic species, is likely critical for the onset and transmissibility of AA amyloidosis.

The persistently high level of SAA during chronic inflammation is an essential prerequisite for AA amyloidosis. However, the relationship between inflammation – unrelated to the high levels of SAA – and amyloid deposition is not easily decoupled. A recent transgenic mouse model study in which the concentration of SAA was increased by doxycycline-inducible expression showed that high expression (>1.0 mg/mL) of SAA is sufficient for the onset of AA amyloidosis independently of inflammation (Simons et al. 2013). In addition, this dose-dependent transgenic model showed that when the concentration of SAA was raised modestly to ~0.12 mg/mL for a few weeks, the still healthy mice were primed for very fast amyloid deposition when the concentration of SAA was subsequently increased to the pathological concentration of >1.0 mg/mL. Furthermore, when the concentration of SAA was decreased after the onset of AA amyloidosis, there was evidence of amyloid clearance, even though the mice remained primed for disease. These remarkable results demonstrate that the structure and biophysical properties of SAA and its fibrils, along with time of exposure to high SAA concentration, are critical factors in the etiology of AA amyloidosis. The apparent competition between AA accumulation and clearance suggests that, unlike other amyloid diseases where a putative toxic species is responsible for pathology (Guerrero-Munoz et al. 2014; Stefani 2012), the pathology in AA amyloidosis arises from the progressive accumulation of amyloid fibrils that interfere with normal organ function.

5.2.2 Intrinsic Factors Affecting the Pathological Accumulation of SAA Fibrils

Although biophysical properties of SAA isoforms, such as nucleation/fibrillation kinetics and nucleus/fibril stability, likely play a major role in the etiology of AA amyloidosis, the relative contributions of these and other endogenous factors remain unclear. In particular, two inherent SAA properties stand out. One is the kinetics of amyloid nucleation and fibril growth; the other is the kinetic stability, i.e. degradation-resistance, of amyloid nuclei and fibrils. The kinetics of amyloid formation depends strongly on the SAA concentration that must exceed a required threshold for nuclei formation and fibril growth. Below a certain concentration, SAA may not fibrillate fast enough to overcome its clearance. In individuals with AA amyloidosis, keeping their SAA levels below 0.01 mg/mL via aggressive treatment of inflammation has been shown to result in beneficial outcome, including regression of amyloid deposits (Gillmore et al. 2001; Lachmann et al. 2007). In contrast, the disease progressed in patients whose SAA levels were persistently above 0.05 mg/mL (Gillmore et al. 2001). Thus, administration of anti-inflammatory medication to keep the SAA concentration at low levels has been effective in reducing the incidence and mortality of AA amyloidosis in developed countries (Gillmore et al. 2001). The persistence of amyloid deposition also depends on nuclei and fibril stability. The inherent ability of SAA to form highly stable amyloid nuclei is also likely a main factor in AA amyloidosis. The persistence of SAA nuclei in mice primed for SAA deposition (Simons et al. 2013) suggests that these precursor aggregates (i.e. seeds) are kinetically stable and, therefore, resistant to proteolysis (Manning and Colon 2004). The stability of SAA seeds would explain the AEF effect observed in mice in vivo, as well as the observation that cheetah exhibits prion-like infectivity of AA amyloidosis (Lundmark et al. 2002; Zhang et al. 2008). Thus, the kinetics of SAA amyloid nucleation and growth, as well as the stability of the precursor aggregates and fibrils are likely the main contributors to the diverse pathogenicity of SAA isoforms.

5.2.3 Endogenous Factors Affecting the Pathological Accumulation of SAA Fibrils

The low incidence of AA amyloidosis indicates that endogenous genetic or environmental factors play an important pathological role. The most important endogenous factor is the expression level of SAA during inflammation. The concentration of SAA may increase up to 1,000-fold during episodes of acute inflammation, but the exact level, as well as the frequency and the duration of the acute episodes, vary for different individuals. Another endogenous factor involves the availability and effect of SAA ligands. Although SAA is an apolipoprotein-related protein that binds HDL, as well as heparin/heparan sulfate (HS) and other ligands, e.g. laminin and various peripheral membrane proteins (Urieli-Shoval et al. 2000), it remains unclear how the endogenous amounts of these ligands impact the etiology of AA amyloidosis. Most SAA that circulates in plasma in HDL-bound form is presumably safe from fibrillation. Therefore, SAA and its fragments seem at the greatest risk for deposition at sites of inflammation, where SAA is released and internalized by macrophages. However, one possibility is that not all SAA may be HDL-bound during acute inflammation and, therefore, it is plausible that free SAA (or SAA bound to some other ligand) may build up in certain tissues and be at risk of aggregating. This is supported by reports of extra-hepatic SAA leading to amyloid deposition in mice brain and chicken joint (Guo et al. 2002; Landman 1999). Therefore, the significance of localized HDL-free SAA may be more pathologically relevant than commonly assumed. Since HDL-free SAA is at risk of aggregation, the endogenous level of HDL may affect the levels of HDL-free SAA in vivo and thereby influence AA amyloidosis.

Interestingly, it has been shown that HS and heparin (Hep) can displace SAA from HDL (Noborn et al. 2012). This observation is significant because of substantial evidence that HS and, perhaps, Hep are important in AA amyloidosis, presumably by accelerating amyloid formation (Elimova et al. 2009; Li et al. 2005; Snow et al. 1991). Thus, it is plausible that the SAA displacement from HDL may occur in the Hep-rich macrophages or in HS-rich extracellular matrix environment, where SAA is then internalized by macrophages. The binding of SAA to HS may also serve as a way to increase the concentration of SAA in amyloidogenic focal points in tissues. In vitro studies have found that HS and

Hep accelerate the fibrillation of many amyloidogenic proteins, including SAA (Elimova et al. 2009; van Horssen et al. 2003; Zhang and Li 2010). Thus, the displacement of SAA from HDL at sites of tissue damage and the consequent conformational changes in SAA are dynamic processes likely modulated by the levels of glycosaminoglycans (GAGs) and other factors.

Another fundamental endogenous factor in SAA amyloid deposition is the efficiency of the protein quality control system, including the degradation of SAA and its amyloid deposits. The term "proteostasis" describes the overall homeostasis of proteins in an organism, and compromised proteostasis may lead to amyloid diseases (Cuanalo-Contreras et al. 2013; Kim et al. 2013). The persistently high concentration of SAA in chronic inflammatory diseases is likely to place a stress on proteostasis, which in some individuals may have pathological consequences. For example, the incomplete proteolysis of SAA may accelerate its fibrillation kinetics or enhance the stability of the resulting AA fibrils. This may explain why AA deposits comprise N-terminal fragments of SAA, although the etiological role of SAA degradation in AA amyloidosis still remains unclear. Compromised proteostasis would also slow down the clearance of amyloid, which is stabilized by its interaction with serum amyloid P and GAGs, in particular HS. Therefore, the endogenous amounts of these "amyloid stabilizers" are also expected to impact the rate of amyloid clearance. In support of this idea, AA amyloidosis was shown to slow down in a transgenic knockout mouse of serum amyloid P (Inoue et al. 2005), further highlighting the precarious balance between amyloid deposition and degradation.

5.3 Structural Properties of SAA

5.3.1 SAA May Form Different Oligomers Upon Refolding In Vitro

SAA expressed in *E. coli* must be purified under denaturing conditions due to its tendency to aggregate. Previous studies have shown that mSAA2.2, mSAA1.1, and hSAA1.1 fold into various oligomeric species in vitro, including hexamer, octamer, tetramer, and dodecamer (Lu et al. 2014; Patke et al. 2013; Srinivasan et al. 2013; Wang et al. 2002, 2011). Interestingly, mSAA2.2 first refolds into a kinetically accessible octamer that at 4 °C converts into a more stable hexamer within 3–4 weeks (Fig. 5.1) (Wang et al. 2011). In contrast, mSAA1.1 refolds into a mixture of dodecamer, tetramer, and monomers that does not change much over time at 4 °C (Srinivasan et al. 2013). In the case of hSAA1.1, size-exclusion chromatography data showed a single peak with a variable elution time consistent with a mixture of octamers and hexamers (Patke et al. 2013).

The biological relevance of the oligomeric structures of SAA observed in vitro at 4 °C is poorly understood. However, it seems unlikely that proteins would evolve to fold and assemble into specific oligomers without selective pressure to do so. Therefore, although SAA oligomers are marginally stable, it is plausible that the very high concentration of SAA during acute inflammation together with ligand binding may modulate the oligomeric structure of HDL-free SAA in vivo. The ability of certain proteins to oligomerize upon ligand binding or to interconvert among oligomeric species is well established. Many proteins exhibit such oligomeric plasticity, including human porphobilinogen synthase (Breinig et al. 2003), geranylgeranyl diphosphate synthase (Miyagi et al. 2007), protective antigen protein of the anthrax toxin (Kintzer et al. 2009), and LIM domain binding proteins (Cross et al. 2010). Thus, the possibility that SAA may form different functional oligomers is intriguing, and further studies are warranted to explore this idea (see Sect. 5.3.3).

5.3.2 SAA Is an Intrinsically Disordered Protein

Although SAA folds into α-helical oligomeric structures at 4 °C, previous studies have shown that mSAA2.2, mSAA1.1, and hSAA1.1 are marginally stable proteins with thermal denaturation

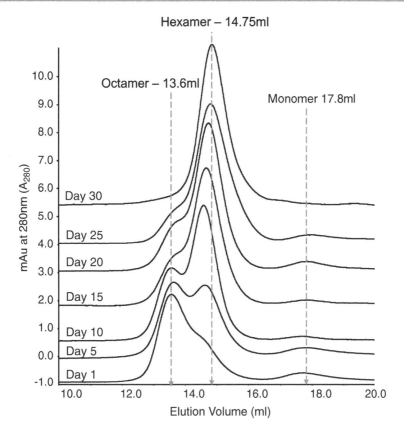

Fig. 5.1 Refolding of murine SAA2.2 monitored by size-exclusion chromatography. Purified and urea-denatured mSAA2.2 was dialyzed against Tris buffer at 4 °C, and protein oligomerization was analyzed by size-exclusion chromatography over time. mSAA2.2 first refolded into an octamer that, over several weeks, converted to a hexamer (Reprinted from (Wang et al. 2011), Copyright (2011), with permission from Elsevier)

mid-points circa 32, 22, and 20 °C, respectively (Patke et al. 2013; Srinivasan et al. 2013; Wang and Colon 2005; Wang et al. 2005). At 37 °C, SAA is largely disordered prior to aggregation, and hence, HDL-free SAA is an intrinsically disordered protein (IDP). IDPs are characterized by their lack of stable secondary and/or tertiary structure in vitro under physiologically-relevant conditions, and by their ability to acquire one or more functional states upon binding to small ligands or macromolecules (Uversky 2013; also see Chap. 2 by Uversky in this volume). Remarkably, many aggregation-prone proteins associated with human neurodegenerative diseases are intrinsically disordered (Uversky 2009). Viewing the structure and function of SAA from the perspective of an IDP, one could explain how

SAA may have multiple functions, depending on the environment and its interactions with ligands. Therefore, in addition to SAA's key function in cholesterol recycling (Kisilevsky and Manley 2012), other putative functions related to lipid metabolism and immune regulation, as well as SAA functions not yet discovered, likely exist within the framework of an IDP. For example, the structure of SAA may be modulated via an induced fit model, in which binding with high specificity – low affinity to various ligands, ranging from small molecules to peripheral membrane proteins, induces the formation of different functionally relevant conformations (Uversky 2013).

The mechanism by which SAA is released from HDL, which is a likely prerequisite for SAA fibrillation, appears to involve HS and Hep

(Noborn et al. 2012). Recent studies have shown that heparin is able to displace SAA from HDL in vitro (Noborn et al. 2012). Since Hep is usually released from mast cells into the blood at sites of tissue injury, one of its physiological roles may involve the release of SAA from HDL at sites of tissue damage. It remains unknown whether turnover of SAA occurs immediately upon release from HDL, or whether Hep/HS binding may modulate the structure of SAA to carry out other functions. In summary, the intrinsic capability of SAA to adopt diverse structures ranging from an intrinsically disordered monomer to HDL-bound form to specific oligomers and higher-order species that are substantially α-helical, is intriguing and may help explain the many putative functions of SAA.

5.3.3 The 3D Structures of Human SAA1.1 and Mouse SAA3 Have Been Solved

Although SAA was discovered over 40 years ago (Levin et al. 1972), the 3D structures of two dif-

ferent SAA proteins were solved only in 2014, providing an unprecedented opportunity to understand how small differences in SAA sequence affect the biochemical and biophysical properties of the protein. The 2.2 Å resolution crystal structure of hSAA1.1 revealed a hexamer comprised of two trimers, with each monomer folding into a cone-shaped four-helix bundle stabilized by the C-terminal tail wrapped around it (Fig. 5.2) (Lu et al. 2014). The structure also showed two clusters of positively charged residues, one around the central pore of the hexamer (involving residues Arg15, Arg19, and Arg47) and another at the trimer apex (involving residues Arg1, Arg62, and His71) (Fig. 5.2). Experiments involving mutations of the relevant Arg residues showed that both charged regions bind to Hep. Subsequent experiments showed that HDL could not bind to hSAA1.1 when Hep was bound to the apex region, indicating that the HDL binding site involves the apex region and can be inhibited by Hep and, presumably, HS (Lu et al. 2014). These results are consistent with the recent observation that HS and Hep can dissociate SAA from HDL (Noborn et al. 2012). The crystal structure of

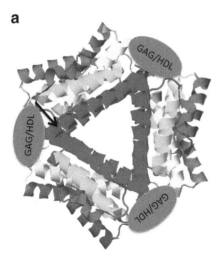
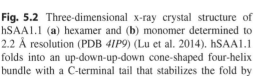
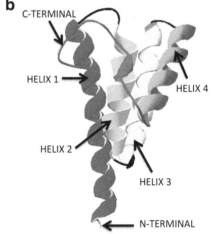

Fig. 5.2 Three-dimensional x-ray crystal structure of hSAA1.1 (**a**) hexamer and (**b**) monomer determined to 2.2 Å resolution (PDB *4IP9*) (Lu et al. 2014). hSAA1.1 folds into an up-down-up-down cone-shaped four-helix bundle with a C-terminal tail that stabilizes the fold by wrapping around the bundle. Each helix is labeled in a different color. The hexamer (viewed down the threefold axis) is comprised of two trimers. Three copies of the N-terminal GAG/HDL binding site located at the trimer apex are indicated

hSAA1.1 also showed that helices 1 and 3, which are the most amyloidogenic regions of SAA, are sequestered within the hexameric structure, in agreement with prior in vitro studies showing that dissociation of the hexameric structure is required for aggregation and formation of fibrils (Patke et al. 2013; Srinivasan et al. 2013; Wang et al. 2005). The 3D structure of hSAA1.1 supports many previous in vitro findings and conclusions regarding the biophysical properties of SAA. These include the hexameric assembly of hSAA1.1 (Patke et al. 2013), the finding that different SAAs are likely to assemble into different oligomers (Srinivasan et al. 2013; Wang et al. 2002, 2011), and the observation that the C-terminal part of SAA is important for α-helical stability and oligomer formation (Patke et al. 2012).

The second crystal structure was that of mouse (m)SAA3, determined to 2.0 Å resolution (Derebe et al. 2014). Comparison of the two atomic structures revealed remarkable similarity in the conformations of hSAA1.1 and mSAA3 molecules. However, unlike hSAA1.1, which was crystallized as a dimer or a hexamer, mSAA3 was crystallized as a tetramer with a central hydrophobic pocket that binds retinol with nanomolar affinity. Retinol, a small lipid-soluble compound plays crucial roles in protecting against bacterial infection, and must be transported by a protein carrier. The remarkable discovery that mSAA3 is a retinol binding protein provides the first direct evidence of a functional oligomeric SAA structure, and more such functions may exist. Interestingly, even though mSAA2.2 forms a hexamer in vitro, electron microscopic images revealed a ring-like assembly where the mSAA2.2 monomers interact side-by-side (Wang et al. 2002), suggesting that more oligomeric architectures of SAA will likely be forthcoming. Thus, the atomic structures of hSAA1.1 and mSAA3 will be major assets in future efforts to understand the breadth of SAA structure-function relationship, including the promising prospect of gaining a structural-level understanding for the diverse pathogenicity of SAA isoforms.

5.4 In Vitro Studies Provide Insights into the Diverse Pathogenicity of SAA Isoforms

5.4.1 Inherent Amyloidogenicity of SAA Does Not Correlate with the Pathogenicity of Different Isoforms

Although many factors likely contribute to the development of AA amyloidosis, the existence of pathogenic and nonpathogenic SAA isoforms suggests that their pathogenic potential correlates with their amyloidogenicity, i.e. with the rate of fibril formation at near-physiologic conditions. However, this is not the case. In fact, nonpathogenic SAA2.2, which is the only isoform present in AA amyloidosis-resistant CE/J mouse, quickly self-assembles into fibrils upon incubation at 37 °C (Wang et al. 2005). In contrast, pathogenic mSAA1.1 exhibits a longer (3–4 days at 0.3 mg/mL SAA1.1) oligomer-rich fibrillation lag phase before forming fibrils (Fig. 5.3).

The lack of correlation between the in vitro rate of SAA fibril formation and pathogenicity is further supported by studies showing that pathogenic hSAA1.1 forms aggregates and protofibril-like species, but does not form mature amyloid fibrils in vitro upon incubation at 37 °C (Fig. 5.4) (Lu et al. 2014; Patke et al. 2013). In contrast, recombinant Met-hSAA1.1 protein expressed in *E. coli*, which contains an additional methionine at position 1, does form fibrils in vitro (Fig. 5.4), suggesting that fibril formation is importantly modulated by the N-terminal methionine (Patke et al. 2013). It is possible the pathologically relevant property is the relative amyloidogenicity of the N-terminal fragments and not of the full-length protein. This would explain the lack of correlation between the in vitro amyloid formation of full-length SAA isoforms and their pathogenicity. Alternatively, it may be that the timescale of in vitro experiments is biologically irrelevant because in vivo formation of the hSAA1.1 amyloid nucleus is probably a much slower process.

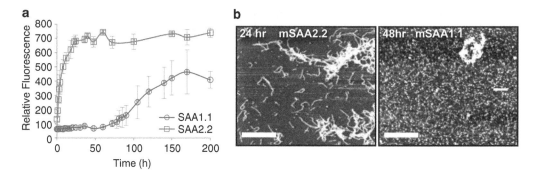

Fig. 5.3 Kinetics of mSAA1.1 and mSAA2.2 fibrillation at 37 °C probed by (**a**) Thioflavin-T fluorescence and (**b**) atomic force microscopy (AFM). Error bars correspond to normalized standard deviation from three independent experiments. AFM images show mSAA2.2 and mSAA1.1 samples incubated at 37 °C for 24 h and 48 h, respectively. The images are shown as height traces, and the scale bar corresponds to 1 μm (The panels in this figure were originally published in *The Journal of Biological Chemistry* (Srinivasan et al. 2013) © the American Society for Biochemistry and Molecular Biology)

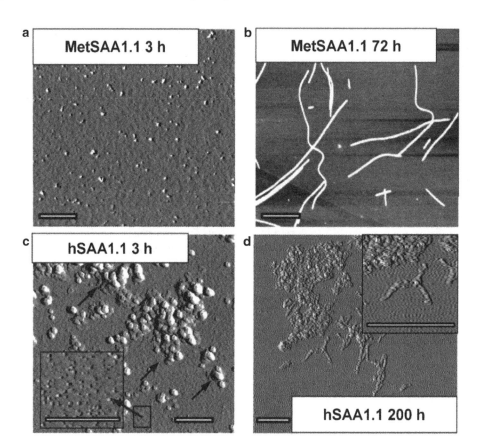

Fig. 5.4 Atomic force microscopy images of fibrils and aggregates formed by MetSAA1.1 and hSAA1.1, respectively. MetSAA1.1 is recombinant hSAA1.1 with an additional methionine at position 1. The samples (0.24 mg/mL) were incubated at 37 °C for the designated amount of time. AFM images correspond to (**a**) MetSAA1.1, 3 h; (**b**) MetSAA1.1, 72 h; (**c**) hSAA1.1, 3 h; (**d**) hSAA1.1, 200 h. The *insets* show expanded regions, and the three *black arrows* in (**c**) outside the inset point to assemblies of spherical aggregates. The images are shown as height traces, and the scale bar corresponds to 1 μm (This figure was originally published in *PLOS One* Patke et al. 2013)

Here, we are using the kinetics of fibril formation to define amyloidogenicity. However, in the years-long timescale of individuals with chronic inflammatory conditions, the biologically relevant definition of amyloidogenicity may not be "how fast", but rather "how persistent/stable". In summary, it may not be the rate of full-length SAA fiber formation in vitro, but perhaps other properties of SAA and other in vivo factors that determine the fibril accumulation in AA amyloidosis.

5.4.2 Stability of SAA Amyloid

The accumulation of SAA deposits is the main histopathological feature in AA amyloidosis. Therefore, the dynamic equilibrium between the accumulation of amyloid and its clearance is of paramount importance. It is clear from mouse model studies that if the fibrillation of SAA is decreased or inhibited, amyloid deposits can be cleared, and AA amyloidosis can be averted (Simons et al. 2013). However, the rate of amyloid clearance depends not only on the robustness of the protein degradation system, but also on the stability of the amyloid fibrils. At one end of the spectrum, marginally stable amyloid fibrils may never accumulate enough to cause the disease, even if they form during acute inflammation. At the other extreme, the formation of kinetically stable, degradation-resistant fibrils will stress the protein degradation system, thereby increasing the risk of disease.

Although most amyloid fibrils formed by different proteins and peptides are known to be very stable, recent studies have shown that amyloid fibrils formed in vitro by non-pathogenic mSAA2.2 are marginally stable (Ye et al. 2011). Although at inflammation-relevant concentration ($\geq$0.3 mg/mL), mSAA2.2 quickly fibrillates upon incubation at 37 °C, the resulting fibrils dissociate upon modest increase in temperature (over 45 °C) and urea concentration (>1.0 M) (Ye et al. 2011). Fibrils formed by pathogenic mSAA1.1 appear to be similarly unstable (unpublished results). Thus, the mSAA1.1 fibrils formed in vitro may not be the same as the pathogenic fibrils formed in vivo.

It is instructive to consider two possible scenarios in which the stability of SAA fibrils may or may not contribute to AA amyloidosis. First, one can envision two distinct types of SAA fibrils differing in stability and morphology. Fibrils of one type may be intrinsically unstable in vivo and be easily degraded, thereby explaining the poorly understood low incidence of AA amyloidosis in humans with chronic inflammatory conditions. Fibrils of the other type, which are more stable and pathogenic, may not be kinetically favored, requiring a long time to form. The latter fibrils may have similar morphology to AEF, thereby explaining the persistence of the AEF and its ability to catalyze the formation of stable pathologically relevant SAA fibrils. Second, the development of AA amyloidosis may require other ligands or factors to either accelerate the formation of stable SAA fibrils or to stabilize the already formed fibrils. Either effect could potentially overwhelm the degradation of SAA fibrils and result in increased amyloid deposition. Thus, understanding the intrinsic stability of amyloid fibrils accessible to different SAA isoforms, and the effect of ligand binding and the cellular environment, will be critical for better understanding the AEF effect, SAA isoform pathogenicity, and species susceptibility in AA amyloidosis.

5.4.3 SAA Can Form Fibrils of Different Morphology: Effects of GAGs

Over the past 15 years there has been an increased appreciation for the morphological differences among amyloid fibrils, including the ability of certain peptides and proteins to form fibrils of different morphologies depending on the solution conditions (Jahn and Radford 2008; Kodali and Wetzel 2007). These morphological disparities may result in (or reflect) different biochemical and biophysical properties of fibrils that may determine the preference for amyloid deposition in certain tissue, its resistance to degradation, and prion-like transmissibility (Meyer-Luehmann et al. 2006; Seilheimer et al. 1997; Tanaka et al. 2006).

Recent studies showed that lipid-free non-pathogenic mSAA2.2 and pathogenic mSAA1.1 form amyloid fibrils with different morphologies (Fig. 5.5). Longitudinal atomic force microscopy (AFM) experiments showed that the morphological differences arose from formation of different pre-fibrillar oligomers and amyloid seeds resulting in different fibrillation pathways (Fig. 5.6) (Srinivasan et al. 2013). This is consistent with earlier studies revealing morphological diversity in ex vivo SAA amyloid deposits (Jimenez et al. 2001). Due to the marginal stability of SAA at 37 °C and its high tendency to quickly aggregate, kinetic factors are likely to play an important role in determining the morphologic outcome of fibril formation (Pellarin et al. 2010). The ability of SAA to form fibrils of different morphologies is consistent with the observation that different amyloid-forming peptides/proteins can serve as

an AEF in mouse models of AA amyloidosis (Westermark et al. 2009).

In addition to the amino acid sequence of SAA, in vivo factors, including the concentration of SAA and binding to ligands, may affect the morphology of SAA fibrils. Of the potential ligands, GAGs are of particular importance because of their role in the etiology of AA amyloidosis (Ancsin and Kisilevsky 1999; Elimova et al. 2004; Li et al. 2005; Snow et al. 1991; Wang et al. 2012). The importance of GAGs, in particular HS, in amyloid diseases is not limited to AA amyloidosis, as it is well known that HS are ubiquitously present in nearly all amyloid deposits (Zhang and Li 2010). In addition, in vitro studies with many proteins have shown that HS can accelerate amyloid formation (see Bourgault et al. 2011; Martin and Ramirez-Alvarado 2011; Solomon et al. 2011). Concerning SAA, it has

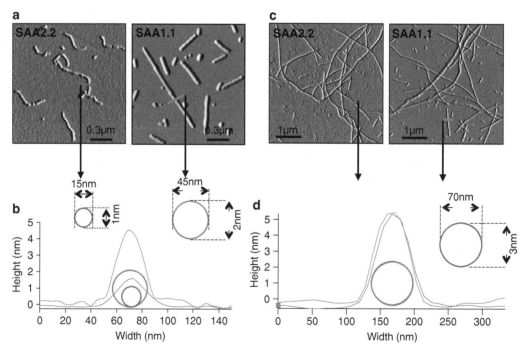

Fig. 5.5 AFM data showing that mSAA1.1 and mSAA2.2 form amyloid fibrils of different morphologies. (**a**) AFM images of mSAA1.1 and mSAA2.2 aggregates after incubation at 37 °C for 70 h and 10 h, respectively. (**b**) Plot of early fibril (i.e. protofibril) cross-section was obtained from the corresponding height traces in (**a**). (**c**) AFM images of mSAA1.1 and mSAA2.2 full-length fibrils formed after incubation at 37 °C for 150 h and 50 h, respectively. (**d**) Plot of the fibril cross-section was obtained from the corresponding height traces in (**c**) (This figure was originally published in *The Journal of Biological Chemistry* (Srinivasan et al. 2013) © the American Society for Biochemistry and Molecular Biology)

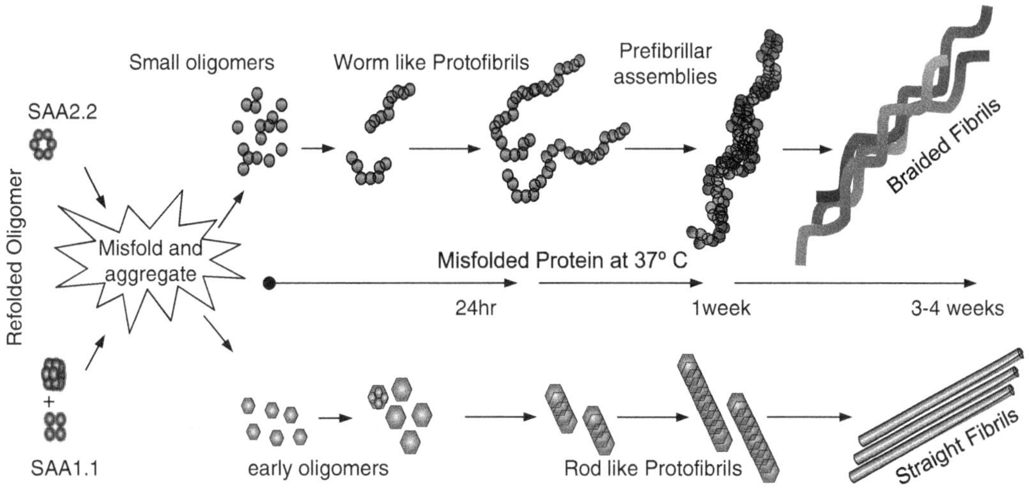

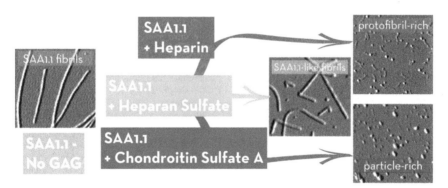

Fig. 5.6 Model of mSAA1.1 and mSAA2.2 fibril formation in vitro. At 37 °C mSAA2.2 misfolds and forms oligomers that self-assemble into curvilinear fibrils. They further grow and intertwine into mature fibrils that may self-assemble into braided bundles. mSAA1.1 quickly forms spherical oligomers larger than those formed by mSAA2.2, which slowly give rise to rod-like protofibrils or short fibrils that assemble into straight fibrils. Mature mSAA1.1 fibrils seem more rigid than those formed by mSAA2.2, and appear to interact laterally with each other rather than intertwine (This figure was originally published in *The Journal of Biological Chemistry* (Srinivasan et al. 2013) © the American Society for Biochemistry and Molecular Biology)

Fig. 5.7 Effect of different GAGs on the aggregation and fibrillation of mSAA1.1 at 37 °C. The central pathways (*green arrows*) show prototypical mSAA1.1-like fibril formation indicating minimal effect of HS on fibril morphology. Hep (*red curve*) caused formation of abundant mSAA1.1 protofibrils that appear to be capped and unable to grow into mature fibrils. mSAA1.1 fibril formation was completely inhibited by interactions with chondroitin sulfate (*blue curve*), which mostly produced spherical species of various size (Reprinted from Aguilera et al. 2014, Copyright (2014), with permission from Elsevier)

been shown that GAGs affect the biophysical properties of SAA, including its aggregation kinetics and aggregate morphology (Aguilera et al. 2014; Elimova et al. 2009). Interestingly, the aggregation of mSAA1.1 in vitro may be modulated by the presence of GAGs (Fig. 5.7) (Aguilera et al. 2014). When HS, hyaluronic acid, and heparosan (these sugars have similar backbone moeities) were individually incubated with refolded mSAA1.1 at 37 °C, SAA1.1 fibril morphology was largely unaffected. In contrast, mono-sulfated chondroitin sulfate A (CSA) blocked SAA fibril formation and enabled the formation of spherical aggregates of various sizes. Heparin – the most sulfated GAG – was not only the most effective GAG in accelerating the aggregation of mSAA1.1, but also resulted in the formation of vast amounts of thin protofibrils that were latent in converting to mature fibrils (Fig. 5.7). Thus, Hep catalyzes the in vitro aggregation of mSAA1.1 along a different pathway (Aguilera et al. 2014).

Numerous studies have probed the effect of GAGs on SAA fibrillation (Elimova et al. 2004, 2009; Kisilevsky and Fraser 1996; Li et al. 2005; Noborn et al. 2012). The results are not in consistent agreement with each other, which is likely a result of differences in experimental design, including in vitro versus in vivo experiments, various SAA isoforms or peptides, differences in the solvent pH, SAA concentrations, etc. Nevertheless, overall these studies suggest that HS and Hep exert a major influence on the aggregation and fibrillation pathway of SAA in vivo.

5.5   Conclusions

A better understanding of the biological functions and pathogenicity of SAA across different species will require a better understanding of the biochemical and biophysical properties of SAA isoforms. In particular, questions related to different pathogenicity of SAA isoforms, the low disease incidence in individuals at risk, the role of ligands (such as HDL and GAGs) in SAA function and in AA disease, and the prion-like transmission potential of SAA are directly related to the biophysical properties of SAA. The current picture emerging from biophysical studies is that, depending on the environmental conditions, SAA proteins have many structural options with distinct functional and pathological implications. At low temperature, SAAs are marginally stable proteins that can fold into various oligomeric structures, whereas at 37 °C SAA behaves like an IDP that is highly prone to aggregation into higher-order species, including fibrils of different morphologies. These morphological differences could be relevant to disease onset and transmissibility and deserve further investigation. In addition, due to the marginal stability

of SAA and its inherent oligomeric and aggregate plasticity, the fibrillation of SAA seems particularly susceptible to modulation by extrinsic factors, in particular HDL and GAGs. Further biophysical studies are needed to explore whether the development of AA amyloidosis is mostly determined by the competition between SAA fibril formation and its degradation. The recently determined atomic structures of SAA have provided much needed biophysical insight, and future structural studies will be essential to understand the biophysical properties of SAA that determine the diverse pathogenicity of SAA isoforms.

Acknowledgments Much of the work was supported by a National Institutes of Health Grant (R01 AG028158) to W.C. The authors are grateful to Ms. Jayeeta Sen for creating Fig. 5.2.

References

Aguilera JJ, Zhang F, Beaudet JM, Linhardt RJ, Colon W (2014) Divergent effect of glycosaminoglycans on the in vitro aggregation of serum amyloid A. Biochimie 104:70–80

Ancsin JB, Kisilevsky R (1999) The heparin/heparan sulfate-binding site on apo-serum amyloid A. Implications for the therapeutic intervention of amyloidosis. J Biol Chem 274:7172–7181

Axelrad MA, Kisilevsky R, Willmer J, Chen SJ, Skinner M (1982) Further characterization of amyloid-enhancing factor. Lab Invest 47:139–146

Benditt EP, Eriksen N (1977) Amyloid protein SAA is associated with high density lipoprotein from human serum. Proc Natl Acad Sci U S A 74:4025–4028

Bourgault S, Solomon JP, Reixach N, Kelly JW (2011) Sulfated glycosaminoglycans accelerate transthyretin amyloidogenesis by quaternary structural conversion. Biochemistry 50:1001–1015

Breinig S, Kervinen J, Stith L, Wasson AS, Fairman R, Wlodawer A, Zdanov A, Jaffe EK (2003) Control of tetrapyrrole biosynthesis by alternate quaternary forms of porphobilinogen synthase. Nat Struct Biol 10:757–763

Chung TF, Sipe JD, McKee A, Fine RE, Schreiber BM, Liang JS, Johnson RJ (2000) Serum amyloid A in Alzheimer's disease brain is predominantly localized to myelin sheaths and axonal membrane. Amyloid 7:105–110

Cross AJ, Jeffries CM, Trewhella J, Matthews JM (2010) LIM domain binding proteins 1 and 2 have different oligomeric states. J Mol Biol 399:133–144

Cuanalo-Contreras K, Mukherjee A, Soto C (2013) Role of protein misfolding and proteostasis deficiency in protein misfolding diseases and aging. Int J Cell Biol 2013:1–10

Cunnane G (2001) Amyloid precursors and amyloidosis in inflammatory arthritis. Curr Opin Rheumatol 13:67–73

de Beer MC, de Beer FC, McCubbin WD, Kay CM, Kindy MS (1993) Structural prerequisites for serum amyloid A fibril formation. J Biol Chem 268:20606–20612

Derebe MG, Zlatkov CM, Gattu S, Ruhn KA, Vaishnava S, Diehl GE, MacMillan JB, Williams NS, Hooper LV (2014) Serum amyloid A is a retinol binding protein that transports retinol during bacterial infection. Elife 3:e03206

Egashira M, Takase H, Yamamoto I, Tanaka M, Saito H (2011) Identification of regions responsible for heparin-induced amyloidogenesis of human serum amyloid A using its fragment peptides. Arch Biochem Biophys 511:101–106

Eklund KK, Niemi K, Kovanen PT (2012) Immune functions of serum amyloid A. Crit Rev Immunol 32:335–348

Elimova E, Kisilevsky R, Szarek WA, Ancsin JB (2004) Amyloidogenesis recapitulated in cell culture: a peptide inhibitor provides direct evidence for the role of heparan sulfate and suggests a new treatment strategy. FASEB J 18:1749–1751

Elimova E, Kisilevsky R, Ancsin JB (2009) Heparan sulfate promotes the aggregation of HDL-associated serum amyloid A: evidence for a proamyloidogenic histidine molecular switch. FASEB J 23:3436–3448

Ericsson LH, Eriksen N, Walsh KA, Benditt EP (1987) Primary structure of duck amyloid protein A. The form deposited in tissues may be identical to its serum precursor. FEBS Lett 218:11–16

Gabay C, Kushner I (1999) Acute-phase proteins and other systemic responses to inflammation. N Engl J Med 340:448–454

Gillmore JD, Lovat LB, Persey MR, Pepys MB, Hawkins PN (2001) Amyloid load and clinical outcome in AA amyloidosis in relation to circulating concentration of serum amyloid A protein. Lancet 358:24–29

Guerrero-Munoz MJ, Castillo-Carranza DL, Kayed R (2014) Therapeutic approaches against common structural features of toxic oligomers shared by multiple amyloidogenic proteins. Biochem Pharmacol 88:468–478

Guo JT, Yu J, Grass D, de Beer FC, Kindy MS (2002) Inflammation-dependent cerebral deposition of serum amyloid a protein in a mouse model of amyloidosis. J Neurosci 22:5900–5909

Hazenberg BP, van Rijswijk MH (2000) Where has secondary amyloid gone? Ann Rheum Dis 59:577–579

Hoffman JS, Ericsson LH, Eriksen N, Walsh KA, Benditt EP (1984) Murine tissue amyloid protein AA. NH2-terminal sequence identity with only one of two serum amyloid protein (ApoSAA) gene products. J Exp Med 159:641–646

Husby G (1998) Treatment of amyloidosis and the rheumatologist. State of the art and perspectives for the future. Scand J Rheumatol 27:161–165

Husebekk A, Skogen B, Husby G, Marhaug G (1985) Transformation of amyloid precursor SAA to protein AA and incorporation in amyloid fibrils in vivo. Scand J Immunol 21:283–287

Immonen K, Finne P, Gronhagen-Riska C, Pettersson T, Klaukka T, Kautiainen H, Hakala M (2011) A marked decline in the incidence of renal replacement therapy for amyloidosis associated with inflammatory rheumatic diseases – data from nationwide registries in Finland. Amyloid 18:25–28

Inoue S, Kawano H, Ishihara T, Maeda S, Ohno S (2005) Formation of experimental murine AA amyloid fibrils in SAP-deficient mice: high resolution ultrastructural study. Amyloid 12:157–163

Ishihara T (1973) Experimental amyloidosis using silver nitrate – electron microscopic study on the relationship between silver granules, amyloid fibrils and reticuloendothelial system. Acta Pathol Jpn 23:439–464

Jahn TR, Radford SE (2008) Folding versus aggregation: polypeptide conformations on competing pathways. Arch Biochem Biophys 469:100–117

Jimenez JL, Tennent G, Pepys M, Saibil HR (2001) Structural diversity of ex vivo amyloid fibrils studied by cryo-electron microscopy. J Mol Biol 311:241–247

Kim YE, Hipp MS, Bracher A, Hayer-Hartl M, Hartl FU (2013) Molecular chaperone functions in protein folding and proteostasis. Annu Rev Biochem 82:323–355

Kintzer AF, Thoren KL, Sterling HJ, Dong KC, Feld GK, Tang II, Zhang TT, Williams ER, Berger JM, Krantz BA (2009) The protective antigen component of anthrax toxin forms functional octameric complexes. J Mol Biol 392:614–629

Kisilevsky R, Boudreau L (1983) Kinetics of amyloid deposition. I. The effects of amyloid-enhancing factor and splenectomy. Lab Invest 48:53–59

Kisilevsky R, Fraser P (1996) Proteoglycans and amyloid fibrillogenesis. Ciba Found Symp 199:58–67 (discussion 68-72, 90-103)

Kisilevsky R, Manley PN (2012) Acute-phase serum amyloid A: perspectives on its physiological and pathological roles. Amyloid 19:5–14

Kisilevsky R, Subrahmanyan L (1992) Serum amyloid A changes high density lipoprotein's cellular affinity. A clue to serum amyloid A's principal function. Lab Invest 66:778–785

Kodali R, Wetzel R (2007) Polymorphism in the intermediates and products of amyloid assembly. Curr Opin Struct Biol 17:48–57

Lachmann HJ, Goodman HJ, Gilbertson JA, Gallimore JR, Sabin CA, Gillmore JD, Hawkins PN (2007) Natural history and outcome in systemic AA amyloidosis. N Engl J Med 356:2361–2371

Landman WJ (1999) Amyloid arthropathy in chickens. Vet Q 21:78–82

Levin M, Franklin EC, Frangione B, Pras M (1972) The amino acid sequence of a major nonimmunoglobulin component of some amyloid fibrils. J Clin Invest 51:2773–2776

Li JP, Galvis ML, Gong F, Zhang X, Zcharia E, Metzger S, Vlodavsky I, Kisilevsky R, Lindahl U (2005) In vivo fragmentation of heparan sulfate by heparanase overexpression renders mice resistant to amyloid protein A amyloidosis. Proc Natl Acad Sci U S A 102:6473–6477

Liang JS, Schreiber BM, Salmona M, Phillip G, Gonnerman WA, de Beer FC, Sipe JD (1996) Amino terminal region of acute phase, but not constitutive, serum amyloid A (apoSAA) specifically binds and transports cholesterol into aortic smooth muscle and HepG2 cells. J Lipid Res 37:2109–2116

Liang JS, Sloane JA, Wells JM, Abraham CR, Fine RE, Sipe JD (1997) Evidence for local production of acute phase response apolipoprotein serum amyloid A in Alzheimer's disease brain. Neurosci Lett 225:73–76

Liepnieks JJ, Kluve-Beckerman B, Benson MD (1995) Characterization of amyloid A protein in human secondary amyloidosis: the predominant deposition of serum amyloid A1. Biochim Biophys Acta 1270:81–86

Lu J, Yu Y, Zhu I, Cheng Y, Sun PD (2014) Structural mechanism of serum amyloid A-mediated inflammatory amyloidosis. Proc Natl Acad Sci U S A 111:5189–5194

Lundmark K, Westermark GT, Nystrom S, Murphy CL, Solomon A, Westermark P (2002) Transmissibility of systemic amyloidosis by a prion-like mechanism. Proc Natl Acad Sci U S A 99:6979–6984

Malle E, Sodin-Semrl S, Kovacevic A (2009) Serum amyloid A: an acute-phase protein involved in tumour pathogenesis. Cell Mol Life Sci 66:9–26

Manning M, Colon W (2004) Structural basis of protein kinetic stability: resistance to sodium dodecyl sulfate suggests a central role for rigidity and a bias toward beta-sheet structure. Biochemistry 43:11248–11254

Martin DJ, Ramirez-Alvarado M (2011) Glycosaminoglycans promote fibril formation by amyloidogenic immunoglobulin light chains through a transient interaction. Biophys Chem 158:81–89

McAdam KP, Sipe JD (1976) Murine model for human secondary amyloidosis: genetic variability of the acute-phase serum protein SAA response to endotoxins and casein. J Exp Med 144:1121–1127

Meek RL, Hoffman JS, Benditt EP (1986) Amyloidogenesis. One serum amyloid A isotype is selectively removed from the circulation. J Exp Med 163:499–510

Meek RL, Urieli-Shoval S, Benditt EP (1994) Expression of apolipoprotein serum amyloid A mRNA in human atherosclerotic lesions and cultured vascular cells: implications for serum amyloid A function. Proc Natl Acad Sci U S A 91:3186–3190

Meyer-Luehmann M, Coomaraswamy J, Bolmont T, Kaeser S, Schaefer C, Kilger E, Neuenschwander A, Abramowski D, Frey P, Jaton AL, Vigouret J-M, Paganetti P, Walsh DM, Mathews PM, Ghiso J, Staufenbiel M, Walker LC, Jucker M (2006) Exogenous induction of cerebral beta-amyloidogenesis is governed by agent and host. Science 313:1781–1784

Miyagi Y, Matsumura Y, Sagami H (2007) Human geranylgeranyl diphosphate synthase is an octamer in solution. J Biochem 142:377–381

Mutru O, Laakso M, Isomaki H, Koota K (1985) Ten year mortality and causes of death in patients with rheumatoid arthritis. Br Med J (Clin Res Ed) 290:1797–1799

Nakamura T, Higashi S, Tomoda K, Tsukano M, Baba S, Shono M (2006) Significance of SAA1.3 allele genotype in Japanese patients with amyloidosis secondary to rheumatoid arthritis. Rheumatology (Oxford) 45:43–49

Noborn F, Ancsin JB, Ubhayasekera W, Kisilevsky R, Li JP (2012) Heparan sulfate dissociates serum amyloid A (SAA) from acute-phase high-density lipoprotein, promoting SAA aggregation. J Biol Chem 287:25669–25677

Obici L, Merlini G (2012) AA amyloidosis: basic knowledge, unmet needs and future treatments. Swiss Med Wkly 142:w13580

Patel H, Bramall J, Waters H, De Beer MC, Woo P (1996) Expression of recombinant human serum amyloid A in mammalian cells and demonstration of the region necessary for high-density lipoprotein binding and amyloid fibril formation by site-directed mutagenesis. Biochem J 318(Pt 3):1041–1049

Patke S, Maheshwari R, Litt J, Srinivasan S, Aguilera JJ, Colon W, Kane RS (2012) Influence of the carboxy terminus of serum amyloid A on protein oligomerization, misfolding, and fibril formation. Biochemistry 51:3092–3099

Patke S, Srinivasan S, Maheshwari R, Srivastava SK, Aguilera JJ, Colon W, Kane RS (2013) Characterization of the oligomerization and aggregation of human serum amyloid a. PLoS One 8:e64974

Pellarin R, Schuetz P, Guarnera E, Caflisch A (2010) Amyloid fibril polymorphism is under kinetic control. J Am Chem Soc 132:14960–14970

Pepys MB (2006) Amyloidosis. Annu Rev Med 57:223–241

Rocken C, Shakespeare A (2002) Pathology, diagnosis and pathogenesis of AA amyloidosis. Virchows Arch 440:111–122

Seilheimer B, Bohrmann B, Bondolfi L, Muller F, Stuber D, Dobeli H (1997) The toxicity of the Alzheimer's beta-amyloid peptide correlates with a distinct fiber morphology. J Struct Biol 119:59–71

Shiroo M, Kawahara E, Nakanishi I, Migita S (1987) Specific deposition of serum amyloid A protein 2 in the mouse. Scand J Immunol 26:709–716

Simms RW, Prout MN, Cohen AS (1994) The epidemiology of AL and AA amyloidosis. Baillieres Clin Rheumatol 8:627–634

Simons JP, Al-Shawi R, Ellmerich S, Speck I, Aslam S, Hutchinson WL, Mangione PP, Disterer P, Gilbertson JA, Hunt T, Millar DJ, Minogue S, Bodin K, Pepys MB, Hawkins PN (2013) Pathogenetic mechanisms of amyloid A amyloidosis. Proc Natl Acad Sci U S A 110:16115–16120

Sipe JD (1992) Amyloidosis. Annu Rev Biochem 61:947–975

Sipe J (1999) Revised nomenclature for serum amyloid A (SAA). Nomenclature committee of the international society of amyloidosis. Part 2. Amyloid 6:67–70

Sipe JD, Carreras I, Gonnerman WA, Cathcart ES, de Beer MC, de Beer FC (1993) Characterization of the inbred CE/J mouse strain as amyloid resistant. Am J Pathol 143:1480–1485

Skinner M, Shirahama T, Benson MD, Cohen AS (1977) Murine amyloid protein AA in casein-induced experimental amyloidosis. Lab Invest 36:420–427

Sletten K, Husby G (1974) The complete amino-acid sequence of non-immunoglobulin amyloid fibril protein AS in rheumatoid arthritis. Eur J Biochem 41:117–125

Snow AD, Bramson R, Mar H, Wight TN, Kisilevsky R (1991) A temporal and ultrastructural relationship between heparan sulfate proteoglycans and AA amyloid in experimental amyloidosis. J Histochem Cytochem 39:1321–1330

Solomon JP, Bourgault S, Powers ET, Kelly JW (2011) Heparin binds 8 kDa gelsolin cross-beta-sheet oligomers and accelerates amyloidogenesis by hastening fibril extension. Biochemistry 50:2486–2498

Srinivasan S, Patke S, Wang Y, Ye Z, Litt J, Srivastava SK, Lopez MM, Kurouski D, Lednev IK, Kane RS, Colon W (2013) Pathogenic serum amyloid A 1.1 shows a long oligomer-rich fibrillation lag phase contrary to the highly amyloidogenic non-pathogenic SAA2.2. J Biol Chem 288:2744–2755

Stefani M (2012) Structural features and cytotoxicity of amyloid oligomers: implications in Alzheimer's disease and other diseases with amyloid deposits. Prog Neurobiol 99:226–245

Steinmetz A, Hocke G, Saile R, Puchois P, Fruchart JC (1989) Influence of serum amyloid A on cholesterol esterification in human plasma. Biochim Biophys Acta 1006:173–178

Tam SP, Flexman A, Hulme J, Kisilevsky R (2002) Promoting export of macrophage cholesterol: the physiological role of a major acute-phase protein, serum amyloid A 2.1. J Lipid Res 43:1410–1420

Tanaka M, Collins SR, Toyama BH, Weissman JS (2006) The physical basis of how prion conformations determine strain phenotypes. Nature 442:585–589

Urieli-Shoval S, Cohen P, Eisenberg S, Matzner Y (1998) Widespread expression of serum amyloid A in histologically normal human tissues. Predominant localization to the epithelium. J Histochem Cytochem 46:1377–1384

Urieli-Shoval S, Linke RP, Matzner Y (2000) Expression and function of serum amyloid A, a major acute-phase protein, in normal and disease states. Curr Opin Hematol 7:64–69

Uversky VN (2009) Intrinsic disorder in proteins associated with neurodegenerative diseases. Front Biosci 14:5188–5238

Uversky VN (2013) Unusual biophysics of intrinsically disordered proteins. Biochim Biophys Acta 1834:932–951

van der Hilst JC (2011) Recent insights into the pathogenesis of type AA amyloidosis. Sci World J 11:641–650

van der Westhuyzen DR, de Beer FC, Webb NR (2007) HDL cholesterol transport during inflammation. Curr Opin Lipidol 18:147–151

van Horssen J, Wesseling P, van den Heuvel LP, de Waal RM, Verbeek MM (2003) Heparan sulphate proteoglycans in Alzheimer's disease and amyloid-related disorders. Lancet Neurol 2:482–492

Wang L, Colon W (2005) Urea-induced denaturation of apolipoprotein serum amyloid A reveals marginal stability of hexamer. Protein Sci 14:1811–1817

Wang L, Lashuel HA, Walz T, Colon W (2002) Murine apolipoprotein serum amyloid A in solution forms a hexamer containing a central channel. Proc Natl Acad Sci U S A 99:15947–15952

Wang L, Lashuel HA, Colon W (2005) From hexamer to amyloid: marginal stability of apolipoprotein SAA2.2 leads to in vitro fibril formation at physiological temperature. Amyloid 12:139–148

Wang Y, Srinivasan S, Ye Z, Javier Aguilera J, Lopez MM, Colon W (2011) Serum amyloid A 2.2 refolds into a octameric oligomer that slowly converts to a more stable hexamer. Biochem Biophys Res Commun 407:725–729

Wang B, Tan YX, Jia J, Digre A, Zhang X, Vlodavsky I, Li JP (2012) Accelerated resolution of AA amyloid in heparanase knockout mice is associated with matrix metalloproteases. PLoS One 7:e39899

Webb NR, de Beer MC, van der Westhuyzen DR, Kindy MS, Banka CL, Tsukamoto K, Rader DL, de Beer FC (1997) Adenoviral vector-mediated overexpression of serum amyloid A in apoA-I-deficient mice. J Lipid Res 38:1583–1590

Westermark GT, Sletten K, Westermark P (1989) Massive vascular AA-amyloidosis: a histologically and biochemically distinctive subtype of reactive systemic amyloidosis. Scand J Immunol 30:605–613

Westermark GT, Engstrom U, Westermark P (1992) The N-terminal segment of protein AA determines its fibrillogenic property. Biochem Biophys Res Commun 182:27–33

Westermark P, Lundmark K, Westermark GT (2009) Fibrils from designed non-amyloid-related synthetic peptides induce AA-amyloidosis during inflammation in an animal model. PLoS One 4:e6041

Westermark GT, Fandrich M, Westermark P (2015) AA amyloidosis: pathogenesis and targeted therapy. Annu Rev Pathol 10:321–344

Woldemeskel M (2012) A concise review of amyloidosis in animals. Vet Med Int 2012:427296

Yamada T, Kluve-Beckerman B, Liepnieks JJ, Benson MD (1995) In vitro degradation of serum amyloid A by cathepsin D and other acid proteases: possible protection against amyloid fibril formation. Scand J Immunol 41:570–574

Ye Z, Bayron Poueymiroy D, Aguilera JJ, Srinivasan S, Wang Y, Serpell LC, Colon W (2011) Inflammation protein SAA2.2 spontaneously forms marginally stable amyloid fibrils at physiological temperature. Biochemistry 50:9184–9191

Zhang X, Li JP (2010) Heparan sulfate proteoglycans in amyloidosis. Prog Mol Biol Transl Sci 93:309–334

Zhang B, Une Y, Fu X, Yan J, Ge F, Yao J, Sawashita J, Mori M, Tomozawa H, Kametani F, Higuchi K (2008) Fecal transmission of AA amyloidosis in the cheetah contributes to high incidence of disease. Proc Natl Acad Sci U S A 105:7263–7268

Interactions of Lipid Membranes with Fibrillar Protein Aggregates

6

Galyna Gorbenko, Valeriya Trusova,
Mykhailo Girych, Emi Adachi, Chiharu Mizuguchi,
and Hiroyuki Saito

Abstract

Amyloid fibrils are an intriguing class of protein aggregates with distinct physicochemical, structural and morphological properties. They display peculiar membrane-binding behavior, thus adding complexity to the problem of protein-lipid interactions. The consensus that emerged during the past decade is that amyloid cytotoxicity arises from a continuum of cross-β-sheet assemblies including mature fibrils. Based on literature survey and our own data, in this chapter we address several aspects of fibril-lipid interactions, including (i) the effects of amyloid assemblies on molecular organization of lipid bilayer; (ii) competition between fibrillar and monomeric membrane-associating proteins for binding to the lipid surface; and (iii) the effects of lipids on the structural morphology of fibrillar aggregates. To illustrate some of the processes occurring in fibril-lipid systems, we present and analyze fluorescence data reporting on lipid bilayer interactions with fibrillar lysozyme and with the N-terminal 83-residue fragment of amyloidogenic mutant apolipoprotein A-I, 1-83/G26R/W@8. The results help understand possible mechanisms of interaction and mutual remodeling of amyloid fibers and lipid membranes, which may contribute to amyloid cytotoxicity.

Keywords

Protein-lipid interactions • Amyloid fibrils • Lysozyme • N-terminal fragment of apolipoprotein A-I • Fluorescence spectroscopy

G. Gorbenko (✉) • V. Trusova • M. Girych
Department of Nuclear and Medical Physics,
V.N. Karazin Kharkiv National University,
4 Svobody Sq., Kharkov 61077, Ukraine
e-mail: galyagor@yahoo.com;
valtrusova@yahoo.com; girichms@gmail.com

E. Adachi • C. Mizuguchi • H. Saito
Institute of Health Biosciences, Graduate School
of Pharmaceutical Sciences, The University
of Tokushima, 1-78-1 Shomachi,
Tokushima 770-8505, Japan
e-mail: c401231014@tokushima-u.ac.jp;
c401331027@tokushima-u.ac.jp; hsaito@
tokushima-u.ac.jp

© Springer International Publishing Switzerland 2015
O. Gursky (ed.), *Lipids in Protein Misfolding*, Advances in Experimental
Medicine and Biology 855, DOI 10.1007/978-3-319-17344-3_6

Abbreviations

1-83/G26R/W@8	N-terminal 1-83 fragment of apoA-I with G26R mutation
AFM	Atomic force microscopy
apoA-I	Apolipoprotein A-I
AV-PC	Anthrylvinyl-labeled PC
Aβ	Amyloid-β peptide
Chol	Cholesterol
CL	Cardiolipin
cyt c	Cytochrome c
FRET	Förster resonance energy transfer
GP	Generalized fluorescence polarization of Laurdan
HR	Helical ribbon
PC	Phosphatidylcholine
PG	Phosphatidylglycerol
PS	Phosphatidylserine
ThT	Thioflavin T
TR	Twisted ribbon

6.1   Introduction

Protein-lipid interactions have long been recognized as an effective modulator of a wide range of membrane processes, including intracellular transport, enzyme function, respiration, antimicrobial defense, signal transduction, motility, etc. (Lee 2003, 2004; Palsdottir and Hunte 2004). Along with continuous expansion, this research area currently undergoes a substantial shift in focus towards ascertaining the role of lipids in modulating the polypeptide self-association and elucidating membrane responses to aggregated proteins. Of paramount interest are amyloid fibrils, a special type of protein aggregates involved in the pathogenesis of numerous conformational disorders, such as Parkinson's, Alzheimer's and Huntington's diseases, type II diabetes, systemic amyloidosis, etc. (Stefani 2004). These aggregates are assembled from proteins or peptides adopting a non-native β-structure-enriched conformation, and have distinct highly ordered structural organization with β-sheets propagating along the fibril axis (Serpell 2000; Kelly 2002). Due to fundamental similarity between intra- and interchain interactions, protein

fibrillization represents an alternative folding pathway of partially unfolded or misfolded proteins that acquire a stable structure by reaching the minimum on the energy landscape (Zbilut et al. 2003). Fibril formation is nucleation-dependent and proceeds through thermodynamically unfavorable monomer association into a critical oligomeric nucleus followed by its subsequent energetically favorable elongation and the exponential fibril growth (Dima and Thirumalai 2002).

Emerging evidence indicates that lipid bilayer can substantially accelerate amyloid nucleation through accumulation of the protein species with specific aggregation-prone conformation, orientation and location at the lipid-water interface (Stefani and Dobson 2003; Gorbenko and Kinnunen 2006; Stefani 2008; Aisenbrey et al. 2008). At the same time, cell membranes are thought to be a primary target for toxic amyloid assemblies (Bucciantini et al. 2014). Membrane damage produced by the early protein oligomers is regarded as the main cause of cytotoxicity (Kinnunen 2009; Relini et al. 2009; Stefani 2010; Butterfield and Lashuel 2010). Cytotoxic action of protein oligomers can be attributed to compromised membrane integrity (Meratan et al. 2011; Huang et al. 2009), formation of non-specific ion channels (Caughey and Lansbury 2003), uptake of lipids into the fibers growing on a membrane template (Sparr et al. 2004; Engel et al. 2008), alterations in the intracellular redox status and free calcium level (Arispe et al. 1993; Squier 2001; Tabner et al. 2002), and impaired functions of membrane proteins (Stefani 2007). Several lines of evidence suggest that oligomeric species display higher membrane-binding affinity compared to mature fibrils. This was demonstrated for several proteins including the N-terminal domain of the hydrogenase maturation factor HypF-N (Relini et al. 2004; Canale et al. 2006, AC), stefin B (Anderluh et al. 2005), α-synuclein (Giannakis et al. 2008; Smith et al. 2008), lysozyme (Meratan et al. 2011) and Abeta (Williams and Serpell 2011). This increased affinity was explained in terms of hydrophobicity-based toxicity mechanism, highlighting the importance of factors such as extensive hydrophobic surfaces and high flexibility of the protein oligomers (Meratan et al. 2011). Nevertheless, it is becoming increasingly clear that, although mature amyloid fibrils bind membrane relatively weakly, they are far from being chemically inert nontoxic species. Numerous studies indicate substantial cytotoxic potential of fibrillar aggregates (Weldon et al. 1998; Forloni 1996; Gharibyan et al. 2007; Matsuzaki 2011; Bucciantini et al. 2012). In particular, Novitskaya et al. showed that mature amyloid fibrils of mammalian prion protein are as toxic as the soluble oligomers to cultured cells and primary neurons (Novitskaya et al. 2006). Different toxicity mechanisms operating through apoptotic and necrotic pathways were revealed for oligomers and fibrils of hen egg white lysozyme in a process leading to neuroblastoma cell death. Mature lysozyme fibrils can induce mitochondrial failure and increase plasma membrane permeability. Hence, cytotoxicity is inherent to a continuum of cross-β-sheet-rich structures rather than to a single uniform species (Gharibyan et al. 2007).

The relationship between the morphology of fibrillar aggregates and their toxicity was reported by Petkova and colleagues who examined the action of mature amyloid fibrils of Alzheimer's amyloid-beta peptide, $A\beta_{1-40}$, on neuronal cell cultures (Petkova et al. 2005). The absence of direct correlation between disease symptoms and total amyloid deposition was considered as the main evidence of the nontoxic nature of mature fibrils. High surface hydrophobicity and solvent accessibility of disulfides in the structure of fibrillar lysozyme were proposed to be responsible for the membrane-disruptive effect of lysozyme fibrils that brought about hemolysis of erythrocytes and their aggregation coupled with intermolecular disulfide cross-linking (Huang et al. 2009). Notably, amyloid fibrils can manifest their toxicity not only nonspecifically, but also via specific pathways involving the activation of cellular receptors, as was demonstrated for transthyretin and Aβ peptide (Bamberger et al. 2003; Sousa et al. 2001). Finally, fragmentation of mature fibrils upon mechanical stress, thermal motion or chaperone activity was reported to enhance cytotoxic potential of amyloid (Carulla

et al. 2005; Smith et al. 2006; Xue et al. 2009). Together, these studies highlight the importance of further in-depth analyses of biological activities of amyloid fibrils in relation to their physicochemical, structural and morphological properties. Evidently, gaining better insights into the membrane-associating and bilayer-modifying behavior of amyloid fibrils, which represent a specific aggregation state of polypeptide chain, adds a new exciting dimension to the problem of protein-lipid interactions.

The present chapter provides a concise overview of the available information concerning structural basis for membrane damage by fibrillar aggregates and presents some of our own pertinent data. Specifically, we scrutinize such processes as (i) changes in membrane structure in response to fibril binding; (ii) competition between fibrillar and monomeric proteins for association with a lipid bilayer; and (iii) the ability of lipid membrane to remodel amyloid fibrils.

6.2 Effects of Amyloid Fibrils on the Membrane

6.2.1 Effects of Protein Fibrils and Oligomers on the Membrane Structure and Dynamics

One of the mechanisms by which amyloid fibril can exert their cytotoxicity involves changes in the structure and dynamics of the cell membrane. It is well established that protein-lipid interactions commonly involve interrelated processes of conformational changes in the protein molecule as well as structural rearrangement of the membrane lipids. However, the conceptual framework developed in this field cannot be directly extrapolated to fibril-lipid interactions. In fact, fibrillar protein aggregates of amyloid type possess unique physical properties, including exceptional rigidity largely arising from highly ordered cross-β array of hydrogen bonds, with Young's modulus on the order of several GPa and persistence length of several tens of micrometers (Adamcik and Mezzenga 2012a). Compared to other biological filaments, amyloid fibrils are extremely stiff, approaching most rigid proteinaceous materials, such as silk, collagen and keratin (Knowles and Buehler 2011). Moreover, regular linear arrangement of the charged, hydrophobic or aromatic amino-acid side chains in fibrillar structure creates reactive surfaces uncharacteristic of natively folded proteins (Adamcik and Mezzenga 2012b). Accordingly, there may be substantial differences in membrane responses to monomeric, oligomeric and fibrillar polypeptides.

This issue has been addressed in a number of recent studies by using a variety of experimental approaches such as atomic force microscopy (AFM) (Giannakis et al. 2008), infrared reflection absorption spectroscopy (Lopes et al. 2007), surface plasmon resonance (Smith et al. 2008), small-angle neutron scattering (Dante et al. 2008), circular dichroism (Quist et al. 2005), fluorescence spectroscopy (Kremer et al. 2001), micropipette manipulation (Kim and Frangos 2008), conductivity measurements (Valincius et al. 2008), and dye release assay (van Rooijen et al. 2009).

To address the problem of fibril-lipid interactions, we used two proteins, hen egg white lysozyme and 1-83/G26R/W@8, which is the N-terminal 1-83 fragment of human apolipoprotein A-I containing a common amyloidogenic mutation G26R and an engineered Trp8 as a reporter group (Adachi et al. 2013). We explored the influence of these two fibrillar proteins on the structural state of model membranes composed of phosphatidylcholine and its mixtures with cholesterol (30 mol%) or cardiolipin (10 mol%). Figure 6.1 shows transmission electron microscopy images of the examined lysozyme (A) and 1-83/G26R/W@8 (B) fibrils.

Lysozyme is a ubiquitous multifunctional protein displaying bactericidal, antitumor and immunomodulatory activities. The mutants of human lysozyme (I56T, F57I, W64R, D67H) undergo pathological fibrillization associated with familial non-neuropathic systemic amyloidosis, a disease affecting kidney, liver and spleen (Pepys et al. 1993). Amyloidogenicity of this protein is thought to arise from the enhanced propensity of its mutants to adopt a partially unfolded aggregation-prone conformation (Frare et al. 2004). Specific segment in hen egg white lysozyme encompassing residues 54–62 was recently proposed to serve as the amyloid core that triggers fibril formation (Tokunaga et al. 2013).

ApoA-I is the major protein component of high-density lipoproteins promoting efflux of phospholipid and cholesterol from plasma membrane (Phillips 2013). Specific naturally occurring variants of human apoA-I, the most common of which is a single substitution mutant G26R, can form amyloid fibrils associated with renal or liver failure in hereditary systemic amyloidosis (Joy et al. 2003). The N-terminal fragments 1-83–1-93 have been identified as the predomi-

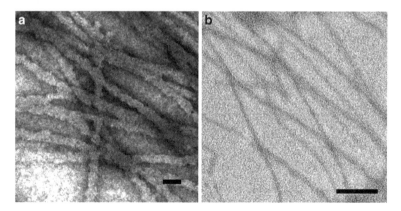

Fig. 6.1 Transmission electron micrographs of negatively stained fibrils of lysozyme (**a**) and 1-83/G26R/W@8 (**b**). Scale bar is 100 nm

nant form of apoA-I in amyloid fibril deposits (Nichols et al. 1990; Obici et al. 2006). One of such fragments is used in our research.

To monitor the changes in the molecular organization of the lipid bilayer induced by fibrillar lysozyme and 1-83/G26R/W@8 peptide, two fluorescent probes differing in their bilayer location have been employed: Laurdan that resides at the lipid-water interface and pyrene that is partitioned in the apolar core of lipid bilayer.

Fluorescent membrane probe Laurdan is characterized by a high sensitivity to variations in membrane hydration and lipid packing density (Parasassi and Gratton 1995; Parasassi et al. 1998). In a lipid bilayer, this amphiphilic fluorophore is localized at the level of the glycerol backbone, with lauric acid tail anchored in the acyl chain region. Laurdan's emission spectrum has two clearly distinct components that are attributed to solvent-unrelaxed (shorter-wavelength band centered *circa* 440 nm) and solvent-relaxed states (longer-wavelength band centered *circa* 490 nm) (Lúcio et al. 2010; Sanchez et al. 2012). This photophysical property is thought to originate from the reorientation of water dipoles around the excited-state dipole of the probe molecule. The environment-dependent spectral changes of Laurdan are generally quantitatively described by the steady-state fluorescence parameter known as the generalized polarization (GP) (Parasassi et al. 1991).

As illustrated in Fig. 6.2, fibrillar lysozyme produced GP increase in the model membranes comprised of phosphatidylcholine (PC) and its mixture with cardiolipin (CL) or cholesterol (Chol). In contrast, amyloid fibrils of the apoA-I fragment 1-83/G26R/W@8 brought about decrease in GP value of PC and PC/Chol bilayers. These findings imply that association of fibrillar apoA-I mutant with the membrane leads to the increase in lipid bilayer hydration and decrease of lipid packing density at the level of glycerol backbone, whereas lysozyme fibrils produce opposite effects. The latter is consistent with the observation that the binding of fibrillar lysozyme to liposomes is followed by the short-wavelength shift of Trp emission maximum, from ~352 to ~343 nm, indicating the transfer Trp62 and Trp108 (which dominate the emission) to interfacial bilayer region containing bound water with restricted mobility (Gorbenko et al. 2012). Therefore, different protein fibrils can produce distinctly different effects on the properties of model liposomes.

Notably, compared to PC vesicles, in similar vesicles of PC:Chol (7:3, mol:mol), both these fibrillar proteins produced much less pronounced changes in GP. Hence, the ability of fibrils to modify physical properties of the interfacial membrane region can be hampered by cholesterol. This observation is in good agreement with numerous studies suggesting that cholesterol can prevent membrane disruption by the aggregated proteins (Sponne et al. 2004; Cecchi et al. 2005; Qiu et al. 2011). In particular, cholesterol was demonstrated to protect primary cortical neurons from neurotoxic effects of soluble oligomeric Aβ

Fig. 6.2 Relative changes in the generalized polarization of Laurdan induced by fibrillar lysozyme and 1-83/G26R/W@8 in the model membranes comprised of PC, PC/CL (9:1, mol:mol) and PC/Chol (7:3, mol:mol). Unilamellar liposomes 100 nm in diameter were used in the lysozyme studies, and 50 nm in diameter in 1-83/G26R/W@8 studies

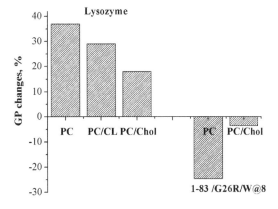

by modulating the physical properties of lipid bilayer. Model membrane studies showed that increase in cholesterol content inhibited aggregation and fusion of liposomes induced by Aβ (1-40) peptide (Sponne et al. 2004). The cell membranes rich in cholesterol were found to strongly resist to the remodelling by prefibrillar aggregates of the N-terminal domain of the prokaryotic hydrogenase maturation factor HypF (Cecchi et al. 2005). The perturbations in the polar part of the model PC membrane by lysozyme oligomers were also suppressed upon cholesterol inclusion in this membrane (Gorbenko and Trusova 2011). Taken together, these studies suggest that cholesterol in lipid membranes prevents their remodeling by pre-fibrillar and fibrillar protein aggregates, perhaps because increased phospholipid headgroup packing in the presence of cholesterol precludes fibril binding to the bilayer.

Next, to monitor lipid bilayer modifications occurring at the level of acyl chains, we employed a classical fluorescent probe pyrene, a polycyclic aromatic compound that is primarily distributed at the level of carbons 4–13 in the hydrocarbon region of the bilayer (Loura et al. 2013). Emission spectrum of this probe has characteristic vibronic structure in the wavelength range 370–400 nm, with relative intensities of vibronic transitions depending on the polarity of the fluorophore microenvironment via the so-called "Ham effect" (Nakajima 1971). Specifically, the intensity of the third vibronic peak (0–2 transition) is significantly enhanced in the hydrophobic environment, while the intensity of the first vibronic peak (0-0 transition) is increased in polar media. For this reason, the intensity ratio of the first-to-third vibronic band, $I1/I3$, has long been employed as an indicator of polarity in the vicinity of pyrene monomers. Accordingly, in water $I1/I3$ was reported to be 1.96, while in the solvents of lower polarity this ratio decreases, reaching the value 0.6 in n-hexane (Karpovich and Blanchard 1995).

Figure 6.3 shows the emission spectra of pyrene in lipid vesicles in the absence and in the presence of fibrillar 1-83/G26R/W@8. Together with similar studies of lysozyme fibers, these results suggest that neither 1-83/G26R/W@8 nor lysozyme fibrils could induce any significant changes in relative intensity of the first-to-third vibronic bands. This suggests that the fibril contacts with the lipid surface do not affect membrane polarity at the level of the initial acyl chain carbons where pyrene monomers are thought to reside.

Moreover, these fibrillar proteins did not markedly influence another pyrene spectral parameter, excimer-to-monomer intensity ratio (E/M). This parameter depends upon the rate of pyrene lateral diffusion and reflects dimerization of the excited- and ground-state probe molecules; such excimer formation manifests itself as a new fluorescent band *circa* 460–470 nm. Since pyrene excimerization is controlled by the frequency of collisions between the probe monomers in a lipid bilayer which, in turn, is a function of molecular packing density, this process is commonly analyzed in terms of the free volume model (Ioffe and Gorbenko 2005). This model considers pyrene diffusion in a lipid phase as a three-step process: (i) formation of dynamic defects (kinks) in the acyl chains followed by opening of the cavities in a lipid monolayer; (ii) jump of diffusing molecules into the cavities coupled with the generation of voids; (iii) sealing the voids by the movement of the packing defects along the adjacent hydrocarbon chains. The appearance of dynamic defects in the membrane interior is associated with *trans-gauche* isomerization of acyl chains initiated by thermal motion and packing constraints. Free volume of the membrane, which is produced by lateral displacements of hydrocarbon chains after kink formation, is defined as the difference between the effective and van der Waals volumes of lipid molecules. Accordingly, the changes in E/M ratio reflect altered rate of *trans-gauche* isomerization of hydrocarbon chains.

The observed invariance of $I1/I3$ and E/M ratios, coupled with pronounced GP changes, strongly suggest that fibrillar lysozyme and 1-83/G26R/W@8 tend to perturb the interfacial bilayer region, with lipid tail order remaining virtually unaffected. This is probably a consequence of

Fig. 6.3 Pyrene emission spectra in PC and PC/Chol unilamellar liposomes in the absence or in the presence of 1-83/G26R/W@8 peptide. Lipid concentration was 16 μM, protein concentration was 1 μM, pyrene concentration was 0.1 μM; liposome diameter was 50 nm

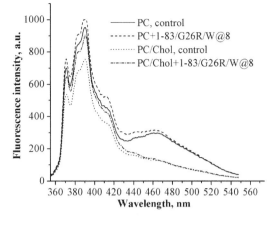

Fig. 6.4 Schematic representation of adsorption of multiple lipid vesicles along the amyloid fibril

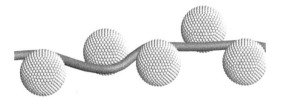

superficial fibril-lipid binding, which may involve adsorption of several liposomes onto a single fibril or wrapping of fibrillar strand around the lipid vesicle (Fig. 6.4). Our data agree with the results of Milanesi and colleagues indicating that fibrillar β_2-microglobulin associates with the membrane surface, as follows from the fibril-induced clustering of liposomes (Milanesi et al. 2012). These studies suggest that protein fibers interact mainly with the surface of the lipid monolayer but cause little or no perturbation in its core.

In contrast to mature fibrils, lysozyme oligomers brought about the decrease in *E/M* ratio which probably reflects lower rate of *trans-gauche* isomerization of hydrocarbon chains (Gorbenko and Trusova 2011). Furthermore, the magnitude of bilayer perturbations induced by oligomeric lysozyme in the nonpolar region of the bilayer appeared to be independent of the membrane charge, suggesting that electrostatic forces do not play determining role in this pertur-

bation. Consequently, hydrophobic interactions are likely to dominate the modification of the membrane structure by lysozyme pre-fibrillar aggregates. In agreement with this finding, various aggregated forms of Aβ amyloid peptide were reported to decrease bilayer fluidity in a fashion that correlated with surface hydrophobicity of the aggregated species (Kremer et al. 2000). Likewise, Aβ aggregates exerted ordering effect on the nonpolar part of the charged and neutral membranes, with little or no perturbation of the headgroup region (Kremer et al. 2001). Furthermore, the accessibility of the hydrocarbon core of the bilayer was demonstrated to modulate the membrane-disruptive effects of synuclein oligomers (van Rooijen et al. 2009). Taken together, these studies suggest that, compared to mature fibers, pre-fibrillar aggregates tend to insert more deeply into lipid bilayers, and that hydrophobic interactions dominate the ensuing structural perturbations in the hydrocarbon core of the bilayer.

6.2.2 Other Mechanisms of Fibril-Membrane Interactions

Evidently, aggregated proteins can influence natural membranes via diverse mechanisms, with lipid bilayer perturbations representing only one of the many effects. For example, lysozyme protofibrils can disintegrate erythrocyte membranes through intermolecular disulfide cross-linking of the membrane proteins (Huang et al. 2009) and induce apoptosis-like death of neurons, fibroblasts and neuroblastoma cells (Malisauskas et al. 2005; Gharibyan et al. 2007). Amyloid Aβ peptide can scramble erythrocyte membrane with subsequent cell death (Nicolay et al. 2007). Similarly to the rigidification of the model membranes (Kremer et al. 2001), Aβ aggregates can also reduce fluidity of hippocampal membranes (Eckert et al. 2000) which results from the peptide influence on the hydrocarbon core. However, other studies reported that aggregated Aβ perturbs mainly the interfacial region of the bilayer and has little effect on its hydrocarbon core (Ma et al. 2002). The main source of such discrepancies probably results from the high heterogeneity of the pre-fibrillar species as well as the structural polymorphism of mature fibrils (Stefani 2010).

Accumulating evidence substantiates the idea that disruption of cellular membrane by amyloid aggregates is one of the principal mechanisms of fibril-induced cellular dysfunction (Gorbenko and Kinnunen 2006; Bucciantini and Cecchi 2010; Bucciantini et al. 2014). The current consensus is that the mechanisms by which amyloidogenic proteins disrupt the membranes resemble those of antimicrobial peptides and include carpeting, toroidal or barrel-stave pore formation, non-specific membrane permeabilization, detergent- and raft-like membrane dissolution, etc. (Demuro et al. 2005; Lashuel and Lansbury 2006; Smith et al. 2009). These mechanisms are not mutually exclusive, so an individual peptide may cause membrane damage via different mechanisms depending on the membrane lipid composition. Changes in membrane integrity, which are initiated by mature fibrils, may result in increased lipid bilayer permeability

(de Planque et al. 2007), lipid loss (Lee et al. 2012), receptor activation (Verdier et al. 2004), membrane fragmentation (Sciacca et al. 2012), and oxidation of membrane lipids (Butterfield et al. 2002).

Recent studies revealed a specific membrane distortion by fibrillar assemblies, which is different from the previously observed mechanisms of lipid bilayer disruption (Milanesi et al. 2012). Using confocal microscopy and cryo-electron tomography, Milanesi and coauthors visualized 3D membrane damage produced by fibrillar β_2-microglobulin. Fibril-lipid interactions were found to result in re-shaping of lipid vesicles, interruptions to the bilayer structure, extraction of lipids from the membrane by removal or blebbing of the outer membrane leaflet, and formation of tiny vesicles from the extracted lipids. Remarkably, the largest distortions were generated by the fragmented fibrils, indicating that fibril ends possess much stronger membrane-modifying propensity than fibril shaft. This was ascribed to the enhanced hydrophobicity of fibril ends, a property shared by the prefibrillar oligomers that are currently regarded as the most toxic species of aggregated proteins (Butterfield and Lashuel 2010; Campioni et al. 2010; Winner et al. 2011; Cremades et al. 2012). This kind of membrane reorganization is thought to be distinct from the membrane breakage characteristic of pore-forming peptides (Tilley and Saibil 2006).

In general, concerted action of factors such as surface topography, hydrophobicity, charge distribution, hydrogen-bonding propensity, conformational flexibility, etc. seems to determine the reactivity of protein aggregates. On the other hand, membrane response to the aggregates of a certain type is likely to be controlled by a range of adjustable parameters of a lipid bilayer, such as surface charge, polarity, acyl chain order, fluidity, lateral pressure, curvature, etc. Therefore, while analyzing the structural basis for amyloid cytotoxicity, a membrane should be considered as a dynamic entity whose properties depend on its interactions with other components such as protein aggregates (Bucciantini et al. 2014).

6.3 Competitive Binding Behavior of Fibrillar Aggregates

The involvement of membrane structures in a wide variety of cellular processes implies multiple potential mechanisms of cell injury by aggregated proteins. In this regard, one cannot rule out competitive relationships between fibrillar species and native proteins within cellular environment. We hypothesized that in vivo amyloid fibrils can impair functionally relevant protein-membrane interactions, thereby initiating deleterious cell responses. As a first step to verify this idea, we determined whether fibrillar lysozyme can compete with the native cytochrome c (cyt c) for binding to the negatively charged model membranes composed of PC mixtures with varying proportions of anionic phospholipids, phosphatidylglycerol (PG), phosphatidylserine (PS) or CL. Biological activities of cytochrome c involve electron transport in the inner mitochondrial membrane and triggering of programmed cell death (Gray and Winkler 2010; Ascenzi et al. 2011). Monomeric lysozyme and cyt c, which are similar in their size (diameter ~3 nm) and charge (~ + 9e at physiological pH), display similar modes of membrane interaction, with relative contributions of electrostatic and hydrophobic interactions depending on the physicochemical characteristics of the membrane and the environmental conditions (Al Kayal et al. 2012).

To track membrane association of cyt c and its displacement with monomeric or polymerized lysozyme, we measured the efficiency of Förster resonance energy transfer (FRET) between anthrylvinyl-labeled PC (AV-PC) as a donor and heme group of cyt c as an acceptor. As shown by ^{1}H-NMR-spectroscopy and AV fluorescence quenching by iodide, anthrylvinyl fluorophore resides at the level of terminal methyl groups preferentially orienting parallel to acyl chains (Molotkovsky et al. 1982). The competition between cyt c and lysozyme for the membrane binding sites manifested itself in the rise of AV quantum yield (Q_r) with increasing lysozyme concentration. Compared to its monomeric counterpart, fibrillar lysozyme caused substantially more detachment of cyt c from the membranes (Fig. 6.5). Moreover, the character of competitive adsorption to lipid bilayers was different for fibrillar and native states of lysozyme. Specifically, upon increasing the content of anionic lipid from 10 to 40 mol% (PG, PS) or from 5 to 25 mol% (CL), desorption curves changed their shape from hyperbolic to sigmoidal, suggesting that desorption of cyt c caused by lysozyme fibrils was a cooperative process. One possible reason for such a cooperativity may lie in simultaneous covering of multiple lipid headgroups in the contact areas of fibrils with liposomes and hence, exclusion of multiple cyt c molecules from these areas.

As follows from the mean-field thermodynamic analysis of surface adsorption in a binary mixture

Fig. 6.5 Relative quantum yield of AV-PC in protein-lipid systems as a function of cytochrome c concentration for PC/PG (3:2, mol:mol) and PC/CL (3:1, mol:mol) liposomes. Lipid concentration was 10 μM, lysozyme concentration was 0.5 μM. Liposome diameter was 100 nm

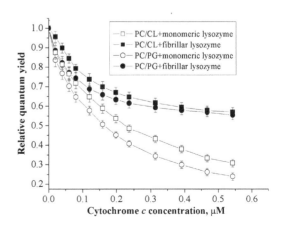

of charged model proteins, competitive behavior of a protein strongly depends upon its molecular parameters such as size, shape, and charge distribution (Fang and Szleifer 2003). The binding of lysozyme and cyt c to negatively charged membranes is largely governed by electrostatic interactions, since these proteins have high positive net charge at physiological pH. At strongly acidic fibrillization conditions (pH 2.0), the net charge on the lysozyme molecule reaches +15e. Notably, the net repulsion among the protein monomers favors fibril growth, while net attraction gives rise to precipitation, as reported by Hill and coauthors (Hill et al. 2011). Moreover, the magnitude of charge repulsion is thought to modulate the assembly pathway and morphology of fibrillar aggregates. Recent studies emphasize the importance of net charge of amyloid fibrils as one of the principal determinants of their cytotoxicity (Hirano et al. 2010, 2012). Higher bilayer-disruptive activity of lysozyme fibrils compared to monomeric protein was attributed to the enhanced electrostatic interactions due to increased charge density upon fibrillization. Likewise, effects on the membranes of the positively charged lysozyme fibrils differed from those of the negatively charged Aβ fibrils (Yoshiike et al. 2007). These observations led Hirano and colleagues to propose that electrostatic interactions dominate the membrane association of fibrillar lysozyme and concomitant bilayer disruption (Hirano et al. 2012). Accordingly, electrostatics is likely to play an essential role in determining the competitive membrane binding of lysozyme fibrils and cyt c discussed here.

Consistent with this idea are the results of Trp fluorescence quenching by acrylamide indicating that decrease in Stern-Volmer constant upon fibril-lipid binding become more pronounced with increasing membrane charge (Gorbenko et al. 2012). For electrostatically-controlled adsorption, the equilibrium association constant can be represented as a combination of an intrinsic (or non-electrostatic) term and an electrostatic component that depends on surface charge density, environmental conditions (pH, ionic strength), and the degree of surface coverage by a protein (Gorbenko et al. 2007). As the proportion of anionic lipid rises, cyt c-bilayer interaction becomes stronger and the number of membrane binding sites increases; therefore, higher lysozyme concentrations are required to reduce cyt c – lipid association constant and surface occupancy to the values at which competition between the proteins becomes significant. This effect may account for the delayed threshold effect in Q_r dependencies on lysozyme concentration observed for liposomes containing 40 mol% PG or PS, or 25 mol% CL, as illustrated in Fig. 6.6 for PC/PG vesicles.

However, there is no unequivocal relation between the membrane charge and the extent of protein competition, since fibrillar lysozyme induced distinct changes in Q_r of the membranes with similar electrostatic surface potentials but different lipid composition. This implies that the amount of desorbed cyt c is not determined exclusively by nonspecific electrostatic factors. The possibility of specific interactions between

Fig. 6.6 Relative quantum yield of AV-PC in protein-lipid systems as a function of lysozyme concentration for PC/PG (3:2, mol:mol) liposomes. Lipid concentration was 10 μM, cytochrome c concentration was 0.08 μM. Liposome diameter was 100 nm

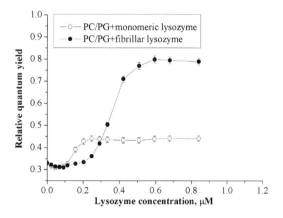

lysozyme fibrils and phospholipid headgroups is also supported by our finding that the degree of Trp solvent exposure differs for lipid bilayers of similar charge but different chemical nature of phospholipid headgroups (Gorbenko et al. 2012).

In sum, our model studies corroborate the idea that pathological fibrillar aggregates forming in vivo can potentially impair proper membrane binding of cellular proteins. The modulating effects of membrane charge and chemical nature of anionic phospholipid headgroups observed in our studies enabled us to propose that lipid-associating and competitive binding properties of lysozyme fibrils are governed by both specific and non-specific protein-lipid interactions. These results suggest that enhanced ability of fibrillar protein aggregates to compete for membrane binding sites are among the possible determinants of amyloid cytotoxicity.

6.4 Fibril Restructuring on a Membrane Template

Specific molecular architecture of amyloid assemblies is stabilized by the enthalpic contribution from the main-chain hydrogen bonding, ionic pairing, aromatic π-π interactions and hydrogen bonds between amino-acid side chains (Nelson et al. 2005; Makin et al. 2005; Petkova et al. 2006), as well as the entropic contribution originating from the release of structured water molecules from the tightly packed amyloid core (Sawaya et al. 2007; Williams et al. 2006). Clearly, all forces stabilizing fibril structure can be modulated by the environmental factors inherent to a lipid bilayer. This notion is supported by the fact that lipids can destabilize and re-solubilize mature fibrils with formation of "reverse oligomers" similar in their cytotoxicity to the oligomers assembled from monomeric proteins. This effect was demonstrated, for instance, for Aβ peptide (Martins et al. 2008). Hence, amyloid plaques can be considered as reservoirs of toxicity in which mature fibrils can be transformed into highly toxic oligomeric species due to alterations in local physicochemical parameters controlled by lipid metabolism.

To further explore fibril transformations induced by the lipid surface, we designed a series of FRET experiments aimed at elucidating the effects of lipids on the structure and morphology of fibrillar aggregates. The aforesaid N-terminal fragment of the amyloidogenic variant of human apoA-I, 1-83/G26R/W@8, was chosen as a model protein for these experiments, while lipid vesicles were formed from PC and its mixture with cholesterol. The idea was to recruit amyloid-specific dye with a defined location within fibril structure as an energy acceptor for a membrane fluorescent probe. To this end, we employed classical amyloid marker Thioflavin T (ThT). Fibril binding of ThT is accompanied by a dramatic fluorescence enhancement arising from restricted torsional oscillations of the benzothiazole and aminobenzoyl rings and nearly planar conformation of the dye molecule incorporated in the solvent-exposed grooves spanning across consecutive β-strands parallel to the fibril axis (Sulatskaya et al. 2010; Stsiapura et al. 2007; Hawe et al. 2008). Another remarkable feature of ThT is its preferential association with the grooves lined with aromatic residues (Biancalana and Koide 2010). Due to these distinct properties, ThT can be used not only for identification of amyloid fibrils, but also for their structural characterization. Our experimental strategy involved: (i) quantitating the dye interaction with fibrillar 1-83/G26R/W@8 peptide and defining possible ThT binding sites through measuring FRET between the engineered Trp8 in this peptide and ThT; (ii) monitoring morphologic changes in 1-83/G26R/W@8 fibrils adsorbing onto the surface of lipid vesicles by examining FRET between Laurdan and ThT.

As a first step, we used the method of double fluorimetric titration to determine quantitative characteristics of ThT-fibril binding (association constant, Ka, and binding stoichiometry, n, in mole of ThT per mole of protein). Scatchard plot obtained from the binding data had a concave-up shape suggesting two types of ThT binding sites. Global fitting based on simultaneous analysis of two-dimensional data arrays acquired by varying both ThT and protein concentration yielded two sets of parameters: $K_{a1} = (6.2 \pm 0.7)\,\mu M^{-1}$,

a

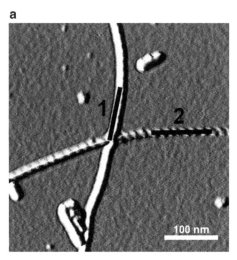

b

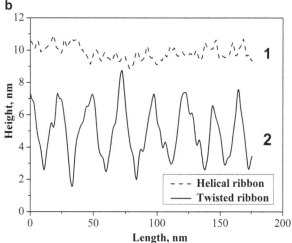

Fig. 6.7 Atomic force microscopy images of 1-83/G26R/W@8 fibrils illustrating the presence of helical and twisted ribbon polymorphs (**a**). Height profiles acquired over contour length for different polymorphs of 1-83/G26R/W@8 fibrils (**b**)

$n_1 = 0.1 \pm 0.02$ (high-affinity sites) and $K_{a2} = (0.14 \pm 0.03)\,\mu M^{-1}$, $n_2 = 0.17 \pm 0.03$ (low-affinity sites), as described in detail in (Girych et al. 2014). The observed heterogeneity of ThT binding sites can stem from: (i) the structurally and compositionally distinct sites within fibrillar assemblies of a certain morphology, and/or (ii) polymorphism inherent to most amyloid preparations. Indeed, AFM shows that 1-83/G26R/W@8 fibrils have either smooth or twisted appearance suggestive of structural polymorphism. The height profiles over contour length analyzed for the two types of fibrils suggested that different polymorphs are represented by the twisted and helical ribbons (Fig. 6.7a, b). Accordingly, we proposed that high- and low-affinity ThT binding sites reside on distinct fibril polymorphs. To better understand these polymorphs, we attempted to develop a tentative structural model of fibrillar 1-83/G26R/W@8 and to ascertain what solvent-exposed grooves within the fibril structure are most likely to bind ThT.

Amino acid sequence analysis of the self-associating properties of apoA-I 1-83 fragment by using AGGRESCAN (Conchillo-Sole et al. 2007), Zyggregator (Tartaglia and Vendruscolo 2008) and TANGO (Linding et al. 2004) algorithms showed that residue segments 14–23 and 50–58 have the highest aggregation tendency (also see Chap. 8 by Das and Gursky in this volume). Recent x-ray crystallographic studies revealed that the apoA-I segment 44–55 forms a native unpaired β-strand in lipid-free protein, suggesting high intrinsic β-sheet propensity by this segment (Gursky et al. 2012; Das et al. 2014). On the other hand, in our model the residues 1–13, 32–40 and 59–83 were excluded from the putative β-strand regions, based on the following considerations: (i) the presence of structure-breaking proline residues at positions 3, 4, 7, 66 and glycine at positions 35, 39, and 65; (ii) the proteolytic accessibility of E34 and F57 (Lagerstedt et al. 2007); and (iii) TANGO prediction that the sequences 14–31 and 41–58 have the highest propensity to form β-sheets in amyloid.

On the basis of these considerations and the observation that the height of 1-83/G26R/W@8 fibrils is ~5–10 nm, while the total length of this fully extended polypeptide is ~28 nm, we hypothesized that 1-83/G26R/W@8 protofilaments have β-strand-loop-β-strand structure in which β-strands from residues 14–31 to 41–58 form a

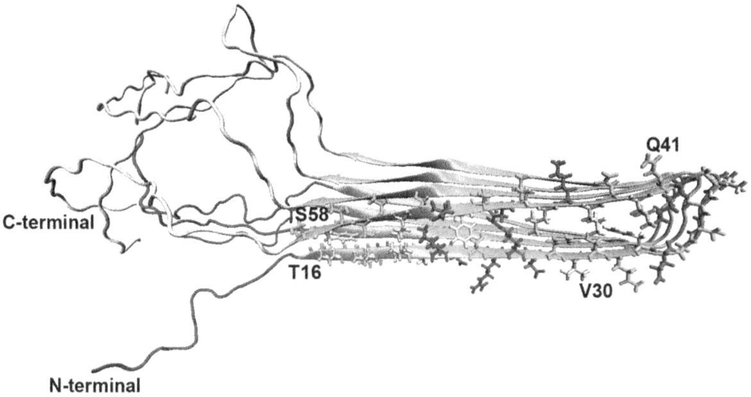

Fig. 6.8 3D model structure of 1-83/G26R/W@8 in a fibrillar state derived from Rosetta calculations. Four individual peptide molecules which were used to build the model are shown

self-complementary steric zipper stabilized by van der Waals and hydrophobic interactions, as well as by the putative salt bridge between R27 and D48 of the neighbouring strands (Girych et al. 2014). Figure 6.8 shows a tentative structure for fibrillar 1-83/G26R/W@8 created for four protein monomers using Rosetta and VMD software (Delano 2005). Notably, the proposed parallel in-register β-sheet structure is consistent with our recent FTIR data recorded of this fibrillar peptide (Adachi et al. 2013).

According to our structural model, surface grooves that can potentially bind ThT sites are formed by the solvent-exposed residues L14_T16_Y18_D20_L22_D24_R26_D28_V30 in the N-terminal β-strand and by the residues Q41_N43_K45_L47_N49_D51, S52_T54_T56_S58 in the C-terminal β-strand. Of those, the grooves lined with aromatic and hydrophobic residues are probably preferred for ThT binding, as demonstrated by Wu and coauthors (Wu et al. 2008, 2009, 2011; Biancalana et al. 2009). These considerations, together with the FRET-based distance estimates between fibril-bound ThT and Trp8 in 1-83/G26R/W@8 (Girych et al. 2014), suggest that the high-affinity ThT binding sites reside within the groove T16_Y18 on the helical ribbon polymorphs. The constant curvature of this region favors nearly planar motionally restricted conformation of the dye, while the low-affinity sites are lined up at the groove

D20_L22 on the twisted ribbon polymorphs with varying curvature.

Next, we used this model as a structural basis to determine whether protein-lipid interactions can alter the structure and morphology of 1-83/G26R/W@8 fibrils. To this end, we employed fluorescent probe Laurdan located at the lipid-water interface as an energy donor for ThT. Fibrillar 1-83/G26R/W@8 peptide was incubated with small (50 nm) PC liposomes doped with Laurdan, followed by titration of ThT. Since in these protein-lipid systems ThT was distributed between the lipid bilayer and the binding sites in the helical and twisted ribbon fibrils, a separate series of FRET experiments was directed towards quantifying the ThT partitioning into the lipid phase and evaluating its membrane location. As illustrated in Fig. 6.9, FRET efficiency in the PC vesicles + fibrillar 1-83/G26R/W@8 system is markedly higher compared to the PC vesicles alone, implying that the number of ThT molecules serving as the energy acceptors increases upon fibril adsorption on the surface of the lipid vesicles. Notably, this effect was observed only for fibril-forming apoA-I variant, while in the case of its non-amyloidogenic counterpart lacking G26R mutation, energy transfer between Laurdan and ThT appeared to be indistinguishable from that in the protein-free PC liposomes. Hence, FRET enhancement

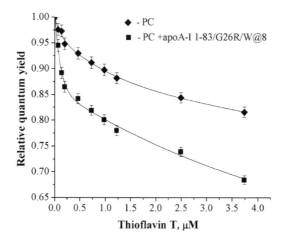

Fig. 6.9 Relative quantum yield of Laurdan, which was incorporated in PC liposomes and PC + 1-83/G26R/W@8 complexes, as a function of Thioflavin T concentration. Lipid concentration was 13 µM, protein concentration was 1.6 µM, Laurdan concentration was 0.5 µM. Liposome diameter was 50 nm. *Solid line* represents the simulation data yielding the best agreement between experiment and theory

could be attributed exclusively to fibril-bound ThT molecules.

Our next goal was to ascertain what kind of spatial distribution of donors and acceptors was consistent with the experimentally measured FRET profiles. In the simulation-based FRET analysis, donor and acceptor coordinates were generated in a virtual square box with edge length l_b and calculating the relative quantum yield averaged over all donors as:

$$Q_r = \frac{1}{N_D} \sum_{j=1}^{N_D} \left[1 + \sum_{i=1}^{N_{AC}} \left(\frac{R_0}{r_{ij}} \right)^6 \right]^{-1} \quad (6.1)$$

Here r_{ij} is the distance between j-th donor and i-th acceptor; N_D and N_{AC} stand for the number of donors and acceptors in a box, respectively; and l_b was taken as 10 R_0, where R_0 is the Förster radius for the Laurdan-ThT donor-acceptor pair estimated to be ~2.9 nm. The NAC value was determined assuming three populations of ThT molecules associated either with lipids or with helical ribbon (HR) or twisted ribbon (TR) fibrils. The respective molar concentrations of these species (B_L, B_{HR} and B_{TR}) were calculated for any given total concentration of ThT by solving the system of the following equations:

$$B_{HR} = \frac{n_1 P \left(Z - B_{HR} - B_{TR} - B_L \right)}{1 / K_{a1} + \left(Z - B_{HR} - B_{TR} - B_L \right)}; \qquad B_{TR} = \frac{n_2 P \left(Z - B_{HR} - B_{TR} - B_L \right)}{1 / K_{a2} + \left(Z - B_{HR} - B_{TR} - B_L \right)}$$

$$B_L = \frac{K_P \left(Z - B_{HR} - B_{TR} - B_L \right) V_L}{V_W} \qquad (6.2)$$

Here, P is the total concentration of the protein monomers, K_P is the dye partition coefficient (630 ± 46 for PC liposomes), and V_L, V_W are the volumes of the lipid and aqueous phases, respectively. Subsequent estimate of the surface density of lipid-bound acceptors and linear density of HR- and TR-associated acceptors (the number of ThT molecules per protein monomer) enabled us to assess the number of various ThT species (N_L, N_{HR} and N_{TR}) in the box.

Our initial assumption was that the fibrils retained their helical or twisted ribbon morphology when interacting with the lipid vesicles. The donors were confined to a plane parallel to the membrane surface and located at a distance r_{FD} from the fibril axis. The lengths of the helical (l_{HR}) and twisted ribbon (l_{TR}) fibrils in the box relative to the box edge l_b were defined by parameters $s_{HR} = l_{HR} / l_b$ and $s_{TR} = l_{TR} / l_b$. The coordinates of ThT molecules distributed over the fiber structure were generated as previously described (Girych et al. 2014). Our goal was to find out if any sets of theoretical parameters $\{r_{FD}, s_{HR}, s_{TR}\}$ were consistent with the experiment. The uncertainty in the mutual orientation of Laurdan emission and ThT absorption transition moments was minimized by setting the lower and upper limits for orientation factor (κ_{min}^2 and κ_{max}^2) using the information on the fluorophore rotational mobility derived from the fluorescence anisotropy measurements (Dale et al. 1979). Although the values of r_{FD}, s_{HR} and s_{TR} were varied in the widest physically plausible range, no good agreement between the experimental FRET data and the simulation results could be achieved. This result suggested that 1-83/G26R/W@8 fibrils undergo lipid-induced morphological changes probably involving the unwinding of the helical and twisted ribbon fibers into more planar ribbons.

Since ThT is thought to reside along the surface side-chain grooves running parallel to the long fibril axis (Krebs et al. 2005; Biancalana et al. 2008; Teoh et al. 2011), in our simulations the fibril-bound acceptors were arranged along the lines parallel to bilayer surface and separated from the donor plane by certain distances, d_{HR} and d_{TR}. Two alternative scenarios were considered in regard to the orientation factor: (i) κ^2 is fixed and varies between κ_{min}^2 and κ_{max}^2, or (ii) κ^2 depends on the donor-acceptor distance suggesting that the orientational behavior of the fibril-associated ThT resembles that of the membrane fluorophores whose transition moments are symmetrically distributed within the cones (Domanov and Gorbenko 2002). Only the latter scenario provided successful theoretical description of the experimental FRET profiles.

Obviously, when a certain fibril fragment attaches to the membrane surface, superficial grooves undergo structural changes due to the rearrangement of the hydrogen-bonded network at the lipid surface. Accordingly, ThT binding sites acquire the properties intermediate between those found on the fibrillar structure and in the lipid phase, and the orientational properties of ThT as an energy acceptor for Laurdan become similar to those of membrane-bound dyes. For this reason, only simulation-based FRET analysis with distance-dependent orientation factor could adequately fit the observed FRET data.

Solid line in Fig. 6.9 represents the simulation results providing the best agreement between experiment and theory, which was achieved by using the following parameter set: l_{HR} ~29 nm, l_{TR} ~6 nm, d_{HR} ~1 nm, d_{TR} ~1 nm. Notably, after converting into planar ribbons, helical and twisted ribbon fibers became structurally indistinguishable, yet the stoichiometries of ThT binding to high-affinity sites in HR and low-affinity sites in TR remained distinct, resulting in different linear densities of the bound dye in HR and TR. Given that Laurdan molecules are located in the planar interface between polar and non-polar parts of lipid bilayer, our estimates for donor-acceptor separation suggest that linear array of the fibril-bound ThT species is located at the water-lipid interface, with ThT-binding grooves facing the aqueous phase. In this orientation, the amyloid core-forming C-terminal segment Q41–S58 must reside within the fibril-membrane contact area.

To verify this idea, we employed the online server HeliQuest to obtain the mean hydrophobicity (<H>), hydrophobic moment (μH), net charge (z) and, eventually, lipid discrimination factor (D) characterizing lipid binding affinity of a given polypeptide fragment (Gautier et al. 2008; Keller 2011). HeliQuest analysis of 1-83/G26R/W@8 sequence showed that this peptide contains four most probable membrane-binding regions, namely R10–R27, K23–K40, L44–R61, and S52–Q69, with the strongest lipid-binding potential in L44–R61. This corroborates the model assumption that the β-sheet in residues 41–58 faces the membrane surface. Since the distance between the β-sheets in our proposed β-strand-

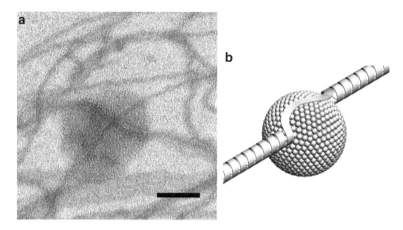

Fig. 6.10 Transmission electron micrographs illustrating adsorption of 1-83/G26R/W@8 fibril on PC vesicle (**a**) and schematic illustration of untwisting of the 1-83/ G26R/W@8 fibrils on a lipid bilayer template drawn approximately to scale (**b**). Scale bar in panel (**a**) is 50 nm

loop-β-strand structure is ~1 nm, our estimates for d_{HR} and d_{TR} are consistent with bilayer penetration by the β-strands to the depth ~1 nm, i.e. to the level of the initial acyl chain carbons.

Furthermore, the values of l_{HR} and l_{TR} allowed us to characterize fibril-lipid binding in terms of the number of lipid molecules per protein monomer. Considering the surface area per lipid headgroup as 0.65 nm², the box with edge length 29 nm contains 1,294 lipid molecules, while the total length of the untwisted fibrils is ~35 nm, corresponding to ~75 protein monomers. Hence, the number of lipid molecules per protein monomer in a planar ribbon configuration of 1-83/ G26R/W@8 fibrils that are associated with PC liposomes is *circa* 17. Similar estimates for PC/ Chol liposomes produced the parameters l_{HR} ~17 nm, $l_{TR}=0$, d_{HR} ~2 nm, with the number of lipid molecules per protein monomer *circa* 34. These results suggest that the presence of 30 mol% cholesterol in the PC bilayer impairs fibril penetration into the polar membrane region and reduces the fibril-lipid contact area.

Notably, helical and twisted ribbon fibers are rather rigid structures, as judged from their persistence length of about 2 μM determined directly from AFM images as described in (Adamcik et al. 2011). Nevertheless, adsorption of lipid vesicles along the fibril length (Fig. 6.10a, b) apparently produces local untwisting of the helical and twisted ribbons, thereby increasing their flexibility and extending the contact area with the membrane surface. In summary, the results reported in this section suggest that lipid membranes can not only trigger amyloid formation, but also modulate the structural morphology of fibrillar assemblies.

6.5 Concluding Remarks

In the past decades, a complex problem of protein-lipid interactions acquired a new intriguing facet concerning the role of cell membranes in initiating the growth of amyloid fibrils and the ensuing cytotoxic action of pre-fibrillar and fibrillar protein aggregates. It became increasingly clear that physicochemical, structural and morphological characteristics of amyloid assemblies are important in determining their membrane-mediated toxicity. On one hand, these characteristics can be fine-tuned by the membrane environment. On the other hand, membrane responses to pathogenic aggregated species are dictated by collective properties of the lipid bilayer, such as the surface charge and curvature (addressed by Uversky, Chap. 2 in this volume), dielectric permeability, viscosity and lateral pressure profiles, elasticity, etc., as well as by the exact chemical nature of individual membrane constituents and their conformational features. In the present chapter we focused mainly on the emerging understanding of the molecular processes involved in the interactions between lipid membranes and mature amyloid fibrils. Our fluorescence studies performed with two model amyloidogenic proteins, lysozyme and 1-83/G26R/W@8 apoA-I peptide, support the following hypotheses. First, superficial fibril binding to lipid bilayers is followed by the changes in lipid packing density and level of hydration in the interfacial bilayer region, with little or no perturbation of the hydrophobic core. Second, fibrillar assemblies can displace membrane-associated proteins. Third, morphological characteristics of protein fibrils can be modulated by the lipid bilayer. Finally, cholesterol can reduce lipid-associating and membrane-modifying abilities of amyloid aggregates. Although extrapolation of these model studies to the in vivo effects of mature amyloid fibrils on the membranes is not straightforward, our ideas provide the basis for a deeper understanding of the membrane-mediated cytotoxicity mechanisms of amyloid assemblies.

Acknowledgements The authors thank Prof. Kenichi Akaji and Dr. Hiroyuki Kawashima (Kyoto Pharmaceutical University) for their help with AFM, and Dr. Rohit Sood (Aalto University) for his aid with transmission electron microscopy. This work was supported by the grant from Fundamental Research State Fund of Ukraine (project number F54.4/015 to G. G.) and Grant-in-Aid for Scientific Research 25293006 (to H. S.) from the Japan Society for the Promotion of Science.

References

Adachi E, Nakajima H, Mizuguchi C, Dhanasekaran P, Kawashima H, Nagao K, Akaji K, Lund-Katz S, Phillips MC, Saito H (2013) Dual role of an N-terminal amyloidogenic mutation in apolipoprotein A-I: destabilization of helix bundle and enhancement of fibril formation. J Biol Chem 288:2848–2856

Adamcik J, Mezzenga R (2012a) Study of amyloid fibrils via atomic force microscopy. Curr Opin Colloid Interface Sci 17:369–376

Adamcik J, Mezzenga R (2012b) Protein fibrils from polymer physics perspective. Macromolecules 45:1137–1150

Adamcik J, Castelletto V, Bolisetty S, Hamley IW, Mezzenga R (2011) Direct observation of time-resolved polymorphic states in the self-assembly of end-capped heptapeptides. Angew Chem 50(24):5495–5498

Aisenbrey C, Borowik T, Byström R, Bokvist M, Lindström F, Misiak H, Sani M, Gröbner G (2008) How is protein aggregation in amyloidogenic diseases modulated by biological membranes? Eur Biophys J 37:247–255

Al Kayal T, Nappini S, Russo E, Berti D, Bucciantini M, Stefani M, Baglioni P (2012) Lysozyme interaction with negatively charged lipid bilayers: protein aggregation and membrane fusion. Soft Matter 8:4524–4534

Anderluh G, Gutierrez-Aguirre I, Rabzelj S, Ceru S, Kopitar-Jerala N, Macek P, Turk V, Zerovnik E (2005) Interaction of human stefin B in the prefibrillar oligomeric form with membranes. Correlation with cellular toxicity. FEBS J 272:3042–3051

Arispe N, Rojas E, Pollard H (1993) Alzheimer's disease amyloid beta protein forms calcium channels in bilayer membranes: blockade by tromethamine and aluminium. Proc Natl Acad Sci U S A 90:567–571

Ascenzi P, Polticelli F, Marino M, Santucci R, Coletta M (2011) Cardiolipin drives cytochrome c proapoptotic and antiapoptotic actions. Life 63:160–165

Bamberger ME, Harris ME, McDonald DR, Husemann J, Landreth GE (2003) A cell surface receptor complex for fibrillar β-amyloid mediates microglial activation. J Neurosci 23:2665–2674

Biancalana M, Koide S (2010) Molecular mechanism of thioflavin-T binding to amyloid fibrils. Biochim Biophys Acta 1804:1405–1412

Biancalana M, Makabe K, Koide A, Koide S (2008) Aromatic cross-strand ladders control the structure and stability of beta-rich peptide self-assembly mimics. J Mol Biol 383:205–213

Biancalana M, Makabe K, Koide A, Koide S (2009) Molecular mechanism of thioflavin-T binding to the surface of beta-rich peptide self-assemblies. J Mol Biol 385:1052–1063

Bucciantini M, Cecchi C (2010) Biological membranes as protein aggregation matrices and targets of amyloid toxicity. Methods Mol Biol 648:231–243

Bucciantini M, Forzan M, Russo E, Martino C, Pieri L, Formigli L, Quercioli F, Soria S, Pavone F, Savistchenko J, Meliki R, Stefani M (2012) Toxic effects of amyloid fibrils on cell membranes: the importance of ganglioside GM1. FASEB J 26:818–831

Bucciantini M, Rigacci S, Stefani M (2014) Amyloid aggregation: role of biological membranes and the aggregate-membrane system. J Phys Chem Lett 5:17–527

Butterfield SM, Lashuel HA (2010) Amyloidogenic protein-membrane interactions: mechanistic insight from model systems. Angew Chem Int Ed 49:5628–5654

Butterfield DA, Castegna A, Lauderback CM, Drake J (2002) Evidence that amyloid beta-peptide-induced lipid peroxidation and its sequelae in Alzheimer's disease brain contribute to neuronal death. Neurobiol Aging 23:655–664

Campioni S, Mannini B, Zampagni M, Pensalfini A, Parrini C, Evangelisti E, Relini A, Stefani M, Dobson CM, Cecchi C, Chiti F (2010) A causative link between the structure of aberrant protein oligomers and their toxicity. Nat Chem Biol 6:140–147

Canale C, Torrassa S, Rispoli P, Relini A, Rolandi R, Bucciantini A, Stefani M, Gliozzi A (2006) Natively folded Hypf-N and its early amyloid aggregates interact with phospholipid monolayers and destabilize supported lipid bilayers. Biophys J 91:1–14

Carulla N, Caddy GL, Hall DR, Zurdo J, Gairi M, Feliz M, Giralt E, Robinson CV, Dobson CM (2005) Molecular recycling within amyloid fibrils. Nature 436:54–558

Caughey B, Lansbury PT (2003) Protofibrils, pores, fibrils, and neurodegeneration: separating the responsible protein aggregates from the innocent bystanders. Annu Rev Neurosci 26:267–298

Cecchi C, Baglioni S, Fiorillo C, Pensalfini A, Liguri G, Nosi D, Rigacci S, Bucciantini M, Stefani M (2005) Insights into the molecular basis of the differing susceptibility of varying cell types to the toxicity of amyloid aggregates. J Cell Sci 118:3459–3470

Conchillo-Sole O, de Groot NS, Avilés F, Vendrell J, Daura X, Ventura S (2007) AGGRESCAN: a server for the prediction and evaluation of "hot spots" of aggregation in polypeptides. BMC Bioinformatics 8:65

Cremades N, Cohen SI, Deas E, Abramov AY, Chen AY, Orte A, Sandal M, Clarke RW, Dunne P, Aprile FA, Bertoncini CW, Wood NW, Knowles TPJ, Dobson CM, Klenerman D (2012) Direct observation of the interconversion of normal and toxic forms of α-synuclein. Cell 149(5):1048–1059

Dale R, Eisinger J, Blumberg W (1979) The orientational freedom of molecular probes. The orientation factor in intramolecular energy transfer. Biophys J 26:161–194

Dante S, Haus T, Brandt A, Dencher NA (2008) Membrane fusogenic activity of the Alzheimer's peptide Abeta(1-42) demonstrated by small-angle neutron scattering. J Mol Biol 376:393–404

Das M, Mei X, Jayaraman S, Atkinson D, Gursky O (2014) Amyloidogenic mutations in human apolipoprotein A-I are not necessarily destabilizing – a common mechanism of apolipoprotein A-I misfolding in familial amyloidosis and atherosclerosis. FEBS J 281:2525–2542

de Planque RR, Raussen V, Contera SA, Rijkers DTS, Liskamp RMJ, Ruysschaert JM, Ryan JF, Separovic F, Watts A (2007) β-sheet structured β-amyloid (1-40) perturbs phosphatidylcholine model membranes. J Mol Biol 368:982–987

Delano WL (2005) The case for open-source software in drug discovery. Drug Discov Today 10:213–217

Demuro A, Mina E, Kayed R, Milton SC, Parker I, Glabe CG (2005) Calcium dysregulation and membrane disruption as a ubiquitous neurotoxic mechanism of soluble amyloid oligomers. J Biol Chem 280:17294–17300

Dima RI, Thirumalai D (2002) Exploring protein aggregation and self-propagation using lattice models: phase diagram and kinetics. Protein Sci 11:1036–1049

Domanov YA, Gorbenko GP (2002) Analysis of resonance energy transfer in model membranes: role of orientational effects. Biophys Chem 99:143–154

Eckert GP, Cairns NJ, Maras A, Gattaz WF, Müllera WE (2000) Cholesterol modulates the membrane-disordering effects of beta-amyloid peptides in the Hippocampus: specific changes in Alzheimer's disease. Dement Geriatr Cogn Disord 11:181–186

Engel MF, Khemtemourian L, Kleijer CC, Meeldijk HJ, Jacobs J, Verkleij AJ, de Kruijff B, Killian JA, Höppener JW (2008) Membrane damage by human islet amyloid polypeptide through fibril growth at the membrane. Proc Natl Acad Sci U S A 105:6033–6038

Fang F, Szleifer I (2003) Competitive adsorption in model charged protein mixtures: equilibrium isotherms and kinetics behavior. J Chem Phys 119:1053–1065

Forloni G (1996) Neurotoxicity of β-amyloid and prion peptides. Curr Opin Neurol 9:492–500

Frare E, Polverino de Laureto P, Zurdo J, Dobson C, Fontana A (2004) A highly amyloidogenic region of hen lysozyme. J Mol Biol 340:1153–1165

Gautier R, Douguet D, Antonny B, Drin G (2008) HeliQuest: a web server to screen sequences with specific alpha-helical properties. Bioinformatics 24:2101–2102

Gharibyan AL, Zamotin V, Yanamandra K, Moskaleva OS, Margulis BA, Kostanyan IA, Morozova-Roche LA (2007) Lysozyme amyloid oligomers and fibrils induce cellular death via different apoptotic/necrotic pathways. J Mol Biol 365:1337–1349

Giannakis E, Pacifico J, Smith DP, Hung LW, Masters CL, Cappai R, Wade JD, Barnham KJ (2008) Dimeric structures of α-synuclein bind preferentially to lipid membranes. Biochim Biophys Acta 1778:1112–1119

Girych M, Gorbenko G, Trusova V, Adachi E, Mizuguchi C, Nagao K, Kawashima H, Akaji K, Phillips M, Saito H (2014) Interaction of thioflavin T with amyloid fibrils of apolipoprotein A-I N-terminal fragment: resonance energy transfer study. J Struct Biol 185:116–124

Gorbenko GP, Kinnunen PKJ (2006) The role of lipid-protein interactions in amyloid-type protein fibril formation. Chem Phys Lipids 141:72–82

Gorbenko G, Trusova V (2011) Effect of oligomeric lysozyme on structural state of model membranes. Biophys Chem 154:73–81

Gorbenko GP, Ioffe VM, Kinnunen PKJ (2007) Binding of lysozyme to phospholipid bilayers: evidence for protein aggregation upon membrane association. Biophys J 93:140–153

Gorbenko G, Trusova V, Sood R, Molotkovsky J, Kinnunen PKJ (2012) The effect of lysozyme amyloid fibrils on cytochrome c–lipid interactions. Chem Phys Lipids 165:769–776

Gray HB, Winkler JR (2010) Electron flow through metalloproteins. Biochim Biophys Acta 1797:1563–1572

Gursky O, Mei X, Atkinson D (2012) The crystal structure of the C-terminal truncated apolipoprotein A-I sheds new light on amyloid formation by the N-terminal fragment. Biochemistry 51:10–18

Hawe A, Sutter M, Jiskoot W (2008) Extrinsic fluorescent dyes as tools for protein characterization. Pharm Res 25:1487–1499

Hill SE, Miti T, Richmond T, Muschol M (2011) Spatial extent of charge repulsion regulates assembly pathways for lysozyme amyloid fibrils. PLoS One 6:e18171

Hirano A, Uda K, Maeda Y, Akasaka T, Shiraki K (2010) One-dimensional protein-based nanoparticles induce lipid bilayer disruption: carbon nanotube conjugates and amyloid fibrils. Langmuir 26:17256–17259

Hirano A, Yoshikawa H, Matsushita S, Yamada Y, Shiraki K (2012) Adsorption and disruption of lipid bilayers by nanoscale protein aggregates. Langmuir 28:3887–3895

Huang B, He J, Ren J, Yan XY, Zeng CM (2009) Cellular membrane disruption by amyloid fibrils involved intermolecular disulfide cross-linking. Biochemistry 48:5794–5800

Ioffe VM, Gorbenko GP (2005) Lysozyme effect on structural state of model membranes as revealed by pyrene excimerization studies. Biophys Chem 114:199–204

Joy T, Wang J, Hahn A, Hegele RA (2003) ApoA-I related amyloidosis: a case report and literature review. Clin Biochem 36:641–645

Karpovich DS, Blanchard GJ (1995) Relating the polarity-dependent fluorescence response to vibronic coupling. Achieving a fundamental understanding of the py polarity scale. J Phys Chem 99:3951–3958

Keller R (2011) New user-friendly approach to obtain an Eisenberg plot and its use as a practical tool in protein sequence analysis. Int J Mol Sci 21:5577–5591

Kelly JW (2002) Towards an understanding of amyloidogenesis. Nat Struct Biol 9:323–325

Kim DH, Frangos JA (2008) Effects of amyloid β-peptides on the lysis tension of lipid bilayer vesicles containing oxysterols. Biophys J 95:620–628

Kinnunen PKJ (2009) Amyloid formation on lipid membrane surfaces. Open Biol J 2:163–175

Knowles TPJ, Buehler MJ (2011) Nanomechanics of functional and pathological amyloid materials. Nature 6:469–479

Krebs MR, Bromley EH, Donald AM (2005) The binding of thioflavin T to amyloid fibrils: localization and implications. J Struct Biol 149:30–37

Kremer JJ, Pallitto MM, Sklansky DJ, Murphy RM (2000) Correlation of beta-amyloid aggregate size and hydrophobicity with decreased bilayer fluidity of model membranes. Biochemistry 39:10309–10318

Kremer JJ, Sklansky DJ, Murphy RM (2001) Profile of changes in lipid bilayer structure caused by β-amyloid peptide. Biochemistry 40:8563–8571

Lagerstedt JO, Cavigiolio G, Roberts LM, Hong HS, Jin LW, Fitzgerald PG, Oda MN, Voss JC (2007) Mapping the structural transition in an amyloidogenic apolipoprotein A-I. Biochemistry 46:9693–9699

Lashuel HA, Lansbury PT (2006) Are amyloid diseases caused by protein aggregates that mimic bacterial pore-forming toxins? Q Rev Biophys 39:167–201

Lee AG (2003) Lipid–protein interactions in biological membranes: a structural perspective. Biochim Biophys Acta 1612:1–40

Lee AG (2004) How lipids affect the activities of integral membrane proteins. Biochim Biophys Acta 1666:62–87

Lee CC, Sun Y, Huang H (2012) How type II diabetes-related islet amyloid polypeptide damages lipid bilayers. Biophys J 102:1059–1068

Linding R, Schymkowitz J, Rousseau F, Diella F, Serrano L (2004) A comparative study of the relationship between protein structure and beta-aggregation in globular and intrinsically disordered proteins. J Mol Biol 342:345–353

Lopes DHJ, Meister A, Gohlke A, Hauser A, Blume A, Winter R (2007) Mechanism of IAPP fibrillation at lipid interfaces studied by infrared reflection absorption spectroscopy (IRRAS). Biophys J 93:3132–3141

Loura LM, Canto AM do., Martins J (2013) Sensing hydration and behavior of pyrene in POPC and POPC/cholesterol bilayers: a molecular dynamics study. Biochim Biophys Acta 1828:1094–1101

Lúcio AD, Vequi-Suplicy CC, Fernandez RM, Lamy MT (2010) Laurdan spectrum decomposition as a tool for the analysis of surface bilayer structure and polarity: a study with DMPG, peptides and cholesterol. J Fluoresc 20:473–482

Ma X, Sha Y, Lin K, Nie S (2002) The effect of fibrillar Aβ1-40 on membrane fluidity and permeability. Protein Pept Lett 9:173–178

Makin OS, Atkins E, Sikorski P, Johansson J, Serpell LC (2005) Molecular basis for amyloid fibril formation and stability. Proc Natl Acad Sci U S A 102:315–320

Malisauskas M, Ostman J, Darinskas A, Zamotin V, Liutkevicius E, Lundgren E, Morozova-Roche LA (2005) Does the cytotoxic effect of transient amyloid oligomers from common equine lysozyme in vitro imply innate amyloid toxicity? J Biol Chem 280:6269–6275

Martins IC, Kuperstein I, Wilkinson H, Maes E, Vanbrabant M, Jonckheere W, Van Gelder P, Hartmann D, D'Hooge R, De Strooper B, Schymkowitz J, Rousseau F (2008) Lipids revert inert Aβ amyloid fibrils to neurotoxic protofibrils that affect learning in mice. EMBO J 27:224–233

Matsuzaki K (2011) Formation of toxic amyloid fibrils by amyloid beta-protein on ganglioside clusters. Int J Alzheimers Dis 956104:1–7

Meratan AA, Ghasemi A, Nemat-Gorgani M (2011) Membrane integrity and amyloid cytotoxicity: a model study involving mitochondria and lysozyme fibrillation products. J Mol Biol 409:826–838

Milanesi L, Sheynis T, Xue WF, Orlova EV, Hellewell AL, Jelinek R, Hewitt EW, Radford SE, Saibil HR (2012) Direct three-dimensional visualization of membrane disruption by amyloid fibrils. Proc Natl Acad Sci U S A 109:20455–20460

Molotkovsky J, Manevich E, Gerasimova E, Molotkovskaya I, Polessky V, Bergelson L (1982) Differential study of phosphatidylcholine and sphingomyelin in human high-density lipoproteins with lipid-specific fluorescent probes. Eur J Biochem 122:573–579

Nakajima A (1971) Solvent effect on the vibrational structure of the fluorescence and absorption spectra of pyrene. Bull Chem Soc Jpn 44:3272–3277

Nelson R, Sawaya MR, Balbirnie M, Madsen A, Riekel C, Grothe R, Eisenberg D (2005) Structure of the cross-β spine of amyloid-like fibrils. Nature 435:773–778

Nichols WC, Gregg RE, Brewer HB, Benson MD (1990) A mutation in apolipoprotein A-I in the Iowa type of familial amyloidotic polyneuropathy. Genomics 8:318–323

Nicolay J, Gatz S, Liebig G, Gulbins E, Lang F (2007) Amyloid induced suicidal erythrocyte death. Cell Physiol Biochem 19:175–184

Novitskaya V, Bocharova OV, Bronstein I, Baskakov IV (2006) Amyloid fibrils of mammalian prion protein are highly toxic to cultured cells and primary neurons. J Biol Chem 281:13828–13836

Obici L, Franceschini G, Calabresi L, Giorgetti S, Stoppini M, Merlini G, Bellotti V (2006) Structure, function and amyloidogenic propensity of apolipoprotein A-I. Amyloid 13:191–205

Palsdottir H, Hunte C (2004) Lipids in membrane protein structures. Biochim Biophys Acta 1666:2–18

Parasassi T, Gratton E (1995) Membrane lipid domains and dynamics as detected by Laurdan fluorescence. J Fluoresc 8:365–373

Parasassi T, De Stasio G, Ravagnan G, Rusch RM, Gratton E (1991) Quantitation of lipid phases in phospholipid vesicles by the generalized polarization of Laurdan fluorescence. Biophys J 60:179–189

Parasassi T, Krasnowska EK, Bagatolli L, Gratton E (1998) Laurdan and Prodan as polarity-sensitive fluorescent membrane probes. J Fluoresc 8:365–373

Pepys MB, Hawkins PN, Booth DR, Vigushin DM, Tennent GA, Souter AK, Totty N, Nguyen O, Blake CC, Terry CJ, Feest TG, Zalin AM, Hsuan JJ (1993) Human lysozyme gene mutations cause hereditary systemic amyloidosis. Nature 362:553–557

Petkova AT, Leapman RD, Guo Z, Yau WM, Mattson MP, Tycko R (2005) Self-propagating, molecular-level polymorphism in Alzheimer's β-amyloid fibrils. Science 307:262–265

Petkova AT, Yau WM, Tycko R (2006) Experimental constraints on quaternary structure in Alzheimer's β-amyloid fibrils. Biochemistry 45:498–512

Phillips MC (2013) New insights into the determination of HDL structure by apolipoproteins. J Lipid Res 54:2034–2048

Qiu L, Buie C, Reay A, Vaughn MW, Cheng KH (2011) Molecular dynamics simulations reveal the protective role of cholesterol in β-amyloid protein-induced membrane disruptions in neuronal membrane mimics. J Phys Chem B 115:9795–9812

Quist A, Doudewski I, Lin H, Azimova R, Ng D, Frangine B, Kagan B, Ghiso J, Lal R (2005) Amyloid ion channels: a common structural link for protein-misfolding disease. Proc Natl Acad Sci U S A 102:10427–10432

Relini A, Torrassa S, Rolandi R, Gliozzi A, Rosano C, Canale C, Bolognesi M, Plakoutsi G, Bucciantini M,

Chiti F, Stefani M (2004) Monitoring the process of HypF fibrillization and liposome permeabilization by protofibrils. J Mol Biol 338:943–957

Relini A, Cavalleri O, Rolandi R, Gliozzi A (2009) The two-fold aspect of the interplay of amyloidogenic proteins with lipid membranes. Chem Phys Lipids 158:1–9

Sanchez SA, Tricerri MA, Gratton E (2012) Laurdan generalized polarization fluctuations measures membrane packing micro-heterogeneity in vivo. Proc Natl Acad Sci U S A 109:7314–7319

Sawaya MR, Sambashivan S, Nelson R, Ivanova M, Sievers S, Apostol M, Thompson M, Balbirnie M, Wiltzius J, McFarlane H, Madsen A, Riekel C, Eisenberg D (2007) Atomic structures of amyloid cross-β spines reveal varied steric zippers. Nature 447:453–457

Sciacca MF, Brender JR, Lee DK, Ramamoorthy A (2012) Phosphatidylethanolamine enhances amyloid-fiber dependent membrane fragmentation. Biochemistry 51:7676–7684

Serpell LC (2000) Alzheimer's amyloid fibrils: structure and assembly. Biochim Biophys Acta 1502:16–30

Smith JF, Knowles TP, Dobson CM, Macphee CE, Welland ME (2006) Characterization of the nanoscale properties of individual amyloid fibrils. Proc Natl Acad Sci U S A 103:15806–15811

Smith DP, Tew DJ, Hill AF, Bottomley SP, Masters CL, Barnham KJ, Cappai R (2008) Formation of a high affinity lipid-binding intermediate during the early aggregation phase of α-synuclein. Biochemistry 47:1425–1434

Smith P, Brender J, Ramamoorthy A (2009) The induction of negative curvature as a mechanism of cell toxicity by amyloidogenic peptides. The case of islet amyloid polypeptide. J Am Chem Soc 131:4470–4478

Sousa MM, Du Yan S, Fernandes R, Guimaraes A, Stern D, Saraiva MJ (2001) Familial amyloid polyneuropathy: receptor for advanced glycation end products-dependent triggering of neuronal inflammatory and apoptotic pathways. J Neurosci 21:7576–7586

Sparr E, Engel MFM, Sakharov DV, Sprong M, Jacobs J, de Kruijf B, Hoppener JWM, Killian JA (2004) Islet amyloid polypeptide-induced membrane leakage involves uptake of lipids by forming amyloid fibers. FEBS Lett 577:117–120

Sponne I, Fifre A, Koziel V, Oster T, Olivier JL, Pillot T (2004) Membrane cholesterol interferes with neuronal apoptosis induced by soluble oligomers but not fibrils of amyloid-β peptide. FASEB J 838:836–838

Squier T (2001) Oxidative stress and protein aggregation during biological aging. Exp Gerontol 36:1539–1550

Stefani M (2004) Protein misfolding and aggregation: new examples in medicine and biology of the dark side of the protein world. Biochim Biophys Acta 1739:5–25

Stefani M (2007) Generic cell dysfunction in neurodegenerative disorders: role of surfaces in early protein misfolding, aggregation, and aggregate cytotoxicity. Neuroscientist 13:519–531

Stefani M (2008) Protein folding and misfolding on surfaces. Int J Mol Sci 9:2515–2542

Stefani M (2010) Biochemical and biophysical features of both oligomer/fibril and cell membrane in amyloid cytotoxicity. FEBS J 277:4602–4613

Stefani M, Dobson C (2003) Protein aggregation and aggregate toxicity: new insights into protein folding, misfolding diseases and biological evolution. J Mol Med 81:678–699

Stsiapura VI, Maskevich AA, Kuznetsova IM (2007) Computational study of thioflavin T torsional relaxation in the excited state. J Phys Chem A 111:4829–4835

Sulatskaya AI, Maskevich AA, Kuznetsova IM, Uversky VN, Turoverov KK (2010) Fluorescence quantum yield of Thioflavin T in rigid isotropic solution and incorporated into the amyloid fibrils. PLoS One 5:e15385

Tabner B, Turnbull S, El-Agnaf O, Allsop D (2002) Formation of hydrogen peroxide and hydroxyl radicals from Aβ and α-synuclein as a possible mechanism of cell death in Alzheimer's disease and Parkinson's disease. Free Radic Biol Med 32:1076–1083

Tartaglia GG, Vendruscolo M (2008) The Zyggregator method for predicting protein aggregation propensities. Chem Soc Rev 37:1395–1401

Teoh CL, Pham CL, Todorova N, Hung A, Lincoln CN, Lees E, Lam YH, Binger KJ, Thomson NH, Radford SE, Smith TA, Müller SA, Engel A, Griffin MD, Yarovsky I, Gooley PR, Howlett GJ (2011) A structural model for apolipoprotein C-II amyloid fibrils: experimental characterization and molecular dynamics simulations. J Mol Biol 4:1246–1266

Tilley SJ, Saibil HR (2006) The mechanism of pore formation by bacterial toxins. Curr Opin Struct Biol 16(2):230–236

Tokunaga Y, Sakakibara Y, Kamada Y, Watanabe K, Sugimoto Y (2013) Analysis of core region from egg white lysozyme forming amyloid fibrils. Int J Biol Sci 9:219–227

Valincius G, Heinrich F, Budvytyte R, Vanderah DJ, McGillivray DJ, Sokolov Y, Hall JE, Losche M (2008) Soluble amyloid β-oligomers affect dielectric membrane properties by bilayer insertion and domain formation: implications for cell toxicity. Biophys J 95:4845–4861

van Rooijen BD, Claessens MMA, Subramaniam V (2009) Lipid bilayer disruption by oligomeric α-synuclein depends on bilayer charge and accessibility of the hydrophobic core. Biochim Biophys Acta 1788:1271–1278

Verdier Y, Zarandi M, Penke B (2004) Amyloid beta-peptide interactions with neuronal and glial cell plasma membrane: binding sites and implications for Alzheimer's disease. J Pept Sci 10:229–248

Weldon DT, Rogers SD, Ghilardi JR, Finke MP, Cleary JP, O'Hare E, Esler WP, Maggio JE, Mantyh PW (1998) Fibrillar β-amyloid induces microglial phagocytosis, expression of inducible nitric oxide synthase, and loss of a select population of neurons in the rat CNS in vivo. J Neurosci 18:2161–2173

Williams TL, Serpell LC (2011) Membrane and surface interactions of Alzheimer's Aβ peptide: insights into the mechanism of cytotoxicity. FEBS J 278:3905–3917

Williams AD, Shivaprasad S, Wetzel R (2006) Alanine scanning mutagenesis of Ab (1–40) amyloid fibril stability. J Mol Biol 357:1283–1294

Winner B, Jappelli R, Maji SK, Desplats PA, Boyer L, Aigner S, Hetzer C, Loher T, Vilar M, Campioni S, Tzitzilonis C, Soragni A, Jessberger S, Mira H, Consiglio A, Pham E, Masliah E, Gage FH, Riek R (2011) In vivo demonstration that alpha-synuclein oligomers are toxic. Proc Natl Acad Sci U S A 108(10):4194–4199

Wu C, Wang Z, Lei H, Duan Y, Bowers MT, Shea JE (2008) The binding of thioflavin T and its neutral analog BTA-1 to protofibrils of the Alzheimer's disease Abeta(16-22) peptide probed by molecular dynamics simulations. J Mol Biol 384:718–729

Wu C, Biancalana M, Koide S, Shea JE (2009) Binding modes of thioflavin-T to the single-layer beta-sheet of the peptide self-assembly mimics. J Mol Biol 394:627–633

Wu C, Bowers MT, Shea JE (2011) On the origin of the stronger binding of PIB over thioflavin T to protofibrils of the Alzheimer amyloid-β peptide: a molecular dynamics study. Biophys J 100:1316–1324

Xue WF, Hellewell AL, Gosal WS, Homans SW, Hewitt EW, Radford SE (2009) Fibril fragmentation enhances amyloid cytotoxicity. J Biol Chem 284:34272–34282

Yoshiike Y, Akagi T, Takashima A (2007) Surface structure of amyloid-β fibrils contributes to cytotoxicity. Biochemistry 46:9805–9812

Zbilut JP, Colosimo A, Conti F, Colafranceschi M, Manetti C, Valerio M, Webber CL Jr, Giuliani A (2003) Protein aggregation/folding: the role of deterministic singularities of sequence hydrophobicity as determined by nonlinear signal analysis of acylphosphatase and Aβ (1–40). Biophys J 85:3544–3557

The Role of Lipid in Misfolding and Amyloid Fibril Formation by Apolipoprotein C-II

7

Timothy M. Ryan, Yee-Foong Mok,
Geoffrey J. Howlett, and Michael D.W. Griffin

Abstract

Apolipoproteins are a key component of lipid transport in the circulatory system and share a number of structural features that facilitate this role. When bound to lipoprotein particles, these proteins are relatively stable. However, in the absence of lipids they display conformational instability and a propensity to aggregate into amyloid fibrils. Apolipoprotein C-II (apoC-II) is a member of the apolipoprotein family that has been well characterised in terms of its misfolding and aggregation. In the absence of lipid, and at physiological ionic strength and pH, apoC-II readily forms amyloid fibrils with a twisted ribbon-like morphology that are amenable to a range of biophysical and structural analyses. Consistent with its lipid binding function, the misfolding and aggregation of apoC-II are substantially affected by the presence of lipid. Short-chain phospholipids at submicellar concentrations significantly accelerate amyloid formation by inducing a tetrameric form of apoC-II that can nucleate fibril aggregation. Conversely, phospholipid micelles and bilayers inhibit the formation of apoC-II ribbon-type fibrils, but induce slow formation of amyloid with a distinct straight fibril morphology. Our studies of the effects of lipid at each stage of amyloid formation, detailed in this chapter, have revealed complex behaviour dependent on the chemical nature of the lipid molecule, its association state, and the protein:lipid ratio.

Keywords

Protein aggregation • Apolipoprotein C-II • Amyloid fibril • Phospholipid • Detergent • Lipid mimetic

Y.-F. Mok • G.J. Howlett • M.D.W. Griffin (⊠)
Department of Biochemistry and Molecular Biology,
Bio21 Molecular Science and Biotechnology
Institute, University of Melbourne,
Parkville, VIC 3010, Australia
e-mail: ymok@unimelb.edu.au; ghowlett@unimelb.
edu.au; mgriffin@unimelb.edu.au

T.M. Ryan
The Florey Institute of Neuroscience and Mental
Health, The University of Melbourne,
Parkville, VIC, Australia
e-mail: tmryan@unimelb.edu.au

© Springer International Publishing Switzerland 2015
O. Gursky (ed.), *Lipids in Protein Misfolding*, Advances in Experimental
Medicine and Biology 855, DOI 10.1007/978-3-319-17344-3_7

Abbreviations

Apo	Apolipoprotein
CD	Circular dichroism
CMC	Critical micelle concentration
DHPC	1,2-dihexanoyl-sn-glycero-3-phosphocholine
DHPS	1,2-dihexanoyl-sn-glycero-3-phospho-L-serine
DMPC	1,2-dimyristoyl-sn-glycero-3-phosphocholine
DnPC	1,2-diacyl-sn-glycero-3-phosphocholine where n=3–9 carbon acyl chains
DPC	Dodecylphosphocholine
FRET	Fluorescence resonance energy transfer
LysoMPC	1-myristoyl-2-hydroxy-sn-glycero-3-phosphocholine
NBD	Nitrobenzoxadiazole
SDS	Sodium dodecyl sulphate
STEM	Scanning transmission electron microscopy
TEM	Transmission electron microscopy
ThT	Thioflavin T

7.1   Introduction

Apolipoproteins belong to a highly conserved family of lipid-binding proteins involved in lipid transport. An inauspicious feature of the members of this family is that they account for a substantial proportion of proteins known to form amyloid in vivo. For instance, apoA-I, apoA-II and apoA-IV form amyloid fibrils that are associated with several hepatic, systemic, and renal amyloid diseases (Bergstrom et al. 2001, 2004; Coriu et al. 2003; Yazaki et al. 2003; Obici et al. 1999, 2006). In addition, apoA-I, apoA-II and apoC-II accumulate in amyloid deposits of atherosclerotic lesions and have been suggested to contribute to the progression of cardiovascular diseases (Westermark et al. 1995; Mucchiano et al. 2001a, b; Medeiros et al. 2004). The high propensity of apolipoproteins to form amyloid fibrils was postulated to stem from their low conformational stability in the absence of bound lipids (Gursky and Atkinson 1998; Hatters and Howlett 2002).

The present review focuses on human apoC-II, a 79 amino-acid apolipoprotein that activates lipoprotein lipase and is involved in the remodelling of very-low density lipoproteins and chylomicrons in the circulation. Immunohistochemical analysis of atherosclerotic plaques showed that apoC-II co-localises with serum amyloid P, an in vivo marker of amyloid (Stewart et al. 2007a). ApoC-II amyloid fibrils also activate macrophages in a CD36 receptor-dependent process that has been proposed as an early step in foam cell formation and the development of atherosclerosis (Medeiros et al. 2004).

ApoC-II shares high sequence similarity and structural homology with other members of the apolipoprotein family. In particular, the apoC-II sequence comprises 11 residue tandem repeats, which form the characteristic amphipathic helical motifs that facilitate lipid surface binding by apolipoproteins (Segrest et al. 1990, 1994; see also Chap. 8 by Das and Gursky in this volume). These shared properties make apoC-II an excellent model for understanding the role of lipids in the folding and misfolding of apolipoproteins in general.

In the presence of micellar lipid and lipid mimetics, apoC-II adopts a predominantly α-helical structure (MacRaild et al. 2004), consistent with its role in binding lipid surfaces. Conversely, lipid-free apoC-II contains relatively little α-helical secondary structure that is not significantly altered in the presence of 5 M guanidine hydrochloride (Hatters and Howlett 2002), suggesting that in the absence of lipid surfaces, apoC-II is natively unfolded. In vitro, this lipid-free form of apoC-II readily self-assembles into homogeneous fibrils with increased levels of β-structure and all the hallmarks of amyloid (Hatters et al. 2000). This process occurs in a reproducible and concentration-dependent manner at physiological ionic strength and pH, resulting in somewhat soluble, "twisted-ribbon" fibrils (Fig. 7.1a) that are amenable to a range of biophysical analyses.

ApoC-II fibrils formed in the absence of lipid have been investigated using a range of structural techniques, including fibre diffraction, scanning transmission electron microscopy (STEM), fluorescence resonance energy transfer (FRET),

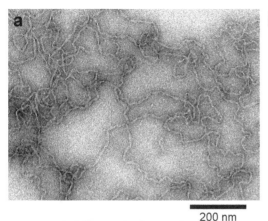

200 nm

Fig. 7.1 The structure of apoC-II fibrils. (**a**) Transmission electron micrograph of apoC-II fibrils formed by incubation of apoC-II at 0.3 mg/mL in 100 mM sodium phosphate, pH 7.4, for 2 days. Note the flat, twisted, ribbon-like morphology. Scale bar represents 200 nm (Figure adapted with permission from Griffin et al. (2008)). (**b**) A structural model for apoC-II amyloid fibrils. ApoC-II monomer assembles into amyloid fibrils with a 'letter-G-like' β-strand-loop-β-strand structure. The model includes residues 21–79 of apoC-II with each β-strand of the monomer contributing to each of the two β-sheets, giving rise to a parallel, in-register structure (Figure adapted with permission from Teoh et al. (2011a))

analytical ultracentrifugation and hydrogen/deuterium exchange NMR. Fibre diffraction experiments show a classical cross-β diffraction pattern with reflections at 4.67 Å and 9.46 Å, indicating the spacing between β-strands in the fibril axis and the average spacing between β-sheets in the fibril cross section, respectively (Teoh et al. 2011b) STEM revealed that the fibrils contain approximately one molecule of apoC-II per 4.7 Å rise in the long axis of the fibril, and hydrogen/deuterium exchange data indicated two protected regions in residues 20–36 and 58–74 (Wilson et al. 2007) that are implicated in formation of the amyloid core. Recently, these and other observations led to the development of a "letter-G-like" β-strand-loop-β-strand structural model for an apoC-II unit within a mature amyloid fibril (Fig. 7.1b) (Teoh et al. 2011b).

Studies of the kinetics of apoC-II fibril formation indicate that it proceeds via a reversible pathway that includes elongation, dissociation, and fibril breaking and rejoining (Binger et al. 2008; Yang et al. 2012). In addition, the concentration dependence of apoC-II aggregation has been analysed (Hatters et al. 2000, 2002a, b). These analyses collectively show that apoC-II forms amyloid fibrils in a time-dependent manner at 0.1–1 mg/mL apoC-II (approximately 10–100 μM), with times for the completion of fibril formation ranging from less than 1 h (at 1 mg/mL apoC-II) to ~7 days (at 0.1 mg/mL). ApoC-II concentrations less than 0.1 mg/mL form amyloid fibrils very slowly (>7 days), facilitating the study of the early events of amyloid fibril formation (Ryan et al. 2008).

Given its role in lipid transport and lipoprotein remodelling, it is perhaps unsurprising that sub-micellar lipids/detergents and lipid surfaces can substantially affect the misfolding of apoC-II and amyloid formation. Lipids alter the rate and pathway of amyloid formation, as well as the structural features of mature fibrils. This chapter reviews our studies of the effects of lipids and lipid-like detergent molecules on amyloid fibril formation by full-length apoC-II and its fibrillogenic peptides.

7.2   The Structure of ApoC-II Bound to Lipid Micelles

Investigating the role of lipids in protein folding is complicated by a range of factors, including, but not limited to, the diverse range of lipids that are present in biological samples, their general insolubility and their tendency to self-associate to form lipid surfaces. These complications render distinguishing the role of individual lipid molecules from the role of lipid surfaces in protein folding difficult, particularly for native lipids that generally form lipid surfaces or complexes. This is compounded by the tendency of molecules that bind to lipid surfaces to alter the properties of the membrane. For example, apoC-II can bind specific phospholipids within lipid membranes, and thereby locally remodel the composition of the lipid surface (Hanson et al. 2003). Conversely, lipid surfaces have a significant effect on apoC-II structure; binding of lipid surfaces almost always results in the stabilisation of a native α-helical structure in this (MacRaild et al. 2001, 2004) and other apolipoproteins [for review see (Wang 2008)]. However, there is evidence that apoC-II can adopt an amyloid structure in atherosclerotic plaques, which are highly enriched in a range of modified lipids (Medeiros et al. 2004; Teoh et al. 2011a). This dichotomy between the effects of lipids in vitro and in in vivo disease lesions suggest that the role of lipids in apoC-II amyloid formation might be much more complex than previously thought.

Initial circular dichroism (CD) and NMR spectroscopy investigations of the structure of apoC-II focused on the protein in complex with micelles of the detergents sodium dodecyl sulphate (SDS) and dodecylphosphocholine (DPC). These lipid-like detergents induced ~60 % α-helical conformation in apoC-II and prevented its aggregation (MacRaild et al. 2001, 2004). NMR structural analyses of apoC-II in complex with SDS micelles showed two broad regions of α-helical structure, corresponding to the N- and C-terminal regions of the molecule. These helices correspond well to the regions of amphipathic helix predicted from primary sequence analysis, linked by a flexible region, which also has some helical character (Fig. 7.2). The N-terminal helical region includes residues 16–36, and forms a curved helix, postulated by MacRaild et al. (2001) to be important for binding to the large-radius surface of chylomicrons and very low density lipoproteins (diameters 50–1,000 nm), while

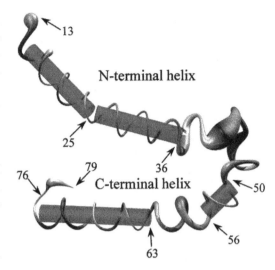

Fig. 7.2 The structure of apoC-II bound to SDS micelles. Averaging of 20 NMR structures results in the main chain trace and reveals two regions of well-defined α-helices (represented by *narrow line* and *cylinders*) connected by a region of less well-defined structure (represented by the thicker C$_\alpha$ trace). *Arrows* and numbers indicate the amino acid position (Adapted with permission from MacRaild et al. (2001); Copyright (2001) American Chemical Society)

the C-terminal helix in residues 50–54 and 63–76 contains the binding site for lipoprotein lipase. Subsequent ^{15}N NMR relaxation measurements coupled with molecular dynamics simulations of apoC-II bound to SDS micelles largely confirmed these findings (Zdunek et al. 2003). ApoC-II in the presence of DPC micelles displayed very similar structural characteristics (MacRaild et al. 2004). While these NMR structures elucidated several conformational features of natively folded apoC-II bound to lipid-like surfaces, they provided little information on the role of individual lipid interactions.

7.3 The Effects of Submicellar Lipid on the Misfolding of ApoC-II

In an attempt to separate the roles of individual phospholipid molecules and lipid surfaces on apoC-II misfolding and aggregation, we have used a range of simple short-chain phospholipids and detergents. These lipid derivatives have the advantage that they exist in two states, a "monomeric" state at low concentrations and a self-associated micellar state at higher concentrations. The critical micelle concentration (CMC) at which these molecules transition from one state to the other depends on structural features of the detergent molecule, including its size and shape, charge distribution on the head group, length and hydrophobicity of the acyl chains and buffer conditions, including pH, salt concentration, and hydrophobicity. By selecting particular lipid derivatives, a range of factors that affect protein folding can be investigated in the absence of lipid surface.

Investigation of the effect of individual lipids on the process of amyloid formation by apoC-II was initially conducted through the use of the short-chain phospholipid 1,2-dihexanoyl-sn-glycero-3-phosphocholine (DHPC), which has a CMC of approximately 10 mM in 100 mM sodium phosphate at pH 7.4 (Hatters et al. 2001). At micellar concentrations this lipid stabilised native structure in apoC-II. However, at submicellar concentrations DHPC enhanced the rate of apoC-II amyloid formation sixfold (Hatters et al. 2001; Griffin et al. 2008; Ryan et al. 2008). Investigation of the effect of acyl chain length, and hence hydrophobicity, was conducted using a series of phosphatidylcholine derivatives with acyl chains containing three to nine carbons at a constant detergent concentration of 10 mM (Griffin et al. 2008) (Fig. 7.3), which is below CMC for short-chain lipids containing three to five carbons. This analysis showed that increasing the length of the acyl chain from three to five carbons increased the rate of fibril formation up to a maximum of sevenfold. This result suggests that higher hydrophobicity increases the rate of fibril formation. Lipids with acyl chains containing six to nine carbons, which approach and surpass their CMC at 10 mM concentration, inhibited amyloid fibril formation by apoC-II. CD spectroscopy indicated that the shorter-chain submicellar lipids induced similar conformation in apoC-II, with a

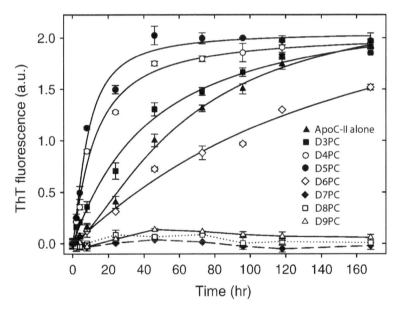

Fig. 7.3 Effect of acyl chain length of short-chain diacyl phospholipids on fibril formation by apoC-II. Diacyl phosphocholine derivatives are designated DnPC, where n refers to the acyl carbon chain length. Fibril formation was monitored using a Thioflavin T (ThT) assay, where solutions of apoC-II (0.3 mg/ml in 100 mM phosphate buffer at pH 7.4) were incubated alone (*filled triangles*) and in the presence of 10 mM D3PC (*filled squares*), D4PC (*open circles*), D5PC (*filled circles*), D6PC (*open diamonds*), D7PC (*filled diamonds, dashed line*), D8PC (*open squares, dotted line*) and D9PC (*open triangles*). ThT fluorescence was measured at the intervals and non-linear best fits to the data using a Hill-3 parameter equation (*continuous lines*) were used to determine the transition rates and maximum ThT fluorescence (Figure adapted with permission from Griffin et al. (2008))

slight trend towards increased negative ellipticity at 222 nm with increasing acyl chain length, suggesting some α-helix formation. The longer-chain micellar lipids invariably induced CD spectra with minima at 208 and 222 nm, characteristic of predominant α-helical structure.

The effect of head group was investigated by using 1,2-dihexanoyl-sn-glycero-3-phospho-L-serine (DHPS), where the choline functionality of the head group is substituted with a serine (Ryan et al. 2008). This lipid has a net negative charge, which allows it to exist in a non-micellar form at significantly higher concentrations (~20 mM) than its zwitterionic DHPC counterpart. When this difference in CMC is taken into account, the effect of submicellar DHPS on amyloid formation by apoC-II is not significantly different from that of DHPC. Investigation of 1,2-dihexanoyl-sn-glycero-3-phospho-L-ethanolamine (zwitterionic) and 1,2-dihexanoyl-sn-glycero-3-phospho-L-glycerol (anionic) at comparable submicellar concentrations indicated very similar effects on the rate of fibril formation to DHPS and DHPC. Taken together, these data indicate that the head group charge has little effect on the misfolding and rate of fibril formation by apoC-II, and that the acyl chain hydrophobicity is the major determinant of the enhancing effects of these submicellar lipids on apoC-II aggregation (Ryan et al. 2008).

7.4 The Effects of Submicellar Lipids on Initial ApoC-II Self-Association

During the studies outlined in the previous section, it became apparent that the short-chain lipids were significantly affecting the lag phase of apoC-II amyloid fibril formation, suggesting that their primary effect was on the initial folding and self-association of apoC-II (Ryan et al. 2008; Griffin et al. 2008). To further investigate this effect, a single cysteine replacement of serine 61 (apoC-II-S61C) was engineered, and this cysteine was labelled with the fluorophore, c_5 maleimide Alexa-488 (Alexa-488-apoC-II)

(Ryan et al. 2008) (Fig. 7.4a), which was used to characterize the molecular environment and the process of amyloid fibril formation. Alexa-488-apoC-II showed a similar rate of amyloid fibril formation to native apoC-II (Ryan et al. 2008), providing a valuable tool for detailed biophysical analysis of this process.

Initial time courses of Alexa-488-apoC-II aggregation indicated that, in the absence of lipids, the protein displayed a time-dependent decrease in fluorescence intensity during fibril formation. Fluorescence anisotropy measurements, which report on the relative rotational motion of a fluorophore, were consistent with this observation, showing a sigmoidal increase to a maximum of approximately 0.1 over 72 h (Fig. 7.4b). This increase in fluorescence anisotropy is consistent with the formation of large aggregates with low tumbling rates, while the decrease in fluorescence intensity can be attributed to homo-fluorescence resonance energy transfer (FRET) occurring when the Alexa-488 fluorophores from different apoC-II molecules are brought into close proximity. Thus, the two fluorescence measures allow analysis of the time course of aggregation and amyloid fibril formation by apoC-II.

Addition of submicellar DHPC or DHPS resulted in significant differences in the time courses of these fluorescence measurements. In both cases, these short-chain lipids at submicellar concentrations induced a rapid increase in the fluorescence intensity and anisotropy of Alexa-488 (Fig. 7.4c). This immediate response was followed by a slower increase in fluorescence intensity during which no significant change in anisotropy was observed. After these initial changes, the quenching of the Alexa-488 fluorescence intensity and further increases in anisotropy due to fibril formation over longer time periods proceeded as for apoC-II alone. The rate of the slower phase of the initial fluorescence changes was independent of protein concentration, and FRET measurements using both Alexa-488 and Alexa-594 labelled apoC-II-S61C indicated significant inter-molecular interactions in the presence of submicellar phospholipids at early time points. These results indicate that sub-

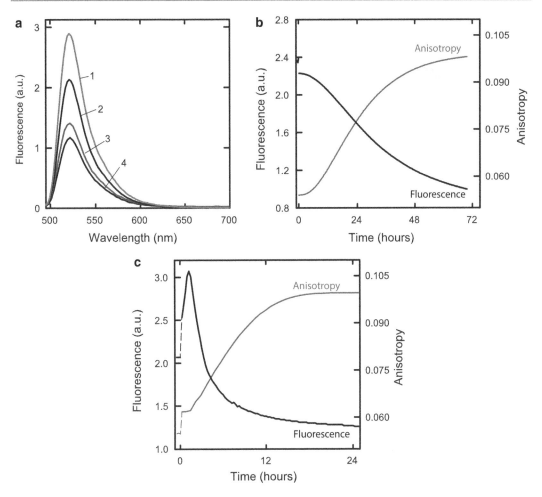

Fig. 7.4 Fluorescence spectroscopy of Alexa488-labeled apoC-II. (**a**) Emission spectra of freshly prepared Alexa488-apoC-II (50 µg/ml in 100 mM phosphate buffer, pH 7.4; excitation at 495 nm) in the absence of lipid (*curve 2*) and in the presence of 10 mM DHPS (*curve 1*); and of fibrillar Alexa488-labeled apoC-II (0.3 mg/ml in 100 mM phosphate buffer, pH 7.4, and diluted to 50 µg/ml) in the absence (*curve 4*) and presence (*curve 3*) of 10 mM DHPS. (**b**) Fluorescence emission intensity (*black line*) and anisotropy (*blue line*) of Alexa488-apoC-II during fibril formation. **c** Fluorescence emission intensity (*black line*) and anisotropy (*blue line*) of Alexa488- apoC-II in the presence of 10 mM DHPS during fibril formation. Data were acquired on a Florolog Tau-2 spectrophotometer using 55° polarization and an emission cutoff of 505 nm. The aggregation kinetics of the labelled and unlabelled apoC-II was similar (Figure adapted with permission from Ryan et al. (2008))

micellar lipids enhance early association events in the apoC-II aggregation pathway.

This hypothesis was further investigated using analytical ultracentrifugation. Sedimentation velocity experiments using Alexa-488-apoC-II enabled the analysis of apoC-II self-association at low protein concentrations, where significant fibril formation does not occur (Ryan et al. 2008). These experiments indicated that addition of sub-micellar phospholipids shifted the modal sedimentation coefficient of apoC-II from 1 S to approximately 3.1 S, equivalent to the formation of an apoC-II tetramer (Fig. 7.5). The size of this larger species was confirmed using sedimentation equilibrium experiments, which showed an approximate molecular weight of 40 kDa, equivalent to four labelled apoC-II molecules. Solutions of this stabilised tetrameric apoC-II were capable of seeding fibril formation, indicating that this oligomeric structure can act as a

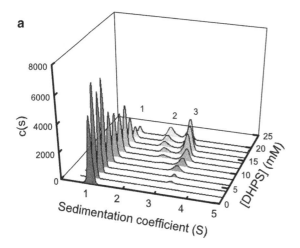
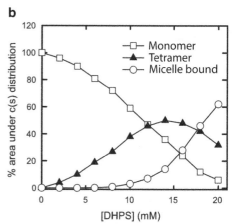

Fig. 7.5 Sedimentation analysis of apoC-II self-association in the presence of the short-chain phospholipid derivative, DHPS. Sedimentation velocity data were acquired for freshly refolded Alexa-488-apoC-II (50 μg/mL, 100 mM phosphate buffer) in the presence of increasing concentrations of DHPS (0–20 mM) using the fluorescence detection system analytical ultracentrifuge. The use of fluorescence provides sensitivity and specificity, allowing analysis of the low apoC-II concentration in the presence of the relatively large lipid concentration. (**a**) Sedimentation velocity data, analysed using the c(S) model, revealed the presence of three peaks that depend on the lipid concentration. The peak *circa* 1 S corresponds to monomeric apoC-II. Increasing submicellar concentrations of DHPS induce a 3.1 S peak corresponding to apoC-II tetramer (Peak 3). Micellar concentrations of DHPS induce a peak at around 2.4 S, consistent with a micelle-associated apoC-II molecule (Peak 2). (**b**) The integrated area under each peak in panel (**a**) is plotted as a function of DHPS concentration, revealing the DHPS concentration-dependence of the monomer (peak 1; *squares*), micelle associated apoC-II (peak 2; *circles*), and tetrameric apoC-II (peak 3; *filled triangles*) (Figure adapted with permission from Ryan et al. (2008))

nucleus for amyloid aggregates (Ryan et al. 2008).

7.5 Kinetics of ApoC-II Amyloid Formation in the Presence of Submicellar Lipids

The results of biophysical studies indicated that lipids affected nucleation of apoC-II amyloid fibrils. To quantify the precise effects of lipids, kinetic modelling of amyloid fibril formation was required. The relatively soluble nature of the apoC-II fibril allows the use of analytical ultracentrifugation (MacRaild et al. 2003) to analyse the time-dependent evolution of the size distribution of these macromolecular structures during aggregation (Binger et al. 2008). This analysis provides both an average aggregate size and the proportion of aggregate over time, which can be used as constraints to mathematically model the aggregation pathway (Binger et al. 2008).

This model showed that apoC-II fibril formation is a reversible process of nucleation, elongation, fibril breaking and rejoining, which resulted in a fairly consistent size distribution for apoC-II amyloid fibrils, and explained the formation of closed loops of apoC-II fibrils (Binger et al. 2008). Application of this method, in conjunction with stopped-flow analysis of the fluorescence changes, indicated that the nucleation phase was increased by several orders of magnitude in the presence of submicellar phospholipid derivatives, while the elongation phase was unaffected (Ryan et al. 2010). This was supported by the observations that the "free pool" of soluble apoC-II in equilibrium with fibrils in the absence and presence of submicellar phospholipids was the same. The size of this free pool is defined solely by the on- and off-rates of elongation.

Further confirmation of the differential effect on nucleation and elongation is provided by two observations. First, submicellar lipids have little effect on the aggregation of synthetic cysteine-

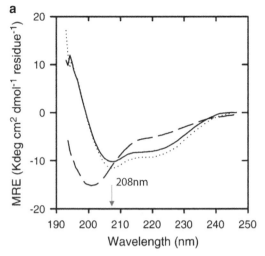

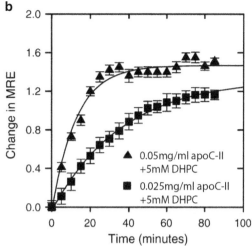

Fig. 7.6 Time-dependent structural changes of the lipid-induced apoC-II tetramer. (**a**) CD spectra of apoC-II (0.05 mg/mL in 10 mM Na phosphate buffer, *dashed line*), immediately after addition of 5 mM DHPC (0 h, *dotted line*), and after 1 h incubation in the presence of 5 mM DHPC (*continuous line*). The CD spectrum of apoC-II alone (*dashed line*) is provided for reference. At this apoC-II concentration, fibril formation is negligible over the time course of the experiment. The data are presented in units of mean residue ellipticity (*MRE*). (**b**) The change in MRE at 208 nm (indicated in panel **a** by *red arrow*) over time for 0.05 mg/mL apoC-II in the presence of 5 mM DHPC (*triangles*), and for 0.025 mg/mL apoC-II in the presence of 5 mM DHPC (*squares*) (Figure adapted with permission from Ryan et al. 2010)

crosslinked dimers of apoC-II-S61C, where the initial association events of the apoC-II fibril formation pathway are already complete. Second, submicellar lipids have no effect on apoC-II aggregation when added midway through the elongation phase. These results suggest that the physicochemical processes underlying the nucleation and the elongation of fibrils are different and can be affected by chemical compounds differentially. This conclusion has potentially important implications for the development of therapeutics designed to modulate fibril formation, but can also provide insight into the basis of in vivo fibril formation.

Another aspect of the nucleation elucidated in this analysis was that the protein tetramerisation in the presence of the submicellar lipid did not constitute the formation of a nucleus, and a second step involving a slow zero-order isomerisation of the tetramer was required to fit the data (Ryan et al. 2010). This isomerisation was confirmed using a combination of stopped-flow fluorescence and CD spectroscopy (Fig. 7.6) showing that this step was rate-limiting for amyloid formation in the presence of submicellar lipids. The nature of the slow isomerisation step is unclear; it could involve a structural rearrangement of the tetramer, a loss of the lipid molecules, or some combination thereof (Ryan et al. 2010). CD time courses indicate that the tetramer structure is subtly altered during this isomerisation step, suggesting that a structural rearrangement does indeed occur.

Notably, this rate-limiting isomerisation step is probably not unique to fibril formation by apoC-II. For example, elegant studies by Singh et al., Chap. 4 in this volume show that the slow rate-limiting step in amyloid fibril formation by a small lipid-binding protein, amylin, also involves re-arrangement and self-association of low-order protein oligomers.

7.6 Lipid Dynamics During ApoC-II Aggregation

To provide information on the dynamics of lipid molecules during apoC-II fibril formation, further analysis was conducted using a fluorescently labelled single-chain phospholipid, 1-dodecyl-[(7-nitro-2,1,3-benzoxadiazol-4-yl)amino]-2-hydroxy-sn-glycero-3-phosphocholine (NBD-lyso-12-PC) (Ryan et al. 2011a). The unlabelled form of this lipid has a CMC near 0.2 mM, and both the labelled and unlabelled lipids at submicellar concentrations induce apoC-II tetramerisation and enhance fibril formation (Griffin et al. 2008; Ryan et al. 2008, 2011a). FRET experiments utilising this labelled lipid and freshly refolded Alexa-594-apoC-II indicated that the lipid and protein were in close proximity, suggesting the formation of a reasonably stable complex (Ryan et al. 2011a). Stopped-flow analysis of this FRET signal indicated rapid pseudo-first-order association with a rate constant of 156,000 s^{-1}.

Fluorescence-detected analytical ultracentrifugation indicated that the lipid associated with both monomeric and tetrameric apoC-II in a concentration-dependent manner. Analysis of the proportion of tetramer, as measured by the weight average sedimentation coefficient of the apoC-II in the presence of NBD-lyso-12-PC, provided the tetramerisation constant of 3.5×10^{-3} μM^{-3}, consistent with that obtained from the kinetic analysis of apoC-II fibril formation in the presence of DHPC. Analysis of the proportion of NBD-lyso-12-PC that co-sedimented with apoC-II provided a stoichiometry of five NBD-lyso-12-PC molecules bound per one apoC-II molecule, and the concentration dependence of this association indicated a $K_d \cong 10$ μM.

The time dependence of the protein-lipid interaction during the fibril formation showed that the FRET signal was lost over 12 h, suggesting that as apoC-II aggregated the lipid affinity decreased. Analysis of the final fibrillar product indicated a complete lack of FRET, suggesting that mature, fibrillar apoC-II did not bind submicellar lipids. This analysis was supported by analytical ultracentrifugation, in which no co-sedimentation between the NBD-lyso-12-PC and mature apoC-II fibrils was observed by fluorescence, and in pelleting assays, where no NBD-lyso-12-PC was found in the pellet fraction of fibrils either added to, or formed in the presence of NBD-lyso-12-PC. Together, these results indicate a catalytic role for submicellar lipids in apolipoprotein misfolding and aggregation, and suggest that the presence of only a small amount of lipid could affect initiation of amyloid fibril formation in vivo (Ryan et al. 2011a).

7.7 The Interaction of ApoC-II Peptide Fragments with Lipids

Analysis of lipid binding by the defined peptide regions of apoC-II has elucidated many biophysical features of apolipoprotein – lipid surface binding. The peptide implicated in forming the primary lipid-binding region, apoC-II$_{19-39}$, was found to adopt a helical structure and bind to lipid surfaces with $K_d = 5$ μM, and to form dimers at very high protein:lipid ratios (MacPhee et al. 1999). Such self-association has been proposed to be important for stabilising apolipoprotein structures and may influence the function of lipoprotein particles (e. g. apoA-I and apoA-IV dimers).

ApoC-II peptides have also been used to elucidate misfolding events and molecular interactions involved in apoC-II aggregation (Wilson et al. 2007; Legge et al. 2007; Griffin et al. 2012). These peptides are derived from the amyloidogenic regions of apoC-II and undergo many of the processes involved in the misfolding and amyloid fibril formation by the full-length protein. However, these peptides do show some distinct differences from the full-length protein, indicating that the amyloid formation by apoC-II involves several regions acting in concert.

Studies of the effect of phospholipids have primarily focused on one of these regions, apoC-II$_{60-70}$, which contains the sequence MSTYTGIFTDQ (Hung et al. 2008, 2009) that is largely hydrophobic and is predicted to be particularly amyloidogenic (Chap. 8 by Das and Gursky in this volume). In contrast to the full-length protein, amyloid fibril formation by this peptide is inhibited by the presence of submicellar lipids, including DHPC (Hung et al. 2009). Interestingly, the peptide responds to the presence of submicellar lipids by forming a range of higher-order oligomers, as evidenced by sedimentation equilibrium experiments, which do not show a significant induction of secondary structure observed by

CD spectroscopy (Hung et al. 2009). Molecular dynamics simulations indicate that the presence of solvated lipid molecules enhanced inter-peptide association, particularly through improved hydrophobic contacts, which act to kinetically trap this apoC-II peptide in oligomeric structures ranging from dimer to much larger species (Hung et al. 2009). This result, while appearing to contradict the results for the full-length protein, is actually complementary. The observation that lipids can drive the peptide self-association correlates to the initial induction of tetrameric species of the full-length protein; it is perhaps other regions of the apoC-II that drive the second phase of nucleation, which ultimately results in the ejection of the lipid and the formation of cross-β structure in amyloid fibrils.

7.8 The Effects of Micellar Lipids and Lipid Surfaces

As discussed above, apoC-II in the presence of large excesses of micellar lipids adopts an α-helical conformation and does not aggregate. One explanation for this effect, in addition to stabilisation of a native-like structure, is that the micelles effectively separate apoC-II molecules from each other. For example, as each DHPC micelle contains approximately 40 lipid molecules (Atcliffe et al. 2001), a concentration of 10 mM DHPC is equivalent to a micelle concentration of approximately 0.25 mM, resulting in a molar ratio of around 7.5 micelles per protein molecule at standard amyloid-forming concentrations of apoC-II. Thus, on average, a micelle is expected to contain fewer than one apoC-II molecule.

Interestingly, the primary lipid-binding region of apoC-II at very high protein:lipid ratios appears to undergo self-association (MacPhee et al. 1999). Under these conditions apoC-II displays some interesting aggregation phenomena (Griffin et al. 2008). The use of low concentrations of micellar lipids, facilitated by using longer-chain phospholipids with lower CMCs, coupled with higher concentrations of apoC-II, results in an initially inhibited aggregation pro-

cess over a period of 200 h. Subsequent sample analysis showed a second phase of aggregation, where the maximal thioflavin T (ThT) fluorescence was significantly greater than in the first phase. This behaviour could be replicated with small unilamellar vesicles comprised of more physiological lipids such as 1,2-dimyristoyl-sn-glycero-3-phosphocholine (DMPC), indicating that it is not specific to a particular lipid surface configuration.

Transmission electron microscopy (TEM) analysis of the apoC-II fibrils formed during the first phase of aggregation showed the common twisted-ribbon morphology. However, TEM analysis after the second phase revealed a previously unidentified apoC-II fibril morphology (Fig. 7.7). These fibrils displayed a straight, cable-like structure 13–14 nm wide, with helical periodicity of 95–100 nm. CD spectroscopy showed that the two fibril types had distinct secondary structure, strongly suggesting a divergent amyloid formation pathway. Seeding experiments supported this observation and indicated that the cable-like fibrils can be seeded and propagated in a lipid-free environment. This shows that phospholipids not only play an important role in the initiation and enhancement of amyloid fibril formation, but also in determining the ultimate fibril structure. This level of polymorphism may be of critical importance in understanding the initiation of amyloid fibril formation in lipid-rich environments in vivo and for understanding the molecular basis of amyloid diseases.

7.9 The Effects of Oxidized Cholesterol

Several studies have demonstrated that amyloid deposits are present in up to 50–60 % of aortic atherosclerotic lesions (Westermark et al. 1995; Mucchiano et al. 2001a). These amyloid deposits contain various proteins, including apoC-II. ApoC-II has also been shown to colocalize in these plaques with a range of amyloid markers, providing indirect evidence that it is part of the amyloid component of atheroma. Atherosclerotic

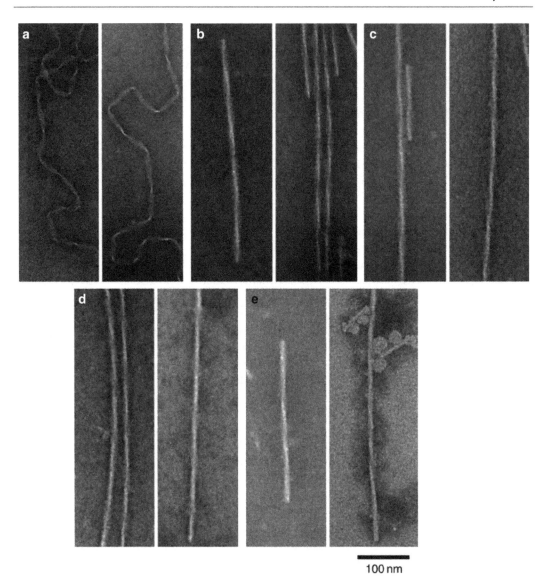

100 nm

Fig. 7.7 Lipids induce a distinct, straight apoC-II fibrillar morphology. Transmission electron micrographs showing the fine detail of (**a**) mature ribbon-type fibrils formed by 1.0 mg/ml apoC-II alone, and straight fibrils formed by 1.0 mg/ml apoC-II in the presence of (**b**) 500 μM D8PC, (**c**) 500 μM D9PC, (**d**) 500 μM Lyso-MPC, and (**e**) 500 μM DMPC. Circular objects adjacent to fibrils in the presence of DMPC are probably small unilamelar lipid vesicles that may interact with the apolipoprotein fibers as described by Gorbenko et al. in Chap. 6 of this volume (Figure adapted with permission from Griffin et al. 2008)

lesions also contain a large component of lipids, including oxidized cholesterol, which has been shown to enhance amyloid formation by α-synuclein and amyloid-β peptide (Bieschke et al. 2005; Zhang et al. 2004; Bosco et al. 2006).

ApoC-II fibril formation is significantly enhanced in the presence of the oxidized cholesterol derivative, 3β-hydroxy-5-oxo-5,6-secocholestan-6-al (Stewart et al. 2007b). Further analysis using HPLC and MALDI-TOF mass spectrometry indicated that this oxidized cholesterol derivative formed a Schiff-base covalent bond with apoC-II lysines 19, 30, 39, 48, 55, and 76 (Stewart et al. 2007b). Isolation of the various oxidized cholesterol adducts and analysis of fibril formation by these modified apoC-II molecules showed that in all cases fibril formation was enhanced, and a variety of alternate fibrillar morphologies were produced (Stewart et al. 2007b).

These observations, in conjunction with the studies of amyloid-β and α-synuclein, suggest that oxidized cholesterol may trigger or enhance the amyloid deposition in various tissues, including atheroma (Stewart et al. 2007b). Another possible implication is a feedback loop with the inflammatory response, which can generate oxidized cholesterol that may worsen amyloid deposition by various proteins. These results provide further evidence that altered lipid metabolism can affect protein folding, and that elucidating the ways in which the diverse lipid molecules are involved in normal and aberrant protein folding may be critical to understanding the basis of amyloid diseases.

7.10 The Effects of Hydrophobic "Lipid-Like" and Detergent Molecules

The effects of lipids and their derivatives on amyloid fibril formation can be altered by the self-association state of the lipid and/or its chemical nature, which varies widely for different lipids. To explore the structural specificity of these effects, a library of 96 diverse amphipathic lipids and detergents was screened at submicellar concentrations for their ability to modulate the rate of amyloid fibril formation by apoC-II (Ryan et al. 2011b) (Fig. 7.8). A combination of ThT fluorescence and light scattering was used to monitor fibril formation by apoC-II (1 mg/mL) in the presence of the 96 compounds at a concentration of one half their respective CMC, ensuring that the majority of the compounds were at submicellar concentrations. Some compounds in the screen, termed non-detergent sulfobetaines, did not form micelles, and others were highly hydrophobic and their CMCs could not be determined. These compounds were screened at concentrations twofold greater than that of apoC-II.

The initial screen showed that the library could be split into three groups of compounds that activated fibril formation, inhibited fibril formation, or had no significant effect. Surprisingly, most lipids fell into the activator or inhibitor groups, with relatively few having little effect on fibril formation. In general, the results suggested that cholesterol-like molecules inhibit apoC-II fibril formation, while compounds that closely resemble fatty acids, phospholipids or sphingolipids were more likely to activate apoC-II fibril formation.

Interestingly, apparently minor structural differences in these compounds caused very significant shifts in their effects on apoC-II amyloid fibril formation. For example, at submicellar concentrations, 1-dodecyl,2-lyso-sn-glycero-phosphocholine activates fibril formation while dodecylphosphocholine inhibits it. The structural difference between these two compounds is a glycerol linker between the phosphocholine head group and the single 12-carbon acyl chain. While a comprehensive analysis of the structural basis for inhibition is challenging, a general defining feature of activators and inhibitors was their average hydrophobicity, as measured by octanol/water partitioning coefficients at pH 7.0 (LogP) calculated from molecular structure. Inhibitors typically had higher overall hydrophobicity (logP $\cong$ +2.2) than activators (logP $\cong$ −0.81). However, examples of both highly hydrophobic activators and hydrophilic inhibitors were also identified (Fig. 7.8).

Structural analysis in the presence of a subset of these lipid-like detergents indicated that, in general, inhibitors induced α-helical structure while activators induced β-sheet structure in apoC-II. Analytical ultracentrifugation analysis indicated induction of tetramers by the activating compounds, further supporting the idea that lipid-induced activation of the amyloidogenic pathway in apoC-II involves tetramer formation. In contrast, the inhibiting compounds induced dimeric, predominantly α-helical apoC-II species, suggesting that the mechanism of inhibition by these molecules involves stabilisation of

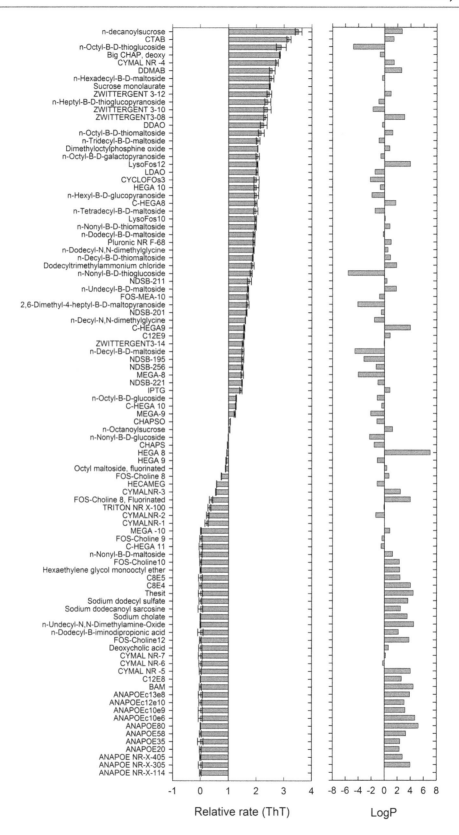

non-amyloidogenic oligomers that may have native-like structure (Ryan et al. 2011b).

Addition of compounds during the nucleation or elongation phases showed differential modulatory effects of some compounds at different stages of fibrillation. These stage-specific effects, whereby molecules were identified that activate nucleation but inhibit elongation of fibrils, and vice versa, emphasise that fibril formation is a multi-step process with different biophysical and structural processes occurring at each stage. This outcome may be important when considering therapeutic intervention in amyloid formation, as screening processes may bias the results towards modulators of nucleation, which may provide little benefit or even have detrimental effects in patients with pre-existing amyloid deposits. Interestingly, a parallel study using the same compound library with amyloid-β peptides displayed a quite different pattern of activators and inhibitors. This indicates that the pattern of activation and inhibition observed with apoC-II is specific to this protein, and should not be generalised to other amyloid-forming proteins (Ryan et al. 2012).

The results of this screen primarily indicate that the response to lipid-like molecules by apoC-II depends on subtle differences in chemical structure of the molecule, and that the vast majority of lipid-like molecules are likely to have significant modulatory effects on apoC-II folding, misfolding and aggregation. This suggests that a balance between the effects of activating and inhibiting molecules governs the proper maintenance of apoC-II folding in vivo, and that subtle shifts in the level or structure of a critical lipid might result in aggregation.

7.11 Conclusions

This chapter presents a comprehensive review of the lipid effects on various stages of apoC-II misfolding and aggregation. The picture that emerges is a complex range of effects that are dictated by the ratios of lipid to protein, the association state of the lipids, and their chemical nature (Fig. 7.9). Lipids are a key component of the pathological site of apoC-II accumulation in atheroma; at the same time, lipids clearly are essential for maintaining lipoprotein stability, and thereby stabilizing the native apolipoprotein structure against misfolding. The results strongly suggest that modified lipids may play a catalytic role in initiating amyloid fibril formation by apolipoproteins, and could play a significant pathological role in the development of amyloid plaques in tissues. Lipids also play a significant role in determining the fibrillar morphology of apoC-II amyloid and, therefore, in the stability and dynamics of the aggregates. Clearly, understanding the role of lipids in the folding and misfolding of proteins is critical to understanding protein folding and aggregation in vivo, both in healthy and disease states.

Acknowledgments T.M.R. is the recipient of the Alzheimer's Australia Dementia Research Foundation Fellowship. M.D.W.G is the recipient of the C.R. Roper Fellowship and an Australian Research Council Post Doctoral Fellowship (project number DP110103528). This work was supported by grants from the Australian Research Council and the National Health and Medical Research Council.

Fig. 7.8 The effects of lipid and lipid-like molecules on apoC-II fibril formation. (**a**) The rate of fibril formation (1.0 mg/mL apoC-II in 100 mM phosphate at pH 7.4, measured using a continuous ThT assay) in the presence of various lipid-like compounds relative to that of apoC-II alone (rate = 1). (**b**) The hydrophobicity scale of the various molecules is expressed as calculated octanol/water partitioning coefficients (Log P), where Log P >0 indicates preferential partitioning into octanol (increased hydrophobicity). Log P values were acquired using MarvinSketch (http://www.chemaxon.com/products/marvin/marvinsketch/) (Figure adapted with permission from Ryan et al. 2011b)

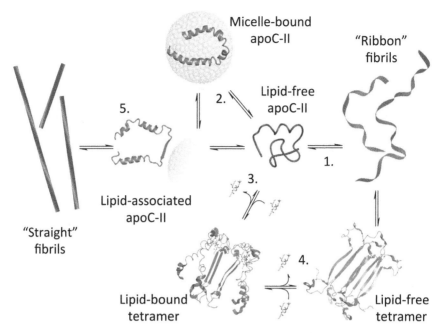

Fig. 7.9 The effects of lipids on apoC-II misfolding and aggregation pathways. In the absence of lipid, apoC-II forms a molten globular structure that can self-associate to form ribbon-type fibrils through a nucleation-dependent process (*1*). At low protein:micelle ratios, apoC-II adopts a stable native-like α-helical structure (*2*). In the presence of submicellar lipids, apoC-II rapidly forms a tetramer that binds individual lipid molecules (*3*). This tetramer undergoes a molecular rearrangement, gaining additional struc-ture and ejecting the bound lipid molecules to form a lipid-free tetramer that rapidly forms ribbon-type fibrils (*4*). At high protein:micelle ratios, apoC-II adopts a non-native conformation that slowly forms fibrils with an alter-nate, straight cable-like morphology (*5*). Structures depicted are schematic models for illustrative purposes, and are not intended to specifically represent the structure of apoC-II in each environment (Figure adapted with per-mission from Griffin et al. 2008)

References

Atcliffe BW, MacRaild CA, Gooley PR, Howlett GJ (2001) The interaction of human apolipoprotein C-I with sub-micellar phospholipid. Eur J Biochem/FEBS 268(10):2838–2846

Bergstrom J, Murphy C, Eulitz M, Weiss DT, Westermark GT, Solomon A, Westermark P (2001) Codeposition of apolipoprotein A-IV and transthyretin in senile systemic (ATTR) amyloidosis. Biochem Biophys Res Commun 285(4):903–908

Bergstrom J, Murphy CL, Weiss DT, Solomon A, Sletten K, Hellman U, Westermark P (2004) Two different types of amyloid deposits–apolipoprotein A-IV and transthyretin–in a patient with systemic amyloidosis. Lab Invest 84(8):981–988

Bieschke J, Zhang Q, Powers ET, Lerner RA, Kelly JW (2005) Oxidative metabolites accelerate Alzheimer's amyloidogenesis by a two-step mechanism, eliminating the requirement for nucleation. Biochemistry 44(13):4977–4983

Binger KJ, Pham CL, Wilson LM, Bailey MF, Lawrence LJ, Schuck P, Howlett GJ (2008) Apolipoprotein C-II amyloid fibrils assemble via a reversible pathway that includes fibril breaking and rejoining. J Mol Biol 376(4):1116–1129

Bosco DA, Fowler DM, Zhang Q, Nieva J, Powers ET, Wentworth P Jr, Lerner RA, Kelly JW (2006) Elevated levels of oxidized cholesterol metabolites in Lewy body disease brains accelerate alpha-synuclein fibrilization. Nat Chem Biol 2(5):249–253

Coriu D, Dispenzieri A, Stevens FJ, Murphy CL, Wang S, Weiss DT, Solomon A (2003) Hepatic amyloidosis resulting from deposition of the apolipoprotein A-I variant Leu75Pro. Amyloid 10(4):215–223

Griffin MD, Mok ML, Wilson LM, Pham CL, Waddington LJ, Perugini MA, Howlett GJ (2008) Phospholipid interaction induces molecular-level polymorphism in apolipoprotein C-II amyloid fibrils via alternative assembly pathways. J Mol Biol 375(1):240–256

Griffin MD, Yeung L, Hung A, Todorova N, Mok YF, Karas JA, Gooley PR, Yarovsky I, Howlett GJ (2012) A cyclic peptide inhibitor of apoC-II peptide fibril formation: mechanistic insight from NMR and molecular dynamics analysis. J Mol Biol 416(5):642–655

Gursky O, Atkinson D (1998) Thermodynamic analysis of human plasma apolipoprotein C-1: high-temperature unfolding and low-temperature oligomer dissociation. Biochemistry 37(5):1283–1291

Hanson CL, Ilag LL, Malo J, Hatters DM, Howlett GJ, Robinson CV (2003) Phospholipid complexation and association with apolipoprotein C-II: insights from mass spectrometry. Biophys J 85(6):3802–3812

Hatters DM, Howlett GJ (2002) The structural basis for amyloid formation by plasma apolipoproteins: a review. Eur Biophys J 31(1):2–8

Hatters DM, MacPhee CE, Lawrence LJ, Sawyer WH, Howlett GJ (2000) Human apolipoprotein C-II forms twisted amyloid ribbons and closed loops. Biochemistry 39(28):8276–8283

Hatters DM, Lawrence LJ, Howlett GJ (2001) Sub-micellar phospholipid accelerates amyloid formation by apolipoprotein C-II. FEBS Lett 494(3):220–224

Hatters DM, Minton AP, Howlett GJ (2002a) Macromolecular crowding accelerates amyloid formation by human apolipoprotein C-II. J Biol Chem 277(10):7824–7830

Hatters DM, Wilson MR, Easterbrook-Smith SB, Howlett GJ (2002b) Suppression of apolipoprotein C-II amyloid formation by the extracellular chaperone, clusterin. Eur J Biochem/FEBS 269(11):2789–2794

Hung A, Griffin MD, Howlett GJ, Yarovsky I (2008) Effects of oxidation, pH and lipids on amyloidogenic peptide structure: implications for fibril formation? Eur Biophys J 38(1):99–110

Hung A, Griffin MD, Howlett GJ, Yarovsky I (2009) Lipids enhance apolipoprotein C-II-derived amyloidogenic peptide oligomerization but inhibit fibril formation. J Phys Chem B 113(28):9447–9453

Legge FS, Treutlein H, Howlett GJ, Yarovsky I (2007) Molecular dynamics simulations of a fibrillogenic peptide derived from apolipoprotein C-II. Biophys Chem 130(3):102–113

MacPhee CE, Howlett GJ, Sawyer WH (1999) Mass spectrometry to characterize the binding of a peptide to a lipid surface. Anal Biochem 275(1):22–29

MacRaild CA, Hatters DM, Howlett GJ, Gooley PR (2001) NMR structure of human apolipoprotein C-II in the presence of sodium dodecyl sulfate. Biochemistry 40(18):5414–5421

MacRaild CA, Hatters DM, Lawrence LJ, Howlett GJ (2003) Sedimentation velocity analysis of flexible macromolecules: self-association and tangling of amyloid fibrils. Biophys J 84(4):2562–2569

MacRaild CA, Howlett GJ, Gooley PR (2004) The structure and interactions of human apolipoprotein C-II in dodecyl phosphocholine. Biochemistry 43(25):8084–8093

Medeiros LA, Khan T, El Khoury JB, Pham CL, Hatters DM, Howlett GJ, Lopez R, O'Brien KD, Moore KJ (2004) Fibrillar amyloid protein present in atheroma activates CD36 signal transduction. J Biol Chem 279(11):10643–10648

Mucchiano GI, Haggqvist B, Sletten K, Westermark P (2001a) Apolipoprotein A-1-derived amyloid in atherosclerotic plaques of the human aorta. J Pathol 193(2):270–275

Mucchiano GI, Jonasson L, Haggqvist B, Einarsson E, Westermark P (2001b) Apolipoprotein A-I-derived amyloid in atherosclerosis. Its association with plasma levels of apolipoprotein A-I and cholesterol. Am J Clin Pathol 115(2):298–303

Obici L, Bellotti V, Mangione P, Stoppini M, Arbustini E, Verga L, Zorzoli I, Anesi E, Zanotti G, Campana C, Vigano M, Merlini G (1999) The new apolipoprotein A-I variant leu(174) → Ser causes hereditary cardiac amyloidosis, and the amyloid fibrils are constituted by the 93-residue N-terminal polypeptide. Am J Pathol 155(3):695–702

Obici L, Franceschini G, Calabresi L, Giorgetti S, Stoppini M, Merlini G, Bellotti V (2006) Structure, function and amyloidogenic propensity of apolipoprotein A-I. Amyloid 13(4):191–205

Ryan TM, Howlett GJ, Bailey MF (2008) Fluorescence detection of a lipid-induced tetrameric intermediate in amyloid fibril formation by apolipoprotein C-II. J Biol Chem 283(50):35118–35128

Ryan TM, Teoh CL, Griffin MD, Bailey MF, Schuck P, Howlett GJ (2010) Phospholipids enhance nucleation but not elongation of apolipoprotein C-II amyloid fibrils. J Mol Biol 399(5):731–740

Ryan TM, Griffin MD, Bailey MF, Schuck P, Howlett GJ (2011a) NBD-labeled phospholipid accelerates apolipoprotein C-II amyloid fibril formation but is not incorporated into mature fibrils. Biochemistry 50(44):9579–9586

Ryan TM, Griffin MD, Teoh CL, Ooi J, Howlett GJ (2011b) High-affinity amphipathic modulators of amyloid fibril nucleation and elongation. J Mol Biol 406(3):416–429

Ryan TM, Friedhuber A, Lind M, Howlett GJ, Masters C, Roberts BR (2012) Small amphipathic molecules modulate secondary structure and amyloid fibril-forming kinetics of Alzheimer disease peptide Abeta(1–42). J Biol Chem 287(20):16947–16954

Segrest JP, De Loof H, Dohlman JG, Brouillette CG, Anantharamaiah GM (1990) Amphipathic helix motif: classes and properties. Proteins 8(2):103–117

Segrest JP, Garber DW, Brouillette CG, Harvey SC, Anantharamaiah GM (1994) The amphipathic alpha helix: a multifunctional structural motif in plasma apolipoproteins. Adv Protein Chem 45:303–369

Stewart CR, Haw A 3rd, Lopez R, McDonald TO, Callaghan JM, McConville MJ, Moore KJ, Howlett GJ, O'Brien KD (2007a) Serum amyloid P colocalizes with apolipoproteins in human atheroma: functional implications. J Lipid Res 48(10):2162–2171

Stewart CR, Wilson LM, Zhang Q, Pham CL, Waddington LJ, Staples MK, Stapleton D, Kelly JW, Howlett GJ (2007b) Oxidized cholesterol metabolites found in human atherosclerotic lesions promote apolipoprotein C-II amyloid fibril formation. Biochemistry 46(18):5552–5561

Teoh CL, Griffin MD, Howlett GJ (2011a) Apolipoproteins and amyloid fibril formation in atherosclerosis. Protein Cell 2(2):116–127

Teoh CL, Pham CL, Todorova N, Hung A, Lincoln CN, Lees E, Lam YH, Binger KJ, Thomson NH, Radford SE, Smith TA, Muller SA, Engel A, Griffin MD, Yarovsky I, Gooley PR, Howlett GJ (2011b) A structural model for apolipoprotein C-II amyloid fibrils: experimental characterization and molecular dynamics simulations. J Mol Biol 405(5):1246–1266

Wang G (2008) NMR of membrane-associated peptides and proteins. Curr Protein Pept Sci 9(1):50–69

Westermark P, Mucchiano G, Marthin T, Johnson KH, Sletten K (1995) Apolipoprotein A1-derived amyloid in human aortic atherosclerotic plaques. Am J Pathol 147(5):1186–1192

Wilson LM, Mok YF, Binger KJ, Griffin MD, Mertens HD, Lin F, Wade JD, Gooley PR, Howlett GJ (2007) A structural core within apolipoprotein C-II amyloid fibrils identified using hydrogen exchange and proteolysis. J Mol Biol 366(5):1639–1651

Yang S, Griffin MD, Binger KJ, Schuck P, Howlett GJ (2012) An equilibrium model for linear and closed-loop amyloid fibril formation. J Mol Biol 421(2–3):364–377

Yazaki M, Liepnieks JJ, Barats MS, Cohen AH, Benson MD (2003) Hereditary systemic amyloidosis associated with a new apolipoprotein AII stop codon mutation Stop78Arg. Kidney Int 64(1):11–16

Zdunek J, Martinez GV, Schleucher J, Lycksell PO, Yin Y, Nilsson S, Shen Y, Olivecrona G, Wijmenga S (2003) Global structure and dynamics of human apolipoprotein CII in complex with micelles: evidence for increased mobility of the helix involved in the activation of lipoprotein lipase. Biochemistry 42(7):1872–1889

Zhang Q, Powers ET, Nieva J, Huff ME, Dendle MA, Bieschke J, Glabe CG, Eschenmoser A, Wentworth P Jr, Lerner RA, Kelly JW (2004) Metabolite-initiated protein misfolding may trigger Alzheimer's disease. Proc Natl Acad Sci U S A 101(14):4752–4757

Amyloid-Forming Properties of Human Apolipoproteins: Sequence Analyses and Structural Insights

Madhurima Das and Olga Gursky

Abstract

Apolipoproteins are protein constituents of lipoproteins that transport cholesterol and fat in circulation and are central to cardiovascular health and disease. Soluble apolipoproteins can transiently dissociate from the lipoprotein surface in a labile free form that can misfold, potentially leading to amyloid disease. Misfolding of apoA-I, apoA-II, and serum amyloid A (SAA) causes systemic amyloidoses, apoE4 is a critical risk factor in Alzheimer's disease, and apolipoprotein misfolding is also implicated in cardiovascular disease. To explain why apolipoproteins are overrepresented in amyloidoses, it was proposed that the amphipathic α-helices, which form the lipid surface-binding motif in this protein family, have high amyloid-forming propensity. Here, we use 12 sequence-based bioinformatics approaches to assess amyloid-forming potential of human apolipoproteins and to identify segments that are likely to initiate β-aggregation. Mapping such segments on the available atomic structures of apolipoproteins helps explain why some of them readily form amyloid while others do not. Our analysis shows that nearly all amyloidogenic segments: (i) are largely hydrophobic, (ii) are located in the lipid-binding amphipathic α-helices in the native structures of soluble apolipoproteins, (iii) are predicted in both native α-helices and β-sheets in the insoluble apoB, and (iv) are predicted to form parallel in-register β-sheet in amyloid. Most of these predictions have been verified experimentally for apoC-II, apoA-I, apoA-II and SAA. Surprisingly, the rank order of the amino acid sequence propensity to form amyloid (apoB > apoA-II > apoC-II ≥ apoA-I, apoC-III, SAA, apoC-I > apoA-IV, apoA-V, apoE) does not correlate with the proteins' involvement in amyloidosis. Rather, it correlates directly with the

M. Das • O. Gursky (✉)
Department of Physiology and Biophysics, Boston
University School of Medicine,
W321, Boston, MA 02118, USA
e-mail: mdas@bu.edu; gursky@bu.edu

© Springer International Publishing Switzerland 2015
O. Gursky (ed.), *Lipids in Protein Misfolding*, Advances in Experimental
Medicine and Biology 855, DOI 10.1007/978-3-319-17344-3_8

strength of the protein-lipid association, which increases with increasing protein hydrophobicity. Therefore, the lipid surface-binding function and the amyloid-forming propensity are both rooted in apolipoproteins' hydrophobicity, suggesting that functional constraints make it difficult to completely eliminate pathogenic apolipoprotein misfolding. We propose that apolipoproteins have evolved protective mechanisms against misfolding, such as the sequestration of the amyloidogenic segments via the native protein-lipid and protein-protein interactions involving amphipathic α-helices and, in case of apoB, β-sheets.

Keywords

Lipid transport • Systemic amyloidosis • Atherosclerosis and Alzheimer's disease • Atomic protein structure • Sequence-based prediction algorithms

Abbreviations

AA	Amyloid A
AApoAI	Familial apoA-I amyloidosis
Apo	Apolipoprotein
Aβ	Amyloid beta peptide
DPC	Dodecylphosphocholine
EPR	Electron paramagnetic resonance
HDL	High-density lipoprotein
LDL	Low-density lipoprotein
NMR	Nuclear magnetic resonance.
SAA	Serum amyloid A
SDS	Sodium dodecyl sulfate
TG	Triacylglycerol
VLDL	Very low-density lipoprotein

8.1 Lipoproteins and Apolipoproteins in Lipid Transport and Metabolism

Lipids in plasma, lymph and cerebrospinal fluid are transported via lipoproteins that are water-soluble heterogeneous nanoparticles comprised of specific proteins, termed apolipoproteins (apos), and hundreds or even thousands of lipid molecules. Most plasma lipoproteins are secreted by the liver or the intestine, while lipoproteins in the central nervous system are secreted largely by the glial cells (Ladu et al. 2000; Huang 2010). Mature lipoproteins contain an apolar lipid core comprised mainly of cholesterol esters and triacylglycerols (TG, or fat) and an amphipathic surface containing a monolayer of polar lipids, mainly phospholipids and unesterified cholesterol, and apolipoproteins (Fig. 8.1). Apolipoproteins form a flexible scaffold on the lipoprotein surface, which is essential for the lipoprotein structure and function. Apolipoproteins confer curvature to the phospholipid monolayer on the particle surface and thereby help encapsulate apolar lipids in its core. Such encapsulation is necessary to transport water-insoluble lipids in the aqueous environment in plasma, lymph and cerebrospinal fluid. In addition, apolipoproteins direct lipoprotein metabolism by interacting with lipophilic enzymes, lipid transfer proteins, lipid transporters, and lipoprotein receptors (Mahley et al. 1984; Goldstein and Brown 1992; Rached et al. 2014; Phillips 2014). Lipid metabolism is an extremely complex process whose defects can manifest themselves as serious diseases, most notably cardiovascular disease that is the leading cause of mortality worldwide (Castelli et al. 1992; Toth et al. 2013). Lipoproteins and lipoprotein receptors also play major roles in normal brain functions and in neurodegenerative diseases (Lane-Donovan et al. 2014).

Plasma lipoproteins form distinct classes differing in the particle size, density, biochemical composition, and function (Fig. 8.1a). Since proteins, which are denser than lipids, are located on the particle surface, the particle density decreases

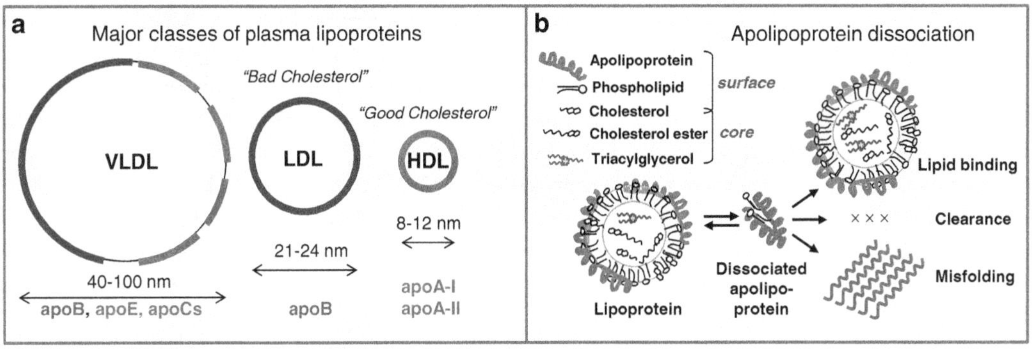

Fig. 8.1 Cartoon representation of lipoproteins and their constituents. (**a**) Major classes of plasma lipoproteins, including very-low, low- and high-density lipoproteins (VLDL, LDL and HDL), are shown in cross-section. Particle diameters are indicated. Intermediate-density lipoproteins (IDL, d=25–35 nm) and chylomicrons (d>100 nm) are not shown. Mature lipoproteins contain a core of apolar lipids (mainly cholesterol esters and triacylglycerols, in *yellow*) surrounded by an amphipathic surface comprised of polar lipids (mainly phospholipids and cholesterol, in *gray*) and apolipoproteins. Exchangeable (water-soluble) apolipoproteins are in *red*, non-exchangeable (water-insoluble) apoB is in *blue*. (**b**) Exchangeable apolipoproteins can reversibly dissociate from the lipoprotein surface. The dissociated free apolipoprotein is structurally labile and can (i) rapidly bind to lipoprotein surface, i.e. associate with another lipoprotein or get recruited to plasma membrane, (ii) get cleared, or (ii) misfold and form amyloid

with increasing size, from high-density lipoproteins (HDL, diameter 8–12 nm) to low-density lipoproteins (LDL, 20–24 nm), intermediate-density lipoproteins (IDL, 25–35 nm), very low-density lipoproteins (VLDL, 40–100 nm), and chylomicrons (>100 nm). Each lipoprotein class has distinct functions (Mahley et al. 1984). Plasma HDL remove excess cholesterol from peripheral cells and deliver it to the liver for excretion via bile or to steroidogenic organs for hormone synthesis. Generally, plasma levels of HDL cholesterol (a. k. a. "good cholesterol") and the major HDL protein, apoA-I, correlate inversely with the risk of developing atherosclerosis (Gordon et al. 1989; Toth et al. 2013), although this relationship is more complex than previously thought. LDL are the major plasma carriers of cholesterol in the form of cholesterol esters, which deliver it to peripheral tissues. Plasma levels of LDL cholesterol ("bad cholesterol") and the major LDL protein, apoB, are the strongest causative risk factors in cardiovascular disease (Castelli et al. 1992). IDL are transient metabolic products of VLDL. VLDL, which are the major plasma carriers of TG, are metabolic precursors of LDL and a risk factor for metabolic syndrome and cardiovascular disease (Krauss 1998). Chylomicrons, which are generated in the gut to transport dietary lipids, contain one copy of a truncated form of apoB, termed B48 (N-terminal 48 %), and various exchangeable proteins.

Each class of plasma lipoprotein has distinct protein composition (Fig. 8.1a). HDL particle contains several copies of exchangeable (water-soluble) apolipoproteins, mainly apoA-I (28 kDa) and apoA-II (9 kDa) that constitute approximately 70 % and 20 % of the total HDL protein mass, respectively, along with minor proteins (apoE, apoCs, etc.) LDL particle contains one copy of the non-exchangeable (water-insoluble) apoB (550 kDa) that constitutes ~95 % of the total LDL protein mass. VLDL particle contain one copy of apoB and multiple copies of exchangeable proteins, mainly apoE (33 kDa) and apoCs (6–9 kDa). Chylomicrons contain one copy of a non-exchangeable protein apoB48, as well as exchangeable proteins (apoE, apoCs, etc.) Other lipophilic proteins including apolipoproteins (apoA-IV, apoA-V, apoD, apoF, apoH, apoJ, etc.) are also found as minor components on lipoproteins. Lipoprotein composition can significantly change during inflammation when most of apoA-I is displaced from plasma HDL by an acute-phase protein, serum amyloid A (SAA) (van der Westhuyzen et al. 2007; Eklund et al. 2012). In contrast to plasma, lipid transport in the central nervous system is mediated solely by

HDL-like particles that contain apoE as their major protein (Huang 2010; Hauser et al. 2011).

Exchangeable apolipoproteins form a multi-gene family that has evolved from a common ancestral gene via the duplication and deletion of 11-mer codon repeats (Li et al. 1988). These proteins have similar gene structure containing 11-mer tandem repeats encoding for amphipathic α-helices that form the major structural and functional motif in this protein family (Segrest et al. 1990, 1992, 1994). Similar 11-mer α-helical repeat motifs are also found in other lipid surface-binding proteins, including parts of the non-exchangeable apoB, as well as in soluble apolipoprotein-related proteins such as SAA and α-synuclein (described by Uversky in Chap. 2 and by Colón and colleagues in Chap. 5 of this volume).

Although over 90 % of all apolipoproteins circulate on lipoproteins, a small fraction of the exchangeable proteins can transiently dissociate from the lipid surface (Fig. 8.1b). This occurs during inflammation, as well as during normal metabolic remodeling of lipoproteins which alters their surface-to-volume ratio and generates excess surface material that can dissociate in the form of apolipoproteins that carry little lipid. Such proteins are termed "lipid-poor" or "lipid-free"; in this chapter, we call them "free" for brevity. Free apolipoproteins can also be generated de novo. They form a metabolically active structurally labile species that can be rapidly recruited to the lipid surface (Rye and Barter 2004; Mulya et al. 2008; Jayaraman et al. 2012) or get degraded or misfolded (Hatters and Howlett 2002) (Fig. 8.1b). The misfolded apolipoproteins can deposit as amyloid fibers, potentially leading to disease.

8.2   Apolipoproteins in Amyloid Diseases

Exchangeable apolipoproteins are prominent players in amyloidoses (Hatters and Howlett 2002). In particular, apoE, which is an important multifunctional apolipoprotein in the central nervous system and in plasma (Getz and Reardon 2009), is found in most systemic and cerebral amyloid deposits in humans and is often co-deposited with other apolipoproteins. Human apoE4 isoform is a major causative risk factor in Alzheimer's and other neurodegenerative diseases (Mahley et al. 2009; Huang 2010; Hauser et al. 2011). Another important plasma apolipoprotein, apoA-I, commonly deposits as fibrils in atherosclerotic plaques in a process that probably contributes to atherosclerosis (Mucchiano et al. 2001; Röcken et al. 2006; Howlett and Moore 2006; Teoh et al. 2011a; Ramella et al. 2011). Moreover, in a rare hereditary form of apoA-I amyloidosis, the N-terminal fragments of apoA-I deposit as fibrils in vital organs (kidneys, liver, heart, etc.) and damage them (Obici et al. 2006; Rowczenio et al. 2011; Ramella et al. 2012; Gursky et al. 2012). Approximately 20 mutations in the apoA-I gene have been linked to this serious disease. A similar disease has been linked to stop codon mutations in human apoA-II, the second-major protein in plasma HDL. These rare mutations produce a C-terminal extension in apoA-II that causes renal amyloidosis and, ultimately, kidney failure in humans (Benson et al. 2001, 2011). In addition, amyloid deposits of a minor apolipoprotein, human apoA-IV, have been found in vivo (Bergström et al. 2004). To-date, no amyloid diseases involving apoCs have been reported. Nevertheless, human apoC-II readily forms amyloid fibrils in vitro in a process that was extensively studied by Howlett and colleagues, providing the most comprehensive model available to-date for understanding the misfolding of larger apos (summarized by Ryan et al. in Chap. 7 of this volume.) Human apoC-III can also form amyloid fibrils in vitro (de Messieres et al. 2014). Moreover, apolipoproteins such as apoC-III, apoA-I and apoE were implicated in binding amyloid-β (Aβ) peptide in circulation and were proposed as potential biomarkers or risk factors of Alzheimer's disease (Lewis et al. 2010; Shih et al. 2014).

The large heavily glycosylated non-exchangeable apoB has also been reported to bind Aβ peptide (Safar et al. 2006). Moreover, apoB has been proposed to form amyloid-like structure in several in vitro studies of LDL

(Stewart et al. 2005; Parasassi et al. 2008; Brunelli et al. 2014), and apoB misfolding in vivo was proposed to contribute to atherosclerosis (Ursini et al. 2002; Asatryan et al. 2005; Brunelli et al. 2014). Although aggregation of modified LDL was observed in these and other studies, including our own work (Jayaraman et al. 2007; Lu et al. 2012; Lu and Gursky 2013), none of these studies provided solid evidence for amyloid formation by apoB. In fact, electron and atomic force microscopy did not show typical amyloid fibers, fiber diffraction has not been reported, spectroscopic studies did not show a large increase in the β-sheet content of apoB upon LDL aggregation, and thioflavine T fluorescence studies were inconclusive since this dye binds not only to amyloid fibrils but also to lipid surfaces (Jayaraman et al. 2007; also see Chap. 6 by Gorbenko and colleagues in this volume). In sum, amyloid formation by apoB awaits experimental verification.

Other lipid surface-binding proteins that contain similar amphipathic α-helices are also prominent in amyloid diseases. In particular, N-terminal fragments of SAA, an acute-phase protein that displaces apoA-I from HDL during inflammation, is deposited in type-A amyloidosis and is also implicated in rheumatoid arthritis and atherosclerosis (van der Westhuyzen et al. 2007; Eklund et al. 2012). SAA in solution is an intrinsically disordered protein that acquires amphipathic α-helical structure on the lipid surface (Segrest et al. 1976) and forms cross-β-sheet in amyloid (see Chap. 5 by Colón et al. in this volume). Another apolipoprotein-related protein, α-synuclein (described by Uversky in Chap. 2 of this volume), is also intrinsically disordered in solution, acquires an amphipathic α-helical structure upon binding to a lipid bilayer, and can form extracellular β-sheet rich amyloid deposits in neurodegenerative diseases such as Parkinson's (Silva et al. 2013).

Why are apolipoproteins and related proteins overrepresented in amyloidoses? Hatters and Howlett (2002) noted that low structural stability and high conformational flexibility of free apolipoproteins, which facilitate their rapid incorporation into the lipid surface, make them labile to misfolding. Importantly, these and other authors have postulated that direct precursors of amyloid are the structurally labile free apolipoproteins, rather than their stable lipoprotein-bound counterparts (Hatters and Howlett 2002; Benson et al. 2011; Gursky et al. 2012; Hauser and Ryan 2013). This concept is supported by studies from several groups including ours showing that free apos in solution adopt a molten globule-like conformation with substantial α-helical content, flexible tertiary structure, low thermodynamic stability, low unfolding cooperativity, substantial solvent exposure of apolar groups, and high aggregating propensity (Gursky and Atkinson 1996a, b; Morrow et al. 2002; Guha et al. 2008). These properties are likely to facilitate β-aggregation.

In contrast to free apolipoproteins, lipoproteins are stabilized by high free energy barriers arising from disruption of extensive protein and lipid interactions during lipoprotein remodeling (Mehta et al. 2003; Jayaraman et al. 2005b, 2006; Guha et al. 2007; Lu et al. 2012). Such remodeling involves partial dissociation of apolipoproteins and changes in lipoprotein morphology such as fusion (Gursky 2005). Free energy barriers decelerate dissociation of apolipoproteins from the lipid surface and thereby delay their misfolding and proteolysis. In sum, strong evidence supports the idea that dissociation from the lipoprotein surface, which generates labile free apolipoproteins, is an obligatory early step in apolipoprotein misfolding and β-aggregation.

The amphipathic α-helix, which is the major lipid surface-binding motif in apolipoproteins and related proteins (Segrest et al. 1990, 1994), was proposed to have high intrinsic propensity for self-association and amyloid formation (Howlett and Hatters 2002). Apolipoprotein α-helices, termed class-A helices, are distinct from the class-G helices found in globular proteins (Segrest et al. 1990). First, class-A helices are comprised of 11-mer tandem repeats (rather than 7-mer repeats typical of globular proteins) that are optimized for lipid surface binding via the apolar helical faces. These large relatively flat apolar faces comprise 30–50 % of the total helical surface area (Fig. 8.2) as compared to ~15 %

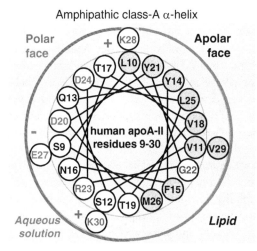

Amphipathic class-A α-helix

Fig. 8.2 Amphipathic class-A α-helix is the major structural and functional motif in apolipoproteins. Helix wheel representation of residues 9–30 of human apoA-II is shown as an example. Class-A α-helix has a large apolar face (*gray filled circles*) flanked by basic residues (+, *blue* letters), with acidic residues (–, *red* letters) located in the middle of the polar face. Polar face of the helix interacts with the aqueous solution (*light-blue*), while the apolar face can bind lipid surface (*beige*) or interact with apolar faces of other helices in free protein

in typical globular proteins. In addition, class-A helices have a relatively large fraction of charged residues (~30 %) with most basic groups flanking the apolar face and most acidic groups located in the middle of the polar face (Fig. 8.2). Such charge distribution facilitates formation of salt bridge networks and is also thought to contribute to lipid surface binding (Segrest et al. 1994). In sum, apolipoprotein α-helices have a distinct distribution of hydrophobic and charged residues and have relatively few polar groups. This contrasts with prion proteins whose misfolding is driven by a domain that is rich in polar residues and poor in charged and hydrophobic groups (reviewed by Shewmaker et al. 2011).

Apolipoproteins readily self-associate in solution in a process that is distinct from β-aggregation. Such self-association is mediated via the apolar helical faces and, similar to lipid surface binding, preserves or even enhances the helical structure (Gursky and Atkinson 1998). Whether or not the apolipoprotein α-helices have high intrinsic propensity to convert into cross-β-sheet in amyloid

is unclear. Recent advances in the bioinformatics approaches enable us to test this idea.

8.3 Recent Advances in Structural and Bioinformatic Studies

Last decade saw important progress in clinical and basic research of apolipoproteins as well as in molecular-level studies of amyloids. In this chapter, we combine recent advances in the structural studies of apolipoproteins with the newly developed amyloid prediction methods to better understand apolipoprotein misfolding.

High-resolution structural studies of apolipoproteins are notoriously difficult because of structural flexibility and high aggregating propensity of these proteins. After years of careful construct optimization, stable non-aggregating apolipoproteins amenable to high-resolution structural analyses have been produced, and atomic structures of several human apolipoproteins have been determined in lipid-free state. These include solution structure of modified full-length apoE determined by NMR (Chen and Wang 2011) and two high-resolution X-ray crystal structures, one of the N-terminal ~75 % residue fragment of apoA-I (Mei and Atkinson 2011) and another of the central domain of apoA-IV (Deng et al. 2012). These structures have provided tremendous insights into the functions of these proteins on and off the lipid surface. In addition, high-resolution structure of human amyloidogenic isoform of serum amyloid A, SAA1.1, in two crystal forms has been determined, providing the much-needed structural basis for understanding amyloid formation by this protein (Lu et al. 2014). Months later, x-ray crystal structure of murine SAA3 has been reported at 2 Å resolution (Derebe et al. 2014). The molecular structures of human SAA1.1 and murine SAA3 turned out to be very similar although the oligomeric states were different, with SAA1.1 crystallized as a dimer or a hexamer (Lu et al. 2014) and SAA3 forming a functional tetramer (Derebe et al. 2014). Notably, all these atomic structures represent free proteins that are direct precursors of amyloid (Howlett and Hatters 2002; Benson et al. 2011; Das et al. 2014).

Another area of recent advances is in the development and improvement of the amyloid prediction software (reviewed by Dovidchenko and Galzitskaya in Chap. 9 of this volume). During the last decade, it became increasingly clear that a wide variety of unrelated proteins and peptides can form amyloid in a process initiated by a limited number of 4- to 10-residue segments, or "hot spots", that have high intrinsic propensity for β-aggregation (Ventura et al. 2004; Thompson et al. 2006; Tzotzos and Doig 2010). Approximately 20 sequence-based algorithms are currently available to predict such segments, including Tango, Waltz, Aggrescan, Zyggregator, 3D Rosetta-based method, FoldAmyloid and PASTA, to name a few. These algorithms use a variety of approaches, from the analysis of side chain distribution to molecular dynamics simulations, and apply a wide range of criteria, such as the side chain hydrophobicity, charge distribution, average packing density, dual α-helix and β-sheet propensity, hidden α-helix to β-sheet switches, β-sheet contiguity, correspondence to known amyloidogenic residue patterns, etc. (Galzitskaya et al. 2006; López de la Paz and Serrano 2004; Fernandez-Escamilla et al. 2004; Conchillo-Solé et al. 2007; Hamodrakas et al. 2007; Trovato et al. 2007; Zhang et al. 2007; Zibaee et al. 2007; Kim et al. 2009; Tian et al. 2009; Maurer-Stroh et al. 2010; Garbuzynskiy et al. 2010; O'Donnell et al. 2011). This multitude of criteria and approaches reflects the fact that β-aggregation and fibril formation is a complex process that can be induced in unrelated proteins by a wide range of factors. Perhaps as a result of this immense complexity, individual algorithms have only limited predictive power. To optimize the selectivity and specificity of the prediction, i.e. minimize the probability of false positives and negatives, a consensus method has been developed by Hamodrakas and colleagues. We use the expanded version of this method, termed AmylPred2 (Tsolis et al. 2013), which combines 11 individual algorithms.

In addition, we use amyloid prediction software PASTA (Trovato et al. 2007; Walsh et al. 2014) and its expanded version, PASTA2.0 (Walsh et al. 2014), that not only identifies the amyloidogenic segments but also estimates free

energy of their pairing and thereby predicts which amino acid register and β-sheet orientation, parallel or antiparallel, is energetically favored. As detailed by Dovidchenko and Galzitskaya in Chap. 9 of this volume, AmylPred2 and PASTA2 are among the best amyloid prediction methods currently available. Following Trovato and colleagues (2007), we assumed that a reliable prediction of the β-sheet pairing corresponds to favorable (negative) PASTA energy of −4 units or lower (1 PASTA energy unit = 1.192 Kcal/mol). This threshold value corresponds to 73.8 % sensitivity and ~15 % selectivity, reflecting the respective rates of false negatives and positives. In this chapter we focus on segment pairings with PASTA2.0 energy circa −4 or lower.

Here, we combine these sequence-based predictions and, whenever possible, their experimental validations, with the available high-resolution structural information to glean new insights into the complex processes of apolipoprotein misfolding. The main focus of this chapter is on the major members of this protein family whose structures and amyloid-forming properties have been explored experimentally. Several minor apolipoproteins are also addressed. Comparison of these proteins leads to surprising new results and helps understand why some of them readily form amyloid while others do not.

8.4 New Insights into Apolipoprotein Misfolding from Sequence and Structural Analyses

8.4.1 Apolipoprotein A-I

Human apolipoprotein A-I (243 a. a.) is the major HDL protein that binds and activates lipid transporters, lipophilic enzymes, and HDL receptor, and thereby orchestrates crucial steps in the HDL biogenesis and functions (Navab et al. 2011; Philips 2013). Although plasma levels of apoA-I and HDL cholesterol (a. k. a. Good Cholesterol) have long been known to correlate inversely with the risk of cardiovascular disease (Gordon et al. 1989), recent studies revealed that this relationship is complex and involves additional factors, particularly functional properties of individual HDL subclasses (Rothblat and Philips 2010; Navab et al. 2011; Banach 2014). The role of these factors is intensely investigated in an effort to develop new HDL-based therapies to complement statins and other lipid-lowering drugs (Kingwell et al. 2014). The prevailing concept is that the protective function of HDL and apoA-I stems mainly from their central role in reverse cholesterol transport, which is the sole pathway of cholesterol removal from the body (Sorci-Thomas and Thomas 2013; Phillips 2014). Additional protection stems from anti-inflammatory, anti-oxidant and antithrombotic effects of apoA-I (Navab et al. 2011).

In plasma, apoA-I is in a dynamic equilibrium between lipid-bound and free forms, both important for reverse cholesterol transport (Rye and Barter 2004). Nearly 95 % of plasma apoA-I circulates in an HDL-bound form that is stabilized by high free energy barriers (Mehta et al. 2003; Jayaraman et al. 2006). The remaining ~5 % forms a transient monomeric lipid-poor/free apoA-I that is generated de novo or through dissociation from HDL (Mulya et al. 2008; Cavigiolio et al. 2010; Jayaraman et al. 2012). This metabolically active marginally stable state probably initiates protein misfolding in amyloid.

Human apoA-I amyloidosis takes on two distinct forms, acquired and familial. Acquired amyloidosis is the most common form where full-length non-variant apoA-I deposits as fibers in arteries and tissues (Röcken et al. 2006; Howlett and Moore 2006; Ramella et al. 2011). Such fibrils were found in aortic plaques of >50 % of patients who underwent atherectomy (Mucciano et al. 2001; Röcken et al. 2006). Although the causes and consequences of apoA-I fibril deposition in vivo are unclear, increasing evidence suggests that this process is pro-atherogenic (Ramella et al. 2011; Teoh et al. 2011a). In fact, apoA-I fibrils reportedly increase vulnerability of atherosclerotic plaques, activate arterial macrophages, and are toxic to cells (Suzuki et al. 2005; Lepedda et al. 2009; Patel et al. 2010; Adachi et al. 2013). Moreover, full-length wild-type human apoA-I can form amyloid in vitro upon methionine oxidation, suggesting a potential mechanism for apoA-I fibrillogenesis in the oxidative environment of atherosclerotic plaques (Wong et al. 2010; Chan et al. 2015).

Familial apoA-I amyloidosis (AApoAI) has entirely different clinical presentation. In this autosomal dominant disorder, fragments 1–83 to 1–93 of mutant apoA-I deposit as fibers in vital organs (kidney, liver, heart, nerves, skin, spleen, testes, etc.) causing organ damage (Obici et al. 2006; Ramella et al. 2012). The only treatment currently available for this potentially lethal disease is end-stage organ transplant (Gillmore et al. 2006). The mechanism of tissue damage and the pathogenic species (apoA-I oligomers or fibrils) are unknown. Moreover, it is unclear whether proteolytic cleavage that generates the N-terminal 9–11 kDa fragments found in amyloid deposits precedes or follows apoA-I misfolding (Wong et al. 2010). It is also unclear why different AApoAI mutations generate N-terminal fragments cleaved at different sites that are distant from the mutation site, and what proteases are responsible for it in vivo. Another conundrum is that symptoms associated with the same AApoAI mutation can range from mild to severe (Obici et al. 2006; Rowczenio et al. 2011), suggesting that additional, still unknown, factors are involved. We hypothesized that such factors include HDL lipids that can prevent or promote apoA-I dissociation, and thereby influence the population of the free protein (Das et al. 2014). Yet another puzzle is that AApoAI patients have lower than normal plasma levels of apoA-I and HDL resulting from reduced secretion or enhanced degradation of the protein (Marchesi et al. 2011; Arciello et al. 2011). Hence, unlike many other amyloid diseases, AApoAI is not due to protein overproduction. Paradoxically, despite apoA-I and HDL deficiency, AApoAI patients are not at a high risk for atherosclerosis (Obici et al. 2006; Rowczenio et al. 2011), which illustrates the complex relationship between plasma levels of HDL and cardiovascular disease.

Nearly 20 naturally occurring human AApoAI mutations have been identified, mainly point substitutions as well as some frame shifts and deletions (Rowczenio et al. 2011; Obici et al. 2006; Gursky et al. 2012, and references therein). Most of these mutations are located within the

N-terminal segment found in amyloid deposits ("inside" mutations), yet several "outside" point mutations have also been found in residues 170–178. This raises a question: how can the "outside" mutations in the C-terminal half of the apoA-I molecule trigger amyloid deposition by its N-terminal part? The answers to this and other questions are suggested by a combined structural and sequence analysis of apoA-I.

Structural analysis of apoA-I has been complicated by its high conformational flexibility and aggregating propensity. Although the atomic structure of full-length apoA-I in solution or on HDL is not available, two x-ray crystal structures of lipid-free truncated human apoA-I have been determined: a low-resolution 4 Å structure of the N-terminally truncated apoA-I, Δ(1–43)apoA-I (Borhani et al. 1997) and a high-resolution 2.2 Å structure of the C-terminally truncated apoA-I,

Δ(185–243)apoA-I (Mei and Atkinson 2011). These important breakthroughs provided tremendous insights into the structure-function relationship of apoA-I (Brouillette et al. 2001; Mei and Atkinson 2011; Gursky 2013; Phillips 2013 and references therein). Notably, the Δ(185–243) apoA-I construct contains all sites of AApoAI mutations and all three apoA-I methionines (M86, M112, and M148), providing a useful model for understanding the effects of these mutations and Met oxidation on apoA-I misfolding (Gursky et al. 2012; Das et al. 2014).

Lipid-free Δ(185–243)apoA-I forms a crystallographic dimer of two antiparallel largely α-helical molecules, stabilized by two four-segment bundles at its ends. The helices are kinked at Pro positions to form an arc 11 nm in diameter, commensurate with that of HDL (Fig. 8.3). This structure provides the basis for

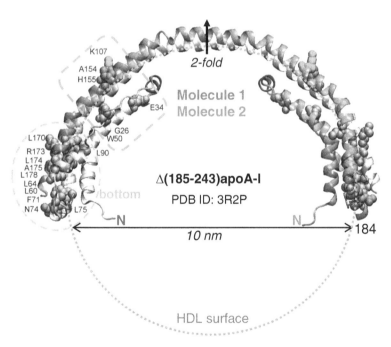

Fig. 8.3 Amyloidogenic mutations mapped on the atomic structure of lipid-free human apoA-I. X-ray crystal structure of the C-terminally truncated protein, Δ(185–243) apoA-I, was determined by Mei and Atkinson (2011). The protein is ~80 % α-helical and forms a crystallographic dimer stabilized by two four-segment bundles. Two antiparallel dimer molecules are color-coded; *arrow* shows twofold symmetry axis. The dimer forms a semi-circular arc with diameter d = 11 nm typical of HDL, suggesting how apoA-I confers curvature to the phospholipid mono-layer on HDL surface (*dotted circle*). Dimer-to-monomer conversion is proposed to occur via the domain swapping of an helical segment around the twofold axis (Mei and Atkinson 2011). All known AApoAI mutations are located in the helix bundle (Gursky et al. 2012). The mutation sites are indicated and are color-coded: point substitutions (*yellow*), deletions (*green*), frame shifts (*pink*). Most mutations are clustered at the bottom of the helix bundle (*dashed oval*), yet some are in the middle/top region (*rectangle*) (Modified from Gursky et al. 2012)

understanding the conformational ensemble of full-length apoA-I, either as a dimer on HDL surface or as a free monomer (Mei and Atkinson 2011; Gursky 2013). Domain swapping around the dimer twofold axis was proposed to convert the dimer into a monomer with a similar helix bundle structure (Mei and Atkinson 2011) (Fig. 8.4c). Mapping AApoAI mutations on this structure showed that most "inside" and all "outside" mutation sites are located in the well-ordered "bottom" half of the helix bundle (Fig. 8.3). This location helped explain the role of the "outside" mutations in structural integrity of free protein and prompted us to propose that these mutations destabilize the helix bundle and thereby promote its misfolding (Gursky et al. 2012).

The crystal structure also revealed that the residue segment 44–55 adopted an unpaired β-strand-like conformation that is unique in apoA-I (Fig. 8.4, dark red), suggesting that this segment helps initiate α-helix to β-sheet conversion (Gursky et al. 2012). Wong et al. (2012) reported that these residues are required for fibril formation. However, segment 44–55 turned out to have only modest predicted sequence propensity to form amyloid (Das et al. 2014). Since in free protein this polar segment is dynamic and solvent-exposed (Chetty et al. 2009; Mei and Atkinson 2011), its relatively low sequence propensity for β-aggregation probably helps protect free apoA-I from misfolding.

Surprisingly, experimental studies of selected naturally occurring mutants of human apoA-I, including common AApoAI variants G26R and W50R, showed that substantial global destabilization of the free protein or its reduced affinity for HDL surface is neither necessary nor sufficient

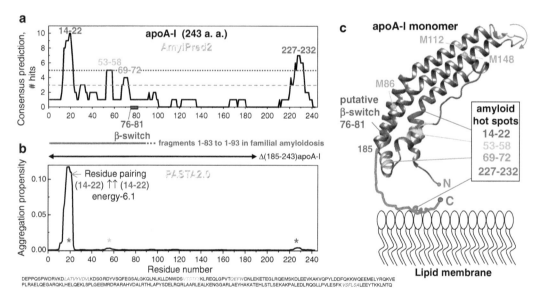

Fig. 8.4 Amyloidogenic segments in human apoA-I. AmylPred2 (**a**) and PASTA2.0 profiles (**b**). Major amyloidogenic segments predicted by five or more methods (at or above the *dotted line*) from the total of 11 methods in AmylPred2 are shown by residue numbers and color-coded: 14–22 (*blue*), 53–58 (*teal*), and 227–232 (*red*). A minor amyloidogenic segment, 69–72 (*green*, above the *dashed line*), is predicted by four methods. *Green bar* shows protein fragments found in AApoAI deposits. *Double arrow* shows the crystallized construct. Amino acid sequence of apoA-I is shown with amyloidogenic segments color-coded. (**c**) Amyloidogenic segments mapped on the proposed structure of free apoA-I mono-mer. The monomer structure was obtained from the crystallographic dimer (Fig. 8.3) by domain swapping (Mei and Atkinson 2011). C-terminal part 185–243 (*gray solid line*), which was missing from the crystallized construct, is partially unfolded in solution, largely α-helical on the lipid, and contains the primary lipid binding site that encompasses the amyloidogenic segment (residues 227–232, in *red*) that overlaps native α-helical structure (Borhani et al. 1997). The N-terminal amyloidogenic segments (in *blue, teal* and *green*) are located in a well-ordered α-helical cluster. The hypothetical pH-dependent switch in residues 76–81, EKETEG, is in *purple* (Modified from Das et al. 2014)

for amyloid formation ((Das et al. 2014) and references therein). This prompted us to search for additional factors driving apoA-I misfolding. We postulated that such factors involve local perturbations in sensitive protein regions, or amyloid "hot spots". To identify such regions, we performed amino acid sequence analyses of human apoA-I by using AmylPred2 and PASTA2.0.

Figure 8.4 summarizes the results of this analysis. AmylPred2 profile shows that most amyloidogenic propensity resides in the N-terminal ~10 kDa segment that deposits in familial amyloidosis. This N-terminal part contains one major amyloid hot spot predicted by up to 10 out of 11 methods in residues 14–22 (LATVYVDVL), and two minor hot spots predicted by four to five methods in residues 53–58 (VTSTFS) and 69–72 (QEFW) (Fig. 8.4a). The only other major amyloidogenic segment is predicted in residues 227–232 (VSFLSA) that form the primary lipid binding site in apoA-I and are largely α-helical on the lipid (Borhani et al. 1997; Lyssenko et al. 2012; Nagao et al. 2014 and references therein). Notably, all predicted amyloidogenic segments are highly hydrophobic. Consistent with AmylPred2, PASTA2 analysis predicts three amyloidogenic segments, with 14–22 being the strongest. These predictions are supported by experimental studies of peptide fragments containing these segments, which readily formed fibrils in vitro (Wong et al. 2012; Adachi et al. 2014; Das et al. 2014; Mendoza-Espinosa et al. 2014).

Moreover, PASTA2 predicts that the major amyloidogenic segment, 14–22, forms parallel in-register intermolecular β-sheet in amyloid (favorable PASTA2 energy –6.1, Fig. 8.4b). This prediction is corroborated by the observation that FTIR spectra of amyloid fibers formed by apoA-I fragment 1–43 lacked the diagnostic peak circa 1,680 cm^{-1} (Adachi et al. 2013) and thus, indicated parallel β-sheet. Moreover, the structural model of 1–83 fragment, which was derived on the basis of fluorescence studies described by Gorbenko et al. in Chap. 6 of this volume, also comprises a parallel in-register intermolecular β-sheet. Similar to the N-terminal segment 14–22, the C-terminal amyloidogenic segment

227–232 is predicted by PASTA to form a parallel in-register β-sheet, albeit with less favorable energy. Such a molecular orientation can facilitate synergy between the two major N- and C-terminal hot spots during amyloid formation by the full-length apoA-I.

Notably, the vast majority of known naturally occurring human apoA-I mutations (Sorci-Thomas and Thomas 2002) are located outside the predicted amyloid hot spots, with very few exceptions such as F71Y located inside a minor hot spot in residues 69–72. Such mutations are not expected to significantly alter the sequence propensity to form amyloid. In fact, our analysis suggests that AApoAI mutations have little or no effect on the proteins' amyloidogenic profiles predicted by AmylPred2 or PASTA2, suggesting that amyloid formation by these mutants in vivo results, at least in part, from factors other than their intrinsic sequence propensity to form amyloid. To identify such factors, we mapped the predicted amyloidogenic segments on the crystal structure of Δ(185–243)apoA-I. The results were quite revealing.

The three N-terminal amyloidogenic segments form a well-ordered α-helical cluster in the "bottom" half of the four-segment bundle (Fig. 8.4c), close to the sites of most AApoAI mutations. In wild type apoA-I, this well-ordered native helical structure is probably instrumental in protecting the amyloidogenic regions from β-aggregation. The C-terminal amyloidogenic segment, 227–238, is probably protected from misfolding by forming an amphipathic α-helix that strongly binds lipid; in free protein, this C-terminal helix probably folds back against the N-terminal bundle to pack against its apolar groove. Therefore, we propose that protein misfolding in AApoAI is triggered by perturbed native helical packing in the N-terminal amyloidogenic segments, which increases their mobility and solvent exposure and thereby facilitates β-aggregation (Das et al. 2014).

This idea is supported by the analysis of local protein structure and dynamics by hydrogen-deuterium exchange mass spectrometry, which was pioneered for apoA-I by Phillips, Englander and colleagues (Chetty et al. 2009, 2012) and is continued by our team. These H-D exchange studies showed increased mobility in the first ~100 residues in G26R mutant, which was proposed to help generate N-terminal fragments found in amyloid deposits (Chetty et al. 2012); however, it was unclear why the proteolysis was limited to just one site from this long mobile helix.

Our proposed mechanism helps explain why methionine oxidation promotes fibril formation in full-length wild-type human apoA-I (Wong et al. 2010). Two of the three apoA-I methionines, M112 and M148, are located at the top of the helix bundle, remote from amyloidogenic segments or the sites of AApoAI mutations (Fig. 8.4c). In contrast, M86 is buried in the hydrophobic cluster at the bottom of the bundle in close proximity to amyloidogenic segments and AApoAI mutation sites. Replacement of the apo-

lar Met side chain with polar Met sulfoxide in position 86 will probably perturb this hydrophobic cluster, particularly the major amyloidogenic segment 14–22. Hence, oxidation of M86 is expected to perturb native packing in this sensitive region (Das et al. 2014). Experimental analyses by Cavigiolio, Jayaraman and colleagues (Chan et al. 2015) helped establish the role of individual Met oxidation in apoA-I fibrillogenesis.

Taken together, the existing studies suggest a unified mechanism of apoA-I misfolding in acquired and familial amyloidosis (Fig. 8.5). We hypothesize that the misfolding is initiated by perturbed amyloidogenic segments in apoA-I, particularly in residues 14–22. On HDL, such segments form amphipathic α-helices that are

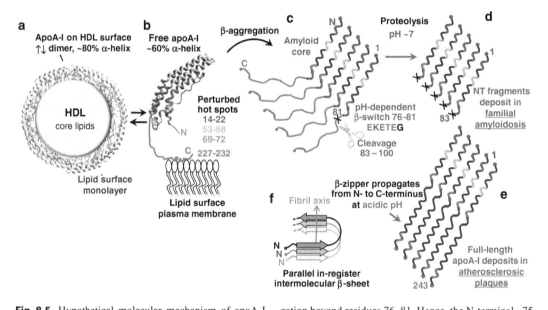

Fig. 8.5 Hypothetical molecular mechanism of apoA-I misfolding in hereditary and in acquired amyloidosis (Modified from Das et al. 2014). (**a**) ApoA-I on HDL surface forms an antiparallel ~80 % α-helical dimer stabilized by high kinetic barriers (Mehta et al. 2003; Jayaraman et al. 2006). (**b**) Dissociation from HDL surface generates a metastable apoA-I monomer that absorbs via its C-terminus to phospholipid surface. (**c**) Perturbed native packing in the N-terminal amyloid hot spots, which can result from AApoAI mutations or Met oxidation, initiates monomer misfolding, from native α-helical structure to parallel in-register β-sheet in amyloid. Residues 76–81, EKETEG (in *purple*), form a putative gate-keeping motif in which three charged Glu are unlikely to be located on the same side of the β-sheet due to their intra- and intermolecular repulsion. (**d**) At pH 7, electrostatic repulsion among these Glu is expected to block the β-zipper propa-

gation beyond residues 76–81. Hence, the N-terminal ~75 residues form amyloid core that is resistant to proteolysis. The polypeptide located C-terminally of residue 81, which looses its helical structure upon misfolding of the N-terminal part, can be readily proteolysed. This mechanism explains why the 9–11 kDa N-terminal fragments are found in systemic apoA-I amyloidosos. (**e**) At acidic pH that approaches pH 5.5 in atherosclerotic plaques, Glu in the EKETEG motif 76–81 are partially protonated and may form carboxyl pairs, facilitating β-zipper propagation from the N- to the C-terminus to encompass all four amyloid hot spots. As a result, full-length apoA-I forms amyloid in atherosclerotic plaques. (**f**) Molecular arrangement in a parallel in-register β-sheet. Different protein molecules are in different shades of *gray*. Main chain hydrogen bonds are aligned along the fibril axis (*arrow*) protected by lipid binding. Transient dissociation from HDL generates the labile free conformation in which these segments are normally protected by their packing in the native helix bundle structure. AApoAI mutations or Met oxidation oxidation perturbs this native packing, which increases the mobility and solvent accessibility of the amyloidogenic segments and triggers their β-aggregation.

This proposed mechanism provides a new explanation why AApoAI deposits contain N-terminal proteolytic fragments cleaved between residues 83 and 93, even though this region is normally protected from proteolysis by its well-ordered native α-helical structure. We hypothesize that the parallel intermolecular β-sheet, which is probably nucleated by the "hot spot" residues 14–22, readily propagates through the N-terminal 75 residues that have high amyloid-forming propensity (Fig. 8.9a). This propagation is probably impeded by the putative β-breaking motif EKETEG in residues 76–81 due to Columbic repulsion among the proximal Glu at pH~7. Hence, residues 1–75 are expected to form the amyloid core; the polypeptide outside this core can be readily proteolysed. According to this mechanism, amyloid core formation precedes apoA-I proteolysis. Still, an alternative (or perhaps an additional) pathway in which proteolytic cleavage of the highly amyloidogenic N-terminal segment happens first is also potentially possible. Which pathway is more relevant to the in vivo conditions is still an open question that can only be resolved once we gain better understanding of apoA-I proteolysis in vivo.

Our hypothetical mechanism also helps explain why arterial deposits in acquired amyloidosis commonly contain full-length apoA-I. In anaerobic regions of deep atherosclerotic plaques, the pH can drop as low as 5.5, potentially leading to partial protonation of Glu and formation of carboxyl pairs in the EKETEG segment. As a result, the repulsion among these carboxyls is diminished or even replaced with attraction, which is expected to facilitate β-zipper propagation from the N- to the C-terminus to encompass all amyloid hot spots in apoA-I in a parallel in-register β-sheet conformation.

This hypothetical mechanism of apoA-I misfolding in various types of human amyloidosis is supported by our on-going studies using hydrogen-deuterium exchange. It illustrates the combined power of structural, bioinformatic and biophysical approaches in dissecting protein misfolding pathways and elucidating the role of protein-lipid interactions and proteolytic degradation in these complex processes.

8.4.2 Apolipoprotein A-II

Apolipoprotein A-II (77 a. a.), which is the second-major HDL protein, is the most hydrophobic exchangeable apolipoprotein. Human apoA-II forms a homodimer disulfide-linked via Cys6, which circulates largely on mature mid-size plasma HDL that also contain apoA-I (Gillard et al. 2009; Gao et al. 2012). In contrast to apoA-I whose cardioprotective role is well-established, the function of apoA-II is less clear. The emerging consensus is that apoA-II restricts the apoA-I conformation on HDL and thereby indirectly influences the interactions of apoA-I with its functional ligands which are critical in HDL metabolism (Silva et al. 2007; Gao et al. 2012; Maiga et al. 2014 and references therein).

Our current knowledge of the apoA-II structure is limited to low resolution. Human apoA-II is proposed to form a highly α-helical antiparallel dimer on HDL (Silva et al. 2007), yet in solution the protein is only partially folded and is marginally stable (Gursky and Atkinson 1996b). The amino acid sequence of apoA-II is predicted to form class-A amphipathic α-helices with particularly large apolar faces spanning ~180° (Fig. 8.2) as compared to ~120° typical of other exchangeable apos (Segrest et al. 1994). As a result, apoA-II α-helices insert more deeply into the phospholipid monolayer and bind more strongly to mature HDL than any other apos (Gao et al. 2012 and references therein). We propose that this strong association with mature HDL, which makes apoA-II practically non-exchangeable, helps protect it from misfolding.

In humans, apoA-II misfolding can cause hereditary amyloidosis, a rare autosomal domi-

nant disorder resulting from stop codon mutations that produce a 21-residue C-terminal
extension in residues 78–98 (Benson et al. 2001).
Full-length mutant apoA-II forms amyloid
deposits primarily in kidneys, causing renal damage and failure. The only existing treatment for
apoA-II amyloidosis is dialysis followed by kidney transplant (Magy et al. 2003).

Amino acid sequence analysis helps explain
why mutant but not wild-type human apoA-II
forms amyloid in vivo (Gursky 2014). In wild-
type human apoA-II, two amyloid hot spots are
predicted strongly in residues 10–18
(LVSQYFQTV) and 60–70 (LVNFLSYFVEL).
These hydrophobic segments are predicted by
8–11 methods in AmylPred2 and have PASTA2
energies ranging from −4.4 to −7.1 (Fig. 8.6).
This is the strongest predicted amyloidogenic
propensity of all exchangeable apolipoproteins,
approaching that of Alzheimer's Aβ(1–42) peptide (Fig. 8.7). Importantly, the two predicted
amyloidogenic segments in apoA-II have been
verified experimentally in peptide fragment studies of murine apoA-II that is homologous to the
human protein (Sawashita et al. 2009). Similar to
apoA-I, the amyloidogenic segments in apoA-II
are located inside the predicted lipid-binding
class-A α-helices (Fig. 8.6), which is expected to
protect them from misfolding.

One likely reason why the C-terminal
extension leads to amyloidosis is because, unlike
the rest of the apoA-II molecule, this extension
cannot form an amphipathic α-helix and hence, is
not well-suited for lipid surface binding (Benson
et al. 2011; Gursky 2014). As a result, the lipid
binding affinity of the mutant protein is expected
to decline, shifting the population distribution
from HDL-bound to free apoA-II that is labile to
misfolding. Another likely reason is that the
C-terminal extension is predicted to form an
additional amyloidogenic segment that can act in
synergy with the major amyloidogenic segment
in residues 60–70. In fact, both these segments
are predicted strongly to form parallel in-register
β-sheet in amyloid (Fig. 8.6c) (Gursky 2014).

In sum, compared to other exchangeable
apos, normal human apoA-II is the most hydrophobic, binds most strongly to mature HDL, and
has the highest sequence propensity to form
amyloid. Two major amyloidogenic segments
predicted in normal apoA-II have been verified
experimentally (Sawashita et al. 2009). These
segments are located inside the native amphipathic α-helices, and the strongest one (residues

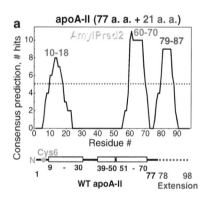

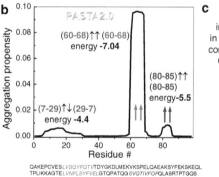

Fig. 8.6 Predicted amyloidogenic segments in human apoA-II and its C-terminal extension. Amyloidogenic pro-files obtained by AmylPred2 (**a**) and PASTA2 (**b**) for human apoA-II with C-terminal extension 78–98 (*dark red*) that causes human amyloidosis. The major amyloido-genic segments predicted by five or more methods in AmylPred2 are shown by residue numbers. *Double arrows* show the orientation of the intermolecular β-sheet predicted by PASTA2.0. The predicted native secondary structure is shown at the *bottom*; *rectangles* and residue numbers indicate α-helices. Cys6 that forms an intermo-lecular disulfide link in human apoA-II dimer is in *orange*. Amino acid sequence of human apoA-II with C-terminal extension is shown; amyloidogenic segments are color-coded. (**c**) Cartoon showing parallel in-register β-sheet predicted for apoA-II with amyloidogenic extension. Different protein molecules are in different shades of *gray* (Modified from Gursky 2014)

Aβ peptide

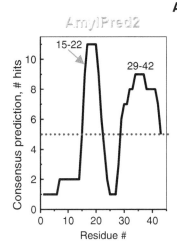
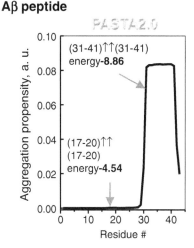

Fig. 8.7 Amyloidogenic profile of Alzheimer's Aβ(1–42) peptide obtained by AmylPred2 and Pasta2.0. Pairing energies exceeding −4 PASTA units are shown; residue numbers indicate amyloid hot spots. The two predicted amyloid hot spots overlap the β-sheet regions observed in the atomic model of Aβ(1–40) fibrils determined by solid-state NMR (Lu et al. 2013), which verifies the prediction

60–70) is predicted to form parallel in-register β-sheet in amyloid. We propose that the reason why normal human apoA-II does not form amyloid is probably due to its particularly strong association with mature HDL, which is further enhanced by the protein dimerization via the disulfide link. This disulfide may also potentially interfere with the structural transition from an antiparallel α-helical dimer (Silva et al. 2007) into a parallel intermolecular β-sheet. C-terminal extension that causes human apoA-II amyloidosis is expected to have a dual effect: it reduces the lipid-binding affinity of the protein and also increases its propensity to form parallel in-register intermolecular β-sheet (Fig. 8.6).

8.4.3 Apolipoprotein A-IV

Human apoA-IV (376 a. a.) is the largest known exchangeable apolipoprotein. It is synthesized in the gut, is secreted on chylomicrons, and circulates in plasma in HDL-bound and free states (Green et al. 1980). This minor apolipoprotein is proposed to regulate chylomicron assembly, modulate reverse cholesterol transport, serve as an anti-oxidant and as a satiety signal, and perform other functions in lipoprotein metabolism

(Deng et al. 2012 and references therein). ApoA-IV was also implicated in clearance of Aβ peptide in the brain (Cui et al. 2011). A case of human cardiac amyloidosis has been reported in which the N-terminal 70-residue fragment of wild-type apoA-IV deposited as fibrils (Bergström et al. 2004).

Similar to other exchangeable apolipoproteins, apoA-IV structure is comprised mainly of amphipathic α-helices. A distinct property of apoA-IV is that it contains a globular core domain (approximately residues 97–311) flanked by the N- and C-terminal segments whose interactions regulate lipid binding (Walker et al. 2014). The x-ray crystal structure of the core domain residues 64–335, which was determined to 2.4 Å resolution by Davidson's team (Deng et al. 2012), resembled that of the C-terminally truncated apoA-I (Mei and Atkinson, 2011). Similar to the N-terminal 75 % of apoA-I, the core domain of apoA-IV forms a four-helix bundle that dimerizes via the domain swapping of a helical segment (Fig. 8.8c) (Deng et al. 2012). In contrast to the highly curved apoA-I dimer that matches the HDL surface curvature, the apoA-IV dimer is nearly linear, which is consistent with its ability to bind much larger lipoproteins with lower surface curvature.

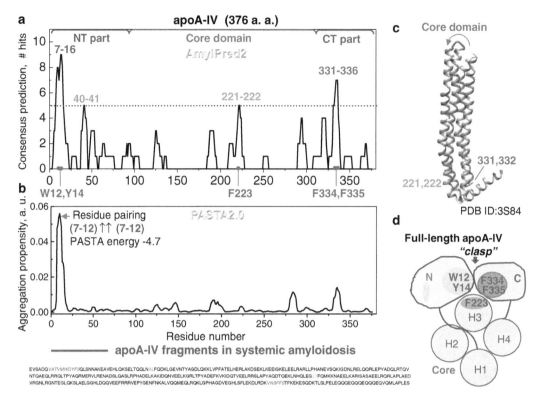

Fig. 8.8 Amyloidogenic profile and the structure of human apoA-IV. AmylPred2 (**a**) and PASTA2.0 (**b**) profiles of full-length apoA-IV are shown. N- and C-terminal regions flanking the core domain are indicated. Residue numbers indicate amyloidogenic segments predicted by five or more methods (above the *dotted line*). Aromatic residues implicated in lipid binding are in *purple*. *Green bar* shows N-terminal fragments found in human apoA-IV amyloidosis (Benson et al. 2001). Predicted residue pairing and PASTA energy in the major amyloidogenic segment is shown. (**c**) Helix bundle structure of the mono- meric core domain of apoA-IV (residues 64–335) obtained via the domain swapping (*circular arrow*) from the X-ray crystal structure of the dimer (Deng et al. 2012). Residues in the amyloidogenic segments predicted by AmylPred2 are indicated. (**d**) Cartoon showing the molecular archi- tecture of free full-length apoA-IV according to (Walker et al. 2014). *Circles* H1–H4 designate α-helices forming four-helix bundle in the core domain. N-terminal segment is in *blue* and C-terminal is in *red*. Filled *ovals* show amy- loid hot spots. Aromatic residues involved in "clasp" for- mation and lipid binding are in *purple* (Walker et al. 2014)

Compared to other exchangeable apos, apoA-IV binds lipid surface relatively poorly, consistent with its near-equal plasma distribution between lipid-bound and free states. Modest lipid binding affinity of apoA-IV is attributed to the strong inter- actions between its N- and C-terminal domains that hold the protein molecule in a closed confor- mation via the so-called "clasp mechanism" that modulates helix bundle opening on the lipid (Deng et al. 2013; Walker et al. 2014). Aromatic residues W12, Y14 and F334, F335 are implicated to be essential in "clasping" and in lipid binding (Walker et al. 2014) (Fig. 8.8d). For example, F334A sub- stitution disrupts the "clasp" and increases lipid binding by apoA-IV (Tubb et al. 2007). Conserved

Pro residues are also implicated in maintaining the "clasp" mechanism (Deng et al. 2015).

We propose that "clasping" of the N- and C-terminal parts helps protect apoA-IV from misfolding. This idea is based upon sequence analysis of human apoA-IV (Fig. 8.8). Both AmylPred2 and PASTA2 predict mild amyloido- genic propensity distributed throughout the poly- peptide chain, with two major aromatic-rich amyloidogenic segments in residues 7–16 (VATVMWDYFS) and 331–336 (VNSFFS) from the N- and C-terminal parts, respectively (Fig. 8.8a, b). PASTA predicts that the N-terminal segment is likely to form parallel in-register β-sheet in amyloid. In addition, AmylPed2 predicts two weaker amyloidogenic segments around residues 40–41 of the N-domain and 221–222 of the core domain (Fig. 8.8a). Three major predicted amyloid hot spots are located in the putative helical regions (Walker et al. 2014) that are outside (residues 7–16 and 40–41) or at the edge (residues 331–336) of the crystallized construct; the fourth hot spot (residues 221–222) is located inside the α-helical region in the crystal structure (Fig. 8.8c). These amyloidogenic segments are probably protected from misfolding by their native helical structure, as well as by the tertiary interactions in free full-length apoA-IV. In fact, aromatic residues W12, Y14 and F334, F335 that are proposed to be essential for lipid binding and for "clasping" (Walker et al. 2014), as well as F223 that modulates lipid binding, are located inside or adjacent to the amyloidogenic segments (Fig. 8.8a, d). Hence, these sensitive regions are probably sequestered via the "clasp" mechanism in free protein or via the lipid binding on HDL.

In sum, we propose that in the native lipid-free apoA-IV, the major predicted amyloidogenic segments are protected from misfolding via the same "clasp" mechanism that regulates protein-lipid interactions (Fig. 8.8d). Further, the presence of two amyloid hot spots in residues 7–16 and 40–41 helps explain the deposition of the N-terminal fragment containing these residues in systemic amyloidosis involving wild type human apoA-IV (Bergström et al. 2004).

8.4.4 Apolipoprotein A-V

ApoA-V (343 a. a.) is a hydrophobic protein that is a potent reducer of plasma TG (Nilsson et al. 2011). Similar to its physiological antagonist apoC-III (described in the next section), apoA-V is secreted by hepatocytes and circulates in plasma as a minor component of VLDL and HDL. To our knowledge, no amyloid diseases resulting from the deposition of apoA-V have been reported. This is not surprising given minute concentration of this protein in plasma (~150 ng/ml). Nevertheless, apoA-V has emerged as an important modulator of TG metabolism, which can influence secretion, processing and clearance

of TG-rich lipoproteins via several mechanisms (Sharma et al. 2013). In an intracellular mechanism, a fraction of apoA-V binds to lipid droplets and thereby influences VLDL secretion. The extracellular mechanisms include synergistic action of apoA-V and lipoprotein lipase, a key enzyme in processing of TG-rich plasma lipoproteins. First, VLDL binding via apoA-V to heparan sulfate proteoglycans in the arterial wall brings together the lipoprotein and the lipase that is anchored to the arterial wall (see Chap. 10 by Leonova and Galzitskaya in this volume). Second, apoA-V can promote clearance of VLDL remnants by binding to a receptor that is essential for clearance of TG-rich lipoproteins (Nilsson et al. 2008). In these reactions, a basic stretch in apoA-V is thought to bind acidic moieties of the receptor and the heparan sulfate proteoglycans. This is reminiscent of apoE and apoB whose basic stretches are critical in binding acidic sites of the lipoprotein receptors or heparan sulfates (Mahley et al. 1984; Wilson et al. 1991).

Structural analysis of apoA-V, pioneered by Ryan and colleagues (Weinberg et al. 2003; Beckstead et al. 2007; Wong et al. 2008; Mauldin et al. 2010; Sharma et al. 2013), has been complicated by the low protein solubility. Most experiments have been performed in solution at pH 3 and have been limited to low-resolution techniques. The protein structure was inferred to contain tandem amphipathic α-helices, which in apoA-V form distinct domains (Weinberg et al. 2003; Beckstead et al. 2007; Wong et al. 2008). The N-terminal residues 1–146 are proposed to form a helix bundle (Wong et al. 2008). The central domain (residues 147–292) contains a basic stretch (residues 186–227) implicated in binding to acidic ligands. Finally, the C-terminal 51 residues are implicated in modulating lipid binding by apoA-V (Beckstead et al. 2007; Mauldin et al. 2010).

Amino acid sequence analysis using AmylPred2 predicts one strong amyloidogenic segment in residues 269–277 (TYLQIAAFT) and three weaker ones in residues 118–120, 150–151 and 204–210 (Fig. 8.9). One predicted amyloidogenic segment, 119–121, is in the N-domain, the other three are in the central

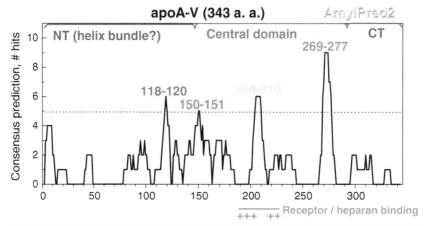

Fig. 8.9 Amyloidogenic profile of human apoA-V obtained by AmylPred2. Residue numbers indicate amyloidogenic segments predicted by five or more methods (above the *dotted line*). *Blue bar* and +++ indicate residue segment that contains eight uncompensated basic residues and three Pro and is implicated in binding to the lipoprotein receptor and to heparan sulfate proteoglycans

domain, and all four are located inside the amphipathic α-helices predicted in the native structure (Weinberg et al. 2003). Notably, a strong amyloidogenic segment predicted in residues 204–210 forms a hydrophobic stretch inside a larger predominantly basic segment 190–225 that is implicated in binding to the receptor and to heparan sulfate proteoglycans (Fig. 8.9). This larger segment contains eight basic residues, no acidic residues, and three structure-breaking Pro. Since stretches of similarly charged residues are disfavored in the parallel in-register amyloid β-sheet due to inter- and intramolecular repulsion, it is unlikely that this residue segment initiates β-aggregation. In fact, PASTA2 does not predict a strong pairing energy for this segment (profile not shown). Further, it predicts parallel in-register β-sheet for apoA-V, with the most favorable pairing energy of −3.9, just below the threshold of −4. Hence, this prediction by PASTA should be taken with a grain of salt.

In sum, apoA-V is a minor hydrophobic protein that is an important modulator of TG metabolism. The atomic structure of apoA-V is unknown. The predicted amyloidogenic potential is localized mainly to the central domain (Fig. 8.9). Similar to other apos, the amyloidogenic segments in apoA-V are hydrophobic, are located in the amphipathic α-helices in the native structure, and are predicted to form parallel in-register β-sheet in amyloid. These segments are expected to be protected from misfolding by their native protein-lipid and/or protein-protein interactions. Amyloidogenic segment 204–210 is probably also protected by its location inside a basic stretch that forms a binding site for VLDL receptor.

8.4.5 Apolipoproteins C-I, C-II and C-III

ApoC-I (6 kDa), C-II (10 kDa) and C-III (10 kDa) are the smallest human apolipoproteins that circulate mainly on VLDL and are also found as minor proteins on HDL and other lipoproteins. ApoCs are particularly important for metabolism of triglyceride-rich lipoproteins, as they modulate the activity of lipophilic enzymes, lipid transfer proteins, and other plasma factors that remodel these lipoproteins in vivo. ApoC-I inhibits cholesterol ester transfer protein, apoC-II is a physiological activator of lipoprotein lipase, while apoC-III is its inhibitor and is also implicated in VLDL production (Jong et al. 1999; Shachter 2001; Yao and Wang 2012; Kei et al. 2012). In addition, apoCs can displace larger proteins such as apoE from VLDL, and thereby inhibit apoE-mediated VLDL receptor binding (Sehayek and Eisenberg 1991).

Free in solution, apoCs have little ordered secondary structure (~30 % α-helix in apoC-I (Gursky and Atkinson 1998), less than that in apoC-II and apoC-II). Hence, similar to other small lipid surface-binding proteins such as apoA-II, SAA, Aβ and α-synuclein (described by Uversky in Chap. 2 of this volume), free apoCs are intrinsically disordered. NMR structures of apoC-I, apoC-II and apoC-III have been determined in lipid-mimetic environment on the micelles of sodium dodecyl sulphate (SDS) or dodecylphosphocholine (DPC) (Rozek et al. 1999; MacRaild et al. 2001, 2004; Gangabadage et al. 2008). These structures showed amphipathic class-A α-helices forming the lipid surface-binding motif in apoCs (Fig. 8.10, bottom panels). Compared to larger apos, apoCs make

fewer contacts on the lipoprotein surface and can dissociate from it relatively easily (Jayaraman et al. 2005a). Such dissociation is expected to occur in vivo, e.g. during VLDL conversion to LDL. Despite marginal structural stability of free apoCs, their amyloid deposition in vivo has not been reported to-date, probably because of the low plasma concentrations of these minor proteins. However, human apoC-II and, as reported recently, apoC-III can form amyloid fibrils in vitro (de Messieres et al. 2014). Extensive studies of apoC-II fibril formation by Howlett's team provided the most detailed model currently available for understanding apolipoprotein misfolding (summarized by Ryan et al. in Chap. 7).

Here, we report amyloidogenic profiles of apoCs (Fig. 8.10). AmylPred2 predicts that apoC-II has substantial amyloid forming potential, especially in the C-terminal part, while the sequences of apoC-I and apoC-III are less amy-

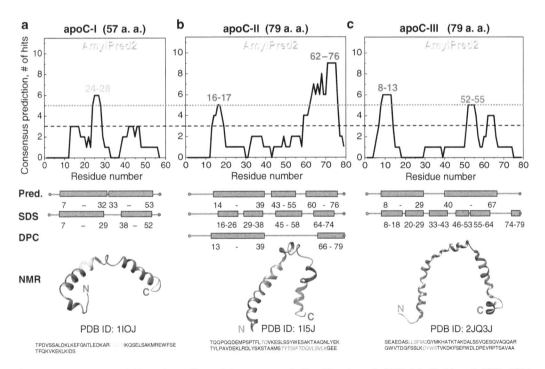

Fig. 8.10 Predicted amyloidogenic profiles and the structures of human apoCs. *Top*: AmylPred2 profiles of human apoC-I (**a**), apoC-II (**b**), and apoC-III (**c**). Residues predicted to form amyloidogenic segments by five or more methods (above the *dotted line*) are indicated. *Middle*: Secondary structures of apoCs that were predicted (Segrest et al. 1992) or observed by NMR on SDS or DPC

micelles (Rozek et al. 1999; MacRaild et al. 2001, 2004; Gangabadage et al. 2008). *Rectangles* and residue numbers indicate helical segments. *Bottom*: NMR structures of human apoCs on SDS micelles. Predicted amyloidogenic segments are mapped on the structures and color-coded. Amino acid sequences of the proteins are shown; the amyloidogenic segments are color-coded loidogenic. Similarly, PASTA2 suggests substantial amyloid-forming potential in the C-terminal part of apoC-II and predicts it to form parallel in-register β-sheet in amyloid (pairing energy of −4.9) (Fig. 8.11). Most favorable pairing energies predicted for apoC-I (−1.9) and apoC-III (−2.7) are well below the threshold of −4 and hence, are not very reliable (PASTA profiles not shown). Nevertheless, human apoC-III can form amyloid fibers in vitro (de Messieres et al. 2014).

The strongest predicted amyloidogenic segment in human apoCs is located in the C-terminal part of apoC-II (residues 62–76, TYTGIFTDQVLSVLK) from the native class-G helix. This prediction has been verified experimentally in careful studies by Wilson et al. (2007) (Fig. 8.11). These studies showed that in fibers of full-length human apoC-II, C-terminal segment is resistant to proteolysis and to H-D exchange, which is characteristic of amyloid core. The authors also showed that peptide fragments corresponding to the C-terminal segments of apoC-II readily form amyloid in vitro. Finally, structural analysis of apoC-II fibrils revealed parallel in-register β-sheets (Wilson et al. 2007), which verifies our prediction by PASTA (Fig. 8.11). In sum, extensive experimental evidence validates the prediction of the major amyloid hot spot in residue

segment 62–76 of apoC-II and suggests that this segment initiates protein misfolding, from the predominantly α-helical native structure to parallel in-register intermolecular β-sheet in amyloid. We speculate that the native α-helical structure in this segment, which forms the major lipid binding site in apoC-II, anchors the protein to the lipid surface and thereby protects it against β-aggregation in vivo. Notably, this segment forms class-G helix that, in contrast to class-A helices, has random radial distribution of charged residues. Hence, amyloidogenic properties of apolipoproteins can be conferred by segments forming other than class-A α-helices in the native structure.

8.4.6 Apolipoprotein E

ApoE (299 a. a.) is arguably the most extensively studied apolipoprotein (~20,000 hits in PubMed) that is an important mediator of lipid transport in plasma and in the central nervous system. ApoE is found in most systemic and cerebral amyloid deposits and is particularly pertinent to the development of Alzheimer's disease (for an excellent recent review see Hauser and Ryan 2013). Plasma apoE circulates mostly on VLDL where it serves as a ligand for LDL-like receptor that

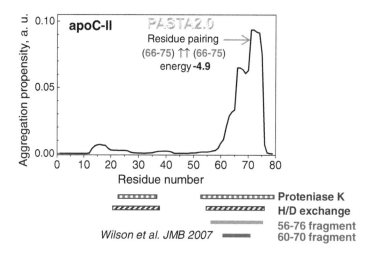

Fig. 8.11 PASTA2.0 profile of human apoC-II and its experimental validation. C-terminal segment 66–75 is predicted to form parallel in-register β-sheet, which is experimentally validated by EPR (Teoh et al. 2011b). Residue segments that are resistant to proteinase digestion

and hydrogen-deuterium exchange and are implicated in amyloid core formation are shown in *patterned bars*. Fragments implicated in amyloid formation in vitro are in *solid bars* (Wilson et al. 2007) mediates VLDL uptake and, ultimately, maturation to LDL (Hauser et al. 2011; Holtzman et al. 2012). During this process, TG in the lipoprotein core are hydrolyzed by lipoprotein lipase, generating excess surface material that can dissociate in the form of free apolipoproteins or small apoE-containing particles that join the plasma pool of HDL. Hence, in addition to VLDL, apoE also circulates as a minor protein on plasma HDL. In contrast to plasma, lipid transport in cerebrospinal fluid is mediated solely by HDL-like particles whose major protein is apoE. This explains why apoE is particularly prominent in neurodegenerative diseases. Notably, apoE4 isoform is the strongest genetic predictor of Alzheimer's disease and is directly associated with the amyloid load in this disease. Although the exact role of apoE in the development of Alzheimer's disease is not fully understood, the consensus is that apoE interacts directly in an isoform-specific manner with Aβ peptide and thereby influences aggregation, accumulation and/or clearance of Aβ from the brain (Strittmatter et al. 1993; Ladu et al. 2000; Hauser and Ryan 2013).

Three major isoforms of human apoE differ in Cys/Arg substitutions at two sites: apoE2 (C112, C158), apoE3 (C112, R158) and apoE4 (R112, R158), resulting in differences in charge, structural stability and functional properties. Compared to apoE3, which is the most common isoform, apoE4 is a strong causative risk factor for Alzheimer's disease, while apoE2 is neuroprotective (Strittmatter et al. 1993; reviewed by Verghese et al. 2011; Suri et al. 2013). ApoE4 is linked strongly to decreased age of the late-onset Alzheimer's disease and more severe Aβ deposition in the brain, while apoE2 has an opposite effect (Verghese et al. 2011; Suri et al. 2013). Thus, the predisposition to Alzheimer's disease increases in the order E2<E3<E4. Although the molecular mechanism via which apoE isoforms differentially influence Aβ accumulation in the brain is subject of debate, the prevailing concept is that this difference resides in the isoform-specific domain interactions in apoE, which influence its interactions with Aβ (reviewed by Hauser and Ryan 2013).

Structurally, apoE is often described as a two-domain protein comprised of a globular 22 kDa N-terminal domain that forms a helix bundle in solution and contains a receptor and heparan binding site in residues 136–150, and a C-terminal 10 kDa domain that is less structured in solution and forms the primary binding site for lipids and Aβ. Four-helix bundle structure of the N-domain of lipid-free apoE3, which was determined by x-ray crystallography to 2.2 Å resolution, was the first known atomic structure of a mammalian apolipoprotein (Wilson et al. 1991). This structure is consistent with the solution conformation of the full-length lipid-free apoE3 determined 20 years later by NMR (Chen and Wang 2011). In the NMR structure, the globular N-domain (residues 1–167) forms a template for folding of the hinge region (168–205) and the C-domain (206–299) (Figs. 8.12 and 8.13).

The two residues, 112 and 158, differing among the apoE isoforms are located in the N-terminal α-helices (orange, Figs. 8.12 and 8.13). Cys to Arg substitution at either site weakens the interhelical interactions and reduces structural stability of the N-domain and of the full-length lipid-free apoE (Acharya et al. 2002; Weers et al. 2003; Nguyen et al. 2014). Hence, the rank order of free protein stability is E4<E3<E2, opposite to that of the pre-disposition to Alzheimer's disease. A similar trend was observed in other proteins whose reduced structural stability often correlates with enhanced amyloid formation and is thought to contribute to it (Hatters et al. 2005, 2006).

Tertiary structural stability of apoE isoforms and other free apolipoproteins in solution correlates inversely with the probability of their helix bundle opening on the lipid, which is a necessary step in lipoprotein formation. Therefore, apoE isoform stability in solution was proposed to correlate inversely with the lipid binding affinity (Morrow et al. 2000; Acharya et al. 2002; Weers et al. 2003). To our knowledge, relative dissociation rates of apoE isoforms from HDL, which determine relative kinetic stabilities of HDL containing these isoforms, have not been reported. Furthermore, no isoform-specific effects have been observed in apoE association with HDL in

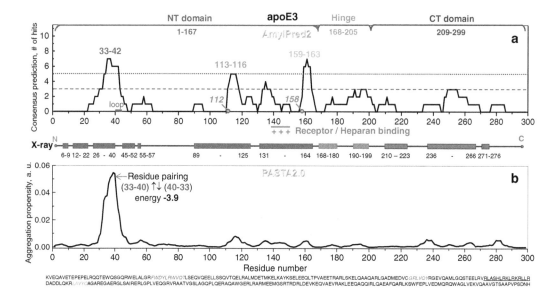

Fig. 8.12 Amyloidogenic profile of human apoE3 predicted by AmylPred2 (**a**) and PASTA2.0 (**b**). Amyloidogenic segments predicted by at least five methods in AmylPred2 are marked. *Orange dots* show locations of Cys/Arg substitutions that differ among the three major isoforms: apoE2 (C112, C158), apoE3 (C112, R158), and apoE4 (R112, R158). Helical segments observed in the NMR structure of the modified full-length human apoE3 (Chen et al. 2011) are shown by *rectangles*. Color code: N-terminal helix bundle domain (*violet*), the connecting hinge (*gray*), and the C-terminal domain (*purple*). + + + show a basic stretch (residues 136–152) implicated in binding to receptor and to heparan sulfate proteoglycans. Protein amino acid sequence with predicted amyloidogenic segments color-coded

cerebrospinal fluid ((Verghese et al. 2011) and references therein). Therefore, it is unclear whether or not differential affinity of apoE isoforms for lipid contributes to the isoform-specific effects in amyloid deposition.

Another potential origin of the isoform-specific effects is in the intrinsic sequence propensity of apoE to form amyloid. In fact, apoE misfolding was proposed to be a primary event that helps nucleate amyloid formation by Aβ (Ma et al. 1994; Wisniewski et al. 1995; Hatters et al. 2006). To test this idea, we analyzed amino acid sequence of apoE by using AmylPred2 and PASTA2 (Fig. 8.12). Surprisingly, both methods predict relatively weak amyloidogenic potential for apoE as compared to other exchangeable apolipoproteins. The major amyloidogenic segment is predicted by both methods in residues 33–42, with the PASTA2 energy of −3.9, just below the threshold of −4 (Fig. 8.12). Interestingly, the preferred orientation of this segment in amyloid is antiparallel β-sheet. Another surprise is that the predicted amyloidogenic propensity of apoE

resides mainly within the N-domain, even though the C-domain is more hydrophobic, binds Aβ, and fragments corresponding to this domain can form fibers in vitro (Cho et al. 2001). This illustrates the ability of most polypeptides to form amyloid under certain conditions. We suggest that the absence of strong amyloidogenic segments in the C-domain of apoE, which adopts a dynamic solvent-exposed conformation in solution, helps protect free protein from misfolding in vivo.

In the N-domain of apoE3, three amyloidogenic segments are predicted in residues 33–42, 113–116 and 159–163 by five to seven methods in AmylPred2 (Fig. 8.12). Mapping these segments on the NMR structure of the modified full-length protein shows that they are located in the amphipathic α-helical regions (Fig. 8.13). The only exception is residues 41 and 42 that are located in a solvent-accessible interhelical loop (blue arrow, Fig. 8.13) and hence, are potentially labile to misfolding. Notably, sites 112 and 158 that distinguish apoE isoforms are located at the

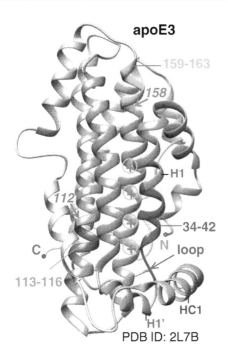

apoE3

159-163

158

H1

112

34-42

N loop

C.

113-116

HC1

H1'

PDB ID: 2L7B

Fig. 8.13 NMR structure of the full-length modified human apoE3, determined by Chen and Wang (2011). Structural domains are color-coded as in Fig. 8.12: N-terminal helix bundle (*violet*), hinge (*gray*), and C-terminal domain (*pink*). Helices H1 and H1′ with their connecting loop in the N-domain, and helix HC1 in the C-domain are indicated. Helical nomenclature follows (Chen et al. 2011). Cys/Arg substitutions at positions 112 and 158 that distinguish the apoE isoforms are in *orange*. The major predicted amyloidogenic segments are color-coded as in Fig. 8.12: residues 34–42 (*blue*), 113–116 (*teal*) and 159–163 (*green*). Residue segment 136–150 from helix H4, which encompasses the receptor and heparan sulfate binding site, is indicated (+ + + +)

edges of the two respective amyloidogenic hot spots, 113–116 and 159–163. Cys to Arg substitutions at these sites slightly reduce the amyloid-forming potential of the adjacent hot spots (profiles not shown). This is not surprising, since the amyloid-forming potential of Arg is lower than that of free Cys in most contexts. Further, amyloidogenic segment 159–163 is close to the basic stretch in residues 136–150 that form the receptor and heparan sulfate binding site in apoE (Wilson et al. 1991). We speculate that, in the absence of polyanionic ligands such as heparan sulfate proteoglycans, this basic stretch may help block the β-zipper propagation along the polypeptide chain.

Since apoE has been previously found to directly nucleate Aβ aggregation in amyloid (Ma et al. 1994; Wisniewski et al. 1995; Hatters et al. 2006), we tested whether a particular segment in apoE is likely to pair with Aβ to nucleate an intermolecular β-sheet co-polymer. To do so, we used AmylPred2 and PASTA2 to analyze a chimeric amino acid sequence comprised of apoE and Aβ$_{42}$ sequences. The two strongest hot spots in AmylPred2 came from Aβ$_{42}$ (Fig. 8.7). Likewise, PASTA2 did not predict any favorable pairings between Aβ and apoE. This analysis suggests that apoE is unlikely to nucleate amyloid formation by Aβ via the segment pairing between the two proteins.

Our analysis can be summarized as follows. First, despite its prominence in amyloid diseases, apoE has weaker sequence propensity to form amyloid than many other apos. Second, this propensity decreases slightly upon Cys to Arg substitutions at sites 112 and 158 that differentiate apoE isoforms. Hence, the amyloid-forming sequence potential increases in order E4<E3<E2, which is reverse of the isoform-specific predisposition to neurodegenerative diseases. Consequently, this pre-disposition must result from factors other than sequence propensity of apoE isoforms for β-aggregation. Together, these findings suggest that apoE misfolding per se is probably not a major factor in amyloid deposition in Alzheimer's disease.

Third, the NMR structure of free full-length apoE shows that a part of the strongest predicted amyloidogenic segment 33–42 is located in the solvent-accessible loop, which makes it potentially labile to β-aggregation. This loop (residues V40, Q41, T42, L43) links N-terminal helices H1 and H1′, and is packed against helix HC1 from the C-domain (Fig. 8.13). Since inter-domain interactions in apoE are isoform-specific, the protection of the major amyloid hot spot by these interactions is also expected to be isoform-specific. We hypothesize that: (i) in lipid-free full-length apoE, the major amyloid hot spot in residues 33–42 is protected from misfolding by secondary and tertiary structure, including inter-domain interactions; (ii) destabilization of the N-domain and changes in its interactions with the C-domain reduce this protection in an isoform-specific manner, from E2 (most stable and most protected) to E4 (least stable and least protected). These differences in protection of amyloid hot spots in the N-domain may potentially contribute to the isoform-specific effects in amyloid deposition.

8.4.7 Serum Amyloid A

Human SAA (104 a. a.) is an intrinsically disordered protein from a family of highly conserved acute-phase proteins whose normal function in immune response is subject of debate (van der Westhuyzen et al. 2007; Eklund et al. 2012; also see Chap. 5 by Colón et al. in this volume). SAA is synthesized predominantly by the liver, circulates mainly on plasma HDL, and is best known for its role in secondary amyloidosis linked to inflammation. In this serious complication, N-terminal fragments of SAA largely containing residues 1–76, termed amyloid A (AA), form extracellular fibrillar deposits in vital organs (kidney, liver, spleen) and damage them. AA amyloidosis can be triggered by chronic infections, inflammation, rheumatoid arthritis, and certain cancers, and is the most prevalent form of human systemic amyloidosis developed by 5–10 % of patients with chronic inflammation. Several SAA isoforms are found in humans, among which SAA 1.1 is most amyloidogenic in vivo. The reasons for the isoform-specific amyloid formation are unclear and were proposed to involve differences in the protein structure, stability and proteolysis (Srinivasan et al. 2013; van der Hilst et al. 2008). It is also unclear whether cleavage of the N-terminal AA fragment precedes or follows SAA misfolding in AA amyloidosis.

During inflammation, plasma levels of SAA can increase up to 1,000-fold, reaching 1 mg/ml or more, and making SAA the most abundant protein in plasma (Gillmore et al. 2001). As a result of mass action, SAA can displace most apoA-I from plasma HDL to form large dense SAA-rich HDL (van der Westhuyzen et al. 2007; Eklund et al. 2012). In addition, in acute phase response, HDL surface is saturated with SAA and a substantial fraction of SAA (up to 25 %) circulates in labile free form (van der Westhuyzen et al. 2007).

SAA on HDL is predominantly α-helical, yet this helical structure is unstable in free protein at near-physiologic conditions, which is thought to promote fibril formation in vivo (Wang et al. 2005). This structural instability of SAA has greatly complicated its atomic-resolution studies that culminated in 2014 with publication of the high-resolution x-ray crystal structure of human SAA1.1 in two crystal forms, one of which contains hexamers and probably depicts the structure of the hexameric SAA found in solution (Lu et al. 2014). Molecular structure of human SAA1 in these crystals is in excellent agreement with the structure of murine SAA3 that was crystallized as a tetramer (Derebe et al. 2014).

Our analysis of human SAA1.1 by using AmylPred2 and PASTA2 shows that the amyloid-forming propensity is modest and is confined largely to the N-terminal ~75 residues (Fig. 8.14a), which is consistent with the deposition of 1–76 residue fragment in AA amyloidosis. AmylPred2 predicts two major amyloid hot spots in residues 2–9 (SFFSFLGE) and 53–55 (WAA), with a minor hot spot in the nearby residues 67–70. Although these consensus predictions are not very strong (4–6 out of 11 methods), they are in excellent agreement with the sequence analysis by using ZipperDB and TANGO algorithms (Lu et al. 2014) as well as PASTA2 (Fig. 8.14b). The latter shows that the major N-terminal amyloid hot spot has weak pairing energy of −2.7 PASTA units which is comparable for the parallel and antiparallel β-sheet. Importantly, the predicted hot spots in SAA have been verified in experimental studies by using site-directed mutagenesis and peptide fragments (Patel et al. 1996; Egashira et al. 2011; Lu et al. 2014).

Compared to other major HDL proteins, the predicted amyloidogenic potential of SAA is comparable to that of apoA-I and much lower than that of apoA-II (Figs. 8.4, 8.6, and 8.14). This unexpected result suggests that AA amyloidosis is probably driven not so much by the high intrinsic propensity of SAA for β-aggregation but mainly by the protein overproduction (de Beer et al. 1982) combined with its structural instability and solvent exposure of the N-terminal amyloidogenic segment in free SAA monomer (Lu et al. 2014).

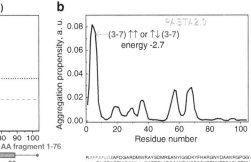
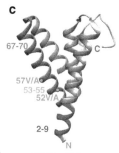

Fig. 8.14 Predicted amyloidogenic profile and the structure of human serum amyloid A. AmylPred2 (**a**) and PASTA profiles (**b**) of SAA1.1 isoform that causes human amyloidosis are shown. AmylPred2 predicts one amyloidogenic segment in residues 2–9 (*blue*) by six methods, and another in residues 65–70 (*green*) by five methods; an additional minor segment in residues 67–70 is predicted by four methods. PASTA2.0 predicts that the major N-terminal amyloidogenic segment can form parallel or antiparallel β-sheet with comparable pairing energy. *Green bar* shows fragment 1–76 that deposits in human AA amyloidosis linked to inflammation. *Rectangles* show helical segments observed in the x-ray crystal structure. Amino acid sequence shows predicted amyloidogenic segments color-coded. (**c**) X-ray crystal structure of human SAA1.1 monomer determined to 2.2 Å resolution by Lu et al. (2014). The predicted amyloidogenic segments are mapped on the structure and color-coded. *Orange dots* show positions 52 and 57 of Ala/Val substitutions that distinguish the major protein isoforms, SAA 1.1 (V52, A57), SAA 1.3 (A52, A57) and SAA 1.5 (A52, V57)

In the crystal structure, this major amyloidogenic segment in residues 2–9 is located at the end of the N-terminal amphipathic α-helix (Fig. 8.14c, blue). This segment is solvent-exposed in free SAA monomer but is sequestered in the hexamer, suggesting that hexamer dissociation is prerequisite for SAA misfolding (Lu et al. 2014). Importantly, this amyloidogenic segment overlaps the primary HDL binding site in SAA (Patel et al. 1996; Ohta et al. 2009), suggesting strongly that SAA binding via its N-terminal helix to HDL surface protects the protein from misfolding (Lu et al. 2014). The second-major amyloidogenic segment is adjacent to residues 52 and 57 that differ among the major human isoforms, SAA 1.1 (V52, A57), SAA 1.3 (A52, A57) and SAA 1.5 (A52, V57) (Fig. 8.14a, c, orange dots). Even though these isoforms are predicted to have comparable sequence propensity for β-aggregation (profiles not shown), one cannot exclude that the location of the isoform-specific mutations at the amyloidogenic segment contributes to the differences in the amyloid formation by SAA1 isoforms (Takase et al. 2014). We speculate that isoform-specific effects in SAA binding to HDL may also be a potential contributing factor in amyloidosis.

In sum, in spite of its name, serum amyloid A has only moderate intrinsic sequence propensity to form amyloid, which is comparable to that of apoA-I and significantly lower than that of apoA-II. The N-terminal end of SAA is predicted and observed to form the major amyloid hot spot that overlaps with the primary HDL binding site and forms a native amphipathic α-helix on the lipid (Ohta et al. 2009). Therefore, this segment is protected against misfolding by binding to lipid surface. In free SAA, this segment is sequestered in the hexamer, yet in the monomer it is solvent exposed and hence, labile to misfolding.

Despite structural instability of free SAA under near-physiologic conditions and its persistently high plasma concentrations in chronic inflammation, only a small percentage of patients develop AA amyloidosis ((Gillmore et al. 2001) and references therein). To explain this observation, SAA has been proposed to develop protective mechanisms against misfolding, such as the proteolysis of its amyloid-forming N-terminal part (Yamada et al. 1995) or hexamer formation (Lu et al. 2014). We hypothesize that modest intrinsic sequence propensity to form amyloid is one of such protective mechanisms.

8.4.8   Apolipoprotein B

ApoB is a 550 kDa glycoprotein synthesized by the liver. Plasma apoB contains a single chain of 4,536 amino acids, which is one of the largest known proteins and the largest apolipoprotein. One copy of full-length apoB, termed apoB100, is present as the major protein on the surface of LDL and VLDL. A shorter variant representing the first 48 % of apoB sequence, termed apoB48, is synthesized in the intestine and forms the major protein on chylomicrons (Nakajima et al. 2014). ApoB forms a flexible structural scaffold on the lipoprotein surface and an important functional ligand for binding to LDL receptor and heparan sulfate proteoglycans (Goldstein and Brown 1976; Camejo et al. 1998). In contrast to other apolipoproteins, apoB is non-exchangeable (water-insoluble) and remains associated with its host particle during metabolism. Also in contrast with the exchangeable apos, apoB is predicted and observed to contain both α-helices and β-sheets (Segrest et al. 2001).

The enormous size and insolubility of apoB have precluded its detailed structural studies. Amino acid sequence analysis suggests five large structural domains (Segrest et al. 2001). Residue segment 1–1,023, termed $\beta\alpha1$ domain, is thought to form an autonomously folded globular domain that protrudes from the LDL surface. Homology model of this domain has been proposed based on the x-ray crystal structure of a related protein, lamprey lipovitellin (Segrest et al. 1999). Epitope labeling combined with cryo-electron microscopy have determined the overall arrangement of human apoB on the mid-size LDL and showed that the protein fully encircles LDL, with the N- and C-terminal domains located near each other to close the circle (Liu and Atkinson 2011; Kumar et al. 2011).

The involvement of apoB in amyloid formation has been suggested in several in vitro studies of modified LDL, including oxidized and electro-negative LDL (Stewart et al. 2005; Parasassi et al. 2008; Brunelli et al. 2014). Aggregation of modified LDL was evident in these and other studies (Bancells et al. 2010), including our own

work (Jayaraman et al. 2007; Lu et al. 2012; Lu and Gursky 2013). However, convincing experimental evidence for amyloid formation by apoB, such as the oriented fiber diffraction and a large increase in cross-β-sheet content measured by spectroscopy, is still lacking.

Our amino acid sequence analysis reveals that apoB has unusually high amyloid-forming potential, comparable to that of Aβ peptide (Figs. 8.7 and 8.15; Table 8.1). Major amyloidogenic segments, which are predicted in all five apoB domains, span regions that are thought to form native α-helices or β-sheets. Figure 8.15 shows amyloidogenic profile in three representative portions of apoB including residues 300–400 (predicted to form native α-helical structure), residues 780–880 (located in a predicted β-sheet region), and residues 2,150–2,250 from the α_2 helical domain that was proposed to be involved in β-aggregation (Parasassi et al. 2008). Many amyloidogenic segments in these and other parts of apoB are predicted strongly by AmylPred2 (8–11 methods) and span 10 or more amino acids, longer than those in the exchangeable apos (Table 8.1). Likewise, PASTA2 predicts unusually strong favorable pairing energies below −9 units for many apoB segments, which is consistent with their high hydrophobicity. In comparison, the lowest PASTA2 energies are about −8.4 for Aβ(1–42), −7 for apoA-II and −5.5 or weaker for other exchangeable apos (Table 8.1). Furthermore, similar to most exchangeable apos, PASTA predicts parallel in-register β-sheet for all major amyloidogenic segments in apoB.

In sum, of all apolipoproteins explored, apoB has by far the strongest sequence propensity to form amyloid, which correlates with high protein hydrophobicity of this water-insoluble protein. Nevertheless, to-date there is no compelling evidence for amyloid formation by apoB in vivo or in vitro. We propose that the reason why apoB does not readily form amyloid is due to its extensive protein-lipid and protein-protein interactions on LDL and VLDL. These native interactions make this enormous protein non-exchangeable and thereby protect it from dissociation, proteolytic degradation and misfolding.

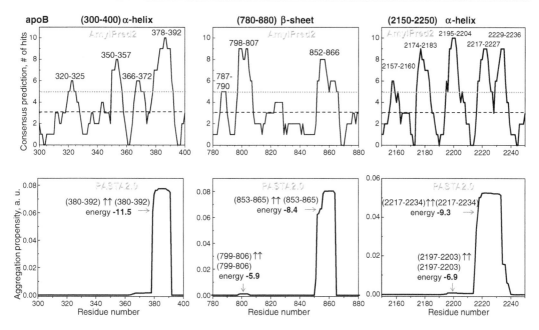

Fig. 8.15 Amyloidogenic propensity of selected portions of human apoB (total protein length 4,536 a. a.) AmylPred2 and PASTA2.0 profiles are shown for 100-residue portions that are predicted to form various native secondary structures: residues 300–400 (*left*) form mainly α-helices that are a part of the so-called A-helix region; residues 780–880 (*middle*) form mainly β-sheets that are a part of the A-sheet region; residues 2,150–2,250 (*right*) form mainly α-helices that are a part of the α₂ region (nomenclature according to Segrest et al. 2001; Wang et al. 2010). Residue numbers in *top panels* indicate amyloidogenic segments predicted by five or more methods in AmylPred2 (at or above the *dotted line*). *Bottom panels* show segments with pairing energy below −4 predicted by PASTA. Other parts of apoB have comparable predicted amyloidogenic propensity (not shown)

8.5   Summary

The studies outlined in this chapter show that amyloidogenic segments in human apolipoproteins (i) are largely hydrophobic, (ii) are located almost entirely in the native amphipathic α-helices in the exchangeable apos while in the non-exchangeable apoB they span both predicted α-helices and β-sheets, and (iii) are predicted to preferentially form parallel in-register β-sheet in amyloid. These amyloidogenic segments are enriched in aromatic residues that often form native interactions stabilizing the helix bundle structures of free proteins such as apoA-I, apoA-IV and apoE (Figs. 8.4, 8.8, and 8.13). Importantly, the predicted amyloidogenic segments have been verified in the experimental studies of apoC-II, apoA-I, apoA-II and SAA (Wilson et al. 2007; Wong et al. 2012; Adachi et al. 2014; Mendoza-Espinosa et al. 2014; Das et al. 2014; Sawashita et al. 2009; Lu et al. 2014). Amyloidogenic segments in many other globular proteins share similar characteristics: they tend to be hydrophobic (with a notable exception of prions), are enriched in aromatic groups, and are often protected from misfolding by forming well-ordered secondary structure such as α-helix (Tzotzos and Doig 2010; Sabate et al. 2012). Furthermore, parallel in-register β-sheet has emerged as the predominant structural motif observed by EPR and solid-state NMR in amyloid fibrils formed by many proteins (Tycko 2006; Margittai and Langen 2008; Shewmaker et al. 2011), including fibrils of apoC-II and α-synuclein (Teoh et al. 2011b; Lv et al. 2012; Der-Sarkissian et al. 2003).

A surprising conclusion emerging from our studies is that the intrinsic sequence propensity to form amyloid decreases in order apoB > apoA-II > apoC-II ≥ apoA-I, apoC-I, apoC-III, SAA > apoA-IV, apoA-V, apoE (Table 8.1).

Table 8.1 Summary of amyloid-forming propensity of human apolipoproteins discussed in this chapter

Protein	Amino acids		AmylPred2		PASTA2
	Total	Hot spots	Hot spot residue segments	Hot spot/total	Min energy
1. apoA-II	77	20 (31)	10–18, 60–70 (78–98)	0.26 (0.38)	−7.0
2. apoC-II	79	17	16–17, 62–76	0.21	−4.9
3. apoC-III	79	10	8–13, 52–55	0.12	−2.7
4. SAA 1.1	104	11	2–9, 53–55	0.10	−2.7
5. apoC-I	57	5	24–28	0.09	−1.9
6. apoA-I	243	22	14–22, 53–58, 227–232	0.09	−6.1
7. apoA-V	343	21	118–120, 150–151, 204–210, 269–277	0.06	−3.9
8. apoE3	299	19	33–42, 113–116, 159–163	0.06	−3.9
9. apoA-IV	376	20	7–16, 40–41, 221–222, 331–336	0.053	−4.7
10. apoB (selected regions from the total of 4,536 a. a.)					
301–400	100	35	320–325, 350–357, 366–372, 378–392	0.35	−11.5
781–880	100	19	787–790, 798–807, 852–866	0.19	−8.4
2,151–2,250	100	43	2,157–60, 2174–83, 2,195–04, 2,217–27, 2,229–36	0.43	−9.3
11. Aβ	42	20	15–22, 30–42	0.46	−8.9

Parameters for the exchangeable apos (*rows 1–9*) and for selected regions of the non-exchangeable apoB (*row 10*) are listed; Aβ(1–42) is included for comparison. Numbers in brackets refer to the naturally occurring C-terminal extension in apoA-II that causes human amyloidosis. Total number of residues in each protein or polypeptide and the number of residues predicted to be in amyloidogenic segments by five or more AmylPred2 methods (major hot spots) are listed. The ratio of these values (hot spot/total) can be used to rank amyloidogenic propensity. This propensity of the exchangeable apos decreases as listed in the top box from 1 to 9. Although other criteria, such as the minimal (most favorable) segment pairing energy by PASTA2, yield somewhat different rank order, various prediction methods consistently show that apoB has the strongest amyloidogenic potential, followed by apoA-II, apoC-II and other apos, with SAA in the middle and apoE near the end of the propensity range

This rank order was unexpected given that apoE and SAA are prominent players in human amyloidoses such as Alzheimer's disease and inflammation-linked AA amyloidosis, while apoB and normal human apoA-II are not. What is the reason for this apparent discrepancy between the predicted and observed amyloid formation in vivo? Clearly, intrinsic sequence propensity is just one of many factors determining amyloid formation in vivo and in vitro. Other important factors include stability of the native protein structure, protein concentration, external conditions such as pH, ligand binding, proteolytic environment, etc. Moreover, the probability of protein misfolding is determined not only by the global stability of its native structure, but also by the local conformation and dynamics in sensitive regions (Chiti and Dobson 2009). We propose that sequestration of such amyloidogenic regions via the native protein-protein and protein-lipid interactions protects apos against misfolding (Das et al. 2014; Gursky 2014; Lu et al. 2014).

Strong apolipoprotein-lipid interactions are expected and observed to be particularly protective. One example is apoB, the largest and most hydrophobic of all human apos, which interacts most strongly with its host lipoprotein. These strong interactions render apoB non-exchangeable and thereby protect it from forming amyloid despite its unusually high sequence propensity to do so (Fig. 8.15), comparable to that of Aβ peptide (Fig. 8.7; Table 8.1). Another example is apoA-II, the most hydrophobic exchangeable apolipoprotein that is tightly bound to the surface of mature HDL. This strong binding is reinforced by the protein dimerization via the Cys6-Cys6 disulfide bond, rendering normal human apoA-II practically non-exchangeable. We propose that this strong association with mature HDL protects wild-type human apoA-II from forming amyloid despite its high sequence propensity to do so (Fig. 8.6, Table 8.1). Notably, naturally occurring mutant apoA-II with C-terminal extension that causes human amyloidosis has high amyloid-forming propensity approaching that of Aβ (Table 8.1).

Interestingly, except for this mutant apoA-II, human apos that commonly form amyloid in vivo are consistently predicted to be in the middle of the amyloid propensity range (SAA1.1) or near its end (apoE) (Table 8.1). These proteins can readily dissociate from the lipid surface and circulate in labile free form. We propose that these proteins have evolved protective strategies against misfolding, which explains why their primary sequences have modest propensity for β-aggregation (Figs. 8.12 and 8.14, Table 8.1). For example, the protective strategy in free apoE is high structural stability of the helix bundle in the N-domain where all major amyloidogenic segments are located, combined with the low amyloidogenic sequence propensity of the flexible hinge and C-domains (Figs. 8.12 and 8.13). The protective strategy in apoA-IV, nearly half of which circulates as free protein, is sequestration of the major amyloidogenic segments via the "clasp" mechanism (Fig. 8.8). Additional protection may also be attained via self-association of free proteins, such as the formation of an antiparallel α-helical dimer by apoA-I (Fig. 8.3) or apoA-IV (Deng et al. 2012), or sequestration of the major amyloidogenic segments in SAA hexamer (Lu et al. 2014). Location of the amyloidogenic segments next to the charged stretches, such as the basic receptor-binding sites in apoA-V and apoE (Figs. 8.9 and 8.12a), is also expected to protect the proteins from β-aggregation. Finally, rapid proteolysis and clearance of free apos can serve as important protective mechanisms against amyloidosis.

We conclude that apolipoproteins' propensity to bind lipid surface and to form amyloid are two sides of the same coin, both rooted in proteins' hydrophobicity. This suggests that functional constraints on apolipoproteins make it difficult to eliminate their pathologic β-aggregation. Nevertheless, the primary sequences that are optimal for lipid surface binding and for amyloid formation are distinctly different, as illustrated in several examples. First, large apolar faces of the amphipathic α-helices or β-sheets found in the exchangeable and non-exchangeable apos are essential for binding to lipid surface but not for amyloid formation. This is exemplified by the naturally occurring C-terminal extension in human apoA-II (Fig. 8.6) which is highly amyloidogenic but cannot form a lipid-binding amphipathic α-helix (Gursky 2014). Other examples include apoA-V and the C-domain of apoE, which are highly lipophilic but not highly amyloidogenic (Figs. 8.9 and 8.12). Furthermore, stretches of similarly charged amino acids, such as the receptor and heparan binding sites in apoE, apoA-V and apoB, are well-tolerated in the polar faces of the lipid-binding class-A α-helices, yet such stretches are expected to disrupt amyloid structure due to their intra- and intermolecular electrostatic repulsion in a parallel in-register β-sheet. These examples illustrate various strategies potentially employed during evolution to optimize reversible binding of apolipoproteins to lipid surface while reducing the risk of amyloid formation.

In sum, the reason why apolipoproteins are overrepresented in amyloid diseases is not due to their native class-A amphipathic α-helices per se. Rather, it is rooted in increased apolipoprotein hydrophobicity that is not only required for lipid binding but also increases protein propensity for β-aggregation. In fact, class-A apolipoprotein α-helices are not necessarily more amyloidogenic than their class-G counterparts, as suggested by the highly amyloidogenic hydrophobic C-terminal segment in apoC-II that forms a native class-G helix (Figs. 8.10b and 8.11, Table 8.1; also see Chap. 7 by Ryan et al. in this volume). Furthermore, the apolipoprotein α-helices do not necessarily have higher amyloid-forming propensity than their β-sheet counterparts. This is evident from the amino acid sequence analysis of apoB, the most amyloidogenic of all apos and the only one with significant β-sheet content. Numerous amyloidogenic segments span the whole length of apoB including predicted α-helices and β-sheets (Fig. 8.15). Importantly, the amyloidogenic segments in the exchangeable apos are located in the amphipathic α-helices and overlap their apolar lipid-binding faces, which probably protects them from misfolding. Likewise, amyloidogenic regions in globular proteins often overlap with their functional interfaces, suggesting a competition between normal protein function and pathogenic β-aggregation (Castillo and Ventura 2009).

This competition implies that increasing apolipoprotein retention on the lipid surface may provide a potential therapeutic strategy against amyloidosis. For example, increased levels of plasma triacylglycerol are well-known to promote apoA-I dissociation from the HDL surface upon enzymatic lipolysis of core TG (Lamarche et al. 1999). Consistent with this observation, clinical studies show that apoA-I amyloid deposits in atherosclerotic arteries correlate directly only with patients' age and with the plasma levels of TG (Röcken et al. 2006). We propose that this correlation may reflect causation, as elevated levels of TG in HDL induce generation of free apoA-I that is the precursor of amyloid. This prompts us to speculate that TG-reducing therapies may help protect apoA-I against dissociation from the lipid, and thereby delay amyloid formation in vivo (Das et al. 2014).

Acknowledgement We thank all members of the Gursky laboratory for useful discussions and support. Special thanks are due to Donald L. Gantz for editorial assistance and to Dr. Oxana Galzitskaya for critical comments and help with apolipoprotein ranking. This work was supported by the National Institutes of Health grant GM067260 to OG.

References

Acharya P, Segall ML, Zaiou M, Morrow J, Weisgraber KH, Phillips MC, Lund-Katz S, Snow J (2002) Comparison of the stabilities and unfolding pathways of human apolipoprotein E isoforms by differential scanning calorimetry and circular dichroism. Biochim Biophys Acta 1584(1):9–19

Adachi E, Nakajima H, Mizuguchi C, Dhanasekaran P, Kawashima H, Nagao K, Akaji K, Lund-Katz S, Phillips MC, Saito H (2013) Dual role of an N-terminal amyloidogenic mutation in apolipoprotein A-I: destabilization of helix bundle and enhancement of fibril formation. J Biol Chem 288:2848–2856

Adachi E, Kosaka A, Tsuji K, Mizuguchi C, Kawashima H, Shigenaga A, Nagao K, Akaji K, Otaka A, Saito H (2014) The extreme N-terminal region of human apolipoprotein A-I has a strong propensity to form amyloid fibrils. FEBS Lett 588(3):389–394

Arciello A, De Marco N, Del Giudice R, Guglielmi F, Pucci P, Relini A, Monti DM, Piccoli R (2011) Insights into the fate of the N-terminal amyloidogenic polypeptide of ApoA-I in cultured target cells. J Cell Mol Med 15:2652–2663

Asatryan L, Hamilton RT, Isas JM, Hwang J, Kayed R, Sevanian A (2005) LDL phospholipid hydrolysis produces modified electronegative particles with an unfolded apoB-100 protein. J Lipid Res 46(1):115–122

Banach M (2014) Editorial: current research, knowledge and controversies on high density lipoprotein. Curr Med Chem 21(25):2853–2854

Bancells C, Villegas S, Blanco FJ, Benitez S, Gallego I, Beloki L, Perez-Cuellar M, Ordonez-Llanos J, Sanchez-Quesada JL (2010) Aggregated electronegative low-density lipoprotein in human plasma shows high tendency to phospholipolysis and particle fusion. J Biol Chem 285(42):32425–32435

Beckstead JA, Wong K, Gupta V, Wan CP, Cook VR, Weinberg RB, Weers PM, Ryan RO (2007) The C terminus of apolipoprotein A-V modulates lipid-binding activity. J Biol Chem 282(21):15484–15489

Benson MD, Liepnieks JJ, Yazaki M, Yamashita T, Hamidi Asl K, Guenther B, Kluve-Beckerman B (2001) A new human hereditary amyloidosis: the result of a stop-codon mutation in the apolipoprotein AII gene. Genomics 72(3):272–277

Benson MD, Kalopissis AD, Charbert M, Liepnieks JJ, Kluve-Beckerman B (2011) A transgenic mouse model of human systemic ApoA-II amyloidosis. Amyloid 18(Suppl 1):32–33

Bergström J, Murphy CL, Weiss DT, Solomon A, Sletten K, Hellman U, Westermark P (2004) Two different types of amyloid deposits – apolipoprotein A-IV and transthyretin – in a patient with systemic amyloidosis. Lab Invest 84(8):981–988

Borhani DW, Rogers DP, Engler JA, Brouillette CG (1997) Crystal structure of truncated human apolipoprotein A-I suggests a lipid-bound conformation. Proc Natl Acad Sci U S A 94(23):12291–12296

Brouillette CG, Anantharamaiah GM, Engler JA, Borhani DW (2001) Structural models of human apolipoprotein A-I: a critical analysis and review. Biochim Biophys Acta 1531(1–2):4–46

Brunelli R, De Spirito M, Mei G, Papi M, Perrone G, Stefanutti C, Parasassi T (2014) Misfolding of apoprotein B-100, LDL aggregation and 17-β-estradiol in atherogenesis. Curr Med Chem 21(20):2276–2283

Camejo G, Hurt-Camejo E, Wiklund O, Bondjers G (1998) Association of apo B lipoproteins with arterial proteoglycans: pathological significance and molecular basis. Atherosclerosis 139(2):205–222

Castelli WP, Anderson K, Wilson PW, Levy D (1992) Lipids and risk of coronary heart disease. The Framingham study. Ann Epidemiol 2(1–2):23–28

Castillo V, Ventura S (2009) Amyloidogenic regions and interaction surfaces overlap in globular proteins related to conformational diseases. PLoS Comput Biol 5(8):e1000476

Caviglio G, Geier EG, Shao B, Heinecke JW, Oda MN (2010) Exchange of apolipoprotein A-I between lipid-associated and lipid-free states: a potential target or oxidative generation of dysfunctional high density lipoproteins. J Biol Chem 285:18847–18857

Chan GK, Witkowski A, Gantz DL, Zhang TO, Zanni MT, Jayaraman S, Caviglio G (2015) Myeloperoxidase-mediated methionine oxidation promotes an amyloidogenic outcome for apolipoprotein A-I. J Biol Chem 290(17):10958–10971

Chen J, Li Q, Wang J (2011) Topology of human apolipoprotein E3 uniquely regulates its diverse biological functions. Proc Natl Acad Sci U S A 108(36):14813–14818

Chetty PS, Mayne L, Lund-Katz S, Stranz D, Englander SW, Phillips MC (2009) Helical structure and stability in human apolipoprotein A-I by hydrogen exchange and mass spectrometry. Proc Natl Acad Sci U S A 106:19005–19010

Chetty PS, Ohshiro M, Saito H, Dhanasekaran P, Lund-Katz S, Mayne L, Englander W, Phillips MC (2012) Effects of the Iowa and Milano mutations on apolipoprotein A-I structure and dynamics determined by hydrogen exchange and mass spectrometry. Biochemistry 51:8993–9001

Chiti F, Dobson CM (2009) Amyloid formation by globular proteins under native conditions. Nat Chem Biol 5(1):15–22

Cho HS, Hyman BT, Greenberg SM, Rebeck GW (2001) Quantitation of apoE domains in Alzheimer disease brain suggests a role for apoE in Aβ aggregation. J Neuropathol Exp Neurol 60:342–349

Conchillo-Solé O, de Groot NS, Avilés FX, Vendrell J, Daura X, Ventura S (2007) AGGRESCAN: a server for the prediction and evaluation of "hot spots" of aggregation in polypeptides. BMC Bioinformatics 8:65

Cui Y, Huang M, He Y, Zhang S, Luo Y (2011) Genetic ablation of apolipoprotein A-IV accelerates Alzheimer's disease pathogenesis in a mouse model. Am J Pathol 178:1298–1308

Das M, Jayaraman S, Mei X, Atkinson D, Gursky O (2014) Amyloidogenic mutations in human apolipoprotein A-I are not necessarily destabilizing: a common mechanism of apoA-I misfolding in familial amyloidosis and atherosclerosis. FEBS J 281(11):2525–2542

De Beer FC, Mallya RK, Fagan EA, Lanham JG, Hughes GR, Pepys MB (1982) Serum amyloid-A protein concentration in inflammatory diseases and its relationship to the incidence of reactive systemic amyloidosis. Lancet 2(8292):231–234

de Messieres M, Huang RK, He Y, Lee JC (2014) Amyloid triangles, squares, and loops of apolipoprotein C-III. Biochemistry 53(20):3261–3263

Deng X, Morris J, Dressmen J, Tubb MR, Tso P, Jerome WG, Davidson WS, Thompson TB (2012) The structure of dimeric apolipoprotein A-IV and its mechanism of self-association. Structure 20(5):767–779

Deng X, Morris J, Chaton C, Schröder GF, Davidson WS, Thompson TB (2013) Small-angle X-ray scattering of apolipoprotein A-IV reveals the importance of its termini for structural stability. J Biol Chem 288(7):4854–4866

Deng X, Walker RG, Morris J, Davidson WS, Thompson TB (2015) Role of conserved proline residues in human apolipoprotein A-IV structure and function. J Biol Chem 290(17):10689–10702

Derebe MG, Zlatkov CM, Gattu S, Ruhn KA, Vaishnava S, Diehl GE, MacMillan JB, Williams NS, Hooper LV (2014) Serum amyloid A is a retinol binding protein that transports retinol during bacterial infection. Elife 3:e03206

Der-Sarkissian A, Jao CC, Chen J, Langen R (2003) Structural organization of α-synuclein fibrils studied by site-directed spin labeling. J Biol Chem 278:37530–37535

Egashira M, Takase H, Yamamoto I, Tanaka M, Saito H (2011) Identification of regions responsible for heparin-induced amyloidogenesis of human serum amyloid A using its fragment peptides. Arch Biochem Biophys 511(1–2):101–1016

Eklund KK, Niemi K, Kovanen PT (2012) Immune functions of serum amyloid A. Crit Rev Immunol 32(4):335–348

Fernandez-Escamilla AM, Rousseaux F, Schymkowitz J, Serrano L (2004) Prediction of sequence-dependent and mutational effects on the aggregation of peptides and proteins. Nat Biotechnol 22:1302–1306

Galzitskaya OV, Garbuzynskiy SG, Lobanov MV (2006) Prediction of amyloidogenic and disordered regions in protein chains. PLoS Comput Biol 2(12):e177

Gangabadage CS, Zdunek J, Tessari M, Nilsson S, Olivecrona G, Wijmenga SS (2008) Structure and dynamics of human apolipoprotein CIII. J Biol Chem 283(25):17416–17427

Gao X, Yuan S, Jayaraman S, Gursky O (2012) Effect of apolipoprotein A-II on the structure and stability of human high-density lipoprotein: implications for the role of apoA-II in HDL metabolism. Biochemistry 51(23):4633–4641

Garbuzynskiy SO, Lobanov MY, Galzitskaya OV (2010) FoldAmyloid: a method of prediction of amyloidogenic regions from protein sequence. Bioinformatics 26(3):326–332

Getz GS, Reardon CA (2009) Apoprotein E as a lipid transport and signaling protein in the blood, liver and artery wall. J Lipid Res 50:S156–S161

Gillard BK, Lin HY, Massey JB, Pownall HJ (2009) Apolipoproteins A-I, A-II and E are independently distributed among intracellular and newly secreted HDL of human hepatoma cells. Biochim Biophys Acta 1791:1125–1132

Gillmore JD, Lovat LB, Persey MR, Pepys MB, Hawkins PN (2001) Amyloid load and clinical outcome in AA amyloidosis in relation to circulating concentration of serum amyloid A protein. Lancet 358(9275): 24–29

Gillmore JD, Stangou AJ, Lachmann HJ, Goodman HJ, Wechalekar AD, Acheson J, Tennent GA, Bybee A, Gilbertson J, Rowczenio D, O'Grady J, Heaton ND, Pepys MB, Hawkins PN (2006) Organ transplantation in hereditary apolipoprotein AI amyloidosis. Am J Transplant 6(10):2342–2347

Goldstein JL, Brown MS (1976) The LDL pathway in human fibroblasts: a receptor-mediated mechanism for the regulation of cholesterol metabolism. Curr Top Cell Regul 11:147–181

Goldstein JL, Brown MS (1992) Lipoprotein receptors and the control of plasma LDL cholesterol levels. Eur Heart J 13(Suppl B):34–36

Gordon DJ, Probstfield JL, Garrison RJ, Neaton JD, Castelli WP, Knoke JD, Jacobs DR Jr, Bangdiwala S, Tyroler HA (1989) High-density lipoprotein cholesterol and cardiovascular disease. Circulation 79:8–15

Green PH, Glickman RM, Riley JW, Quinet E (1980) Human apolipoprotein A-IV. Intestinal origin and distribution in plasma. J Clin Invest 65:911–919

Guha M, England CO, Herscovitz H, Gursky O (2007) Thermal transitions in human very low-density lipoprotein: fusion, rupture and dissociation of HDL-like particles. Biochemistry 46(20):6043–6049

Guha M, Gao X, Jayaraman S, Gursky O (2008) Structural stability and functional remodeling of high-density lipoproteins: the importance of being disordered. Biochemistry 47(44):11393–11397

Gursky O (2005) Apolipoprotein structure and dynamics. Curr Opin Lipidol 16(3):287–294

Gursky O (2013) Crystal structure of Δ(185–243)apoA-I suggests a mechanistic framework for the protein adaptation to the changing lipid load in good cholesterol: from flatland to sphereland via double belt, belt buckle, double hairpin and trefoil/tetrafoil. J Mol Biol 425:1–16

Gursky O (2014) Hot spots in apolipoprotein A-II misfolding and amyloidosis in mice and men. FEBS Lett 588(6):845–850

Gursky O, Atkinson D (1996a) Thermal unfolding of human high-density apolipoprotein A-1: implications for a lipid-free molten globular state. Proc Natl Acad Sci U S A 93:2991–2995

Gursky O, Atkinson D (1996b) High- and low-temperature unfolding of human high-density apolipoprotein A-2. Protein Sci 5(9):1874–1882

Gursky O, Atkinson D (1998) Thermodynamic analysis of human plasma apolipoprotein C-1: high-temperature unfolding and low-temperature oligomer dissociation. Biochemistry 37(5):1283–1291

Gursky O, Mei X, Atkinson D (2012) Crystal structure of the C-terminal truncated apolipoprotein A-I sheds new light on the amyloid formation by the N-terminal segment. Biochemistry 51(1):10–18

Hamodrakas SJ, Liappa C, Iconomidou VA (2007) Consensus prediction of amyloidogenic determinants in amyloid-forming proteins. Int J Biol Macromol 41:295–300

Hatters DM, Howlett GJ (2002) The structural basis for amyloid formation by plasma apolipoproteins: a review. Eur Biophys J 31(1):2–8

Hatters DM, Budamagunta MS, Voss JC, Weisgraber KH (2005) Modulation of apolipoprotein E structure by domain interaction: differences in lipid-bound and lipid-free forms. J Biol Chem 280(40):34288–34295

Hatters DM, Zhong N, Rutenber E, Weisgraber KH (2006) Amino-terminal domain stability mediates apolipoprotein E aggregation into neurotoxic fibrils. J Mol Biol 361(5):932–944

Hauser PS, Ryan RO (2013) Impact of apolipoprotein E on Alzheimer's disease. Curr Alzheimer Res 10(8): 809–817

Hauser PS, Narayanaswami V, Ryan RO (2011) Apolipoprotein E: from lipid transport to neurobiology. Prog Lipid Res 50(1):62–74

Holtzman DM, Herz J, Bu G (2012) Apolipoprotein E and apolipoprotein E receptors: normal biology and roles in Alzheimer disease. Cold Spring Harb Perspect Med 2(3):a006312

Howlett GJ, Moore KJ (2006) Untangling the role of amyloid in atherosclerosis. Curr Opin Lipidol 17(5):541–547

Huang Y (2010) Mechanisms linking apolipoprotein E isoforms with cardiovascular and neurological diseases. Curr Opin Lipidol 21(4):337–345

Jayaraman S, Gantz DL, Gursky O (2005a) Kinetic stabilization and fusion of discoidal lipoproteins containing human apoA-2 and DMPC: comparison with apoA-1 and apoC-1. Biophys J 88(4):2907–2918

Jayaraman S, Gantz DL, Gursky O (2005b) Structural basis for thermal stability of human low-density lipoprotein. Biochemistry 44(10):3965–3971

Jayaraman S, Gantz DL, Gursky O (2006) Effects of salt on the thermal stability of human plasma high-density lipoprotein. Biochemistry 45(14):4620–4628

Jayaraman S, Gantz DL, Gursky O (2007) Effects of oxidation on the structure and stability of human low-density lipoprotein. Biochemistry 46(19):5790–5797

Jayaraman S, Cavigiolio G, Gursky O (2012) Folded functional lipid-poor apolipoprotein A-I obtained by heating of high-density lipoproteins: relevance to HDL biogenesis. Biochem J 442(3):703–712

Jong MC, Hofker MH, Havekes LM (1999) Role of ApoCs in lipoprotein metabolism: functional differences between ApoC1, ApoC2, and ApoC3. Arterioscler Thromb Vasc Biol 19(3):472–484

Kei AA, Filippatos TD, Tsimihodimos V, Elisaf MS (2012) A review of the role of apolipoprotein C-II in lipoprotein metabolism and cardiovascular disease. Metabolism 61(7):906–921

Kim C, Choi J, Lee SJ, Welsh WJ, Yoon S (2009) NetCSSP: web application for predicting chameleon sequences and amyloid fibril formation. Nucleic Acids Res 37:W469–W473

Kingwell BA, Chapman MJ, Kontush A, Miller NE (2014) HDL-targeted therapies: progress, failures and future. Nat Rev Drug Discov 13(6):445–464

Krauss RM (1998) Atherogenicity of triglyceride-rich lipoproteins. Am J Cardiol 81(4A):13B–17B

Kumar V, Butcher SJ, Öörni K, Engelhardt P, Heikkonen J, Kaski K, Ala-Korpela M, Kovanen PT (2011) Three-dimensional cryoEM reconstruction of native LDL particles to 16Å resolution at physiological body temperature. PLoS ONE 6:e18841

Ladu MJ, Reardon C, Van Eldik L, Fagan AM, Bu G, Holtzman D, Getz GS (2000) Lipoproteins in the central nervous system. Ann N Y Acad Sci 903:167–175

Lamarche B, Rashid S, Lewis GF (1999) HDL metabolism in hypertriglyceridemic states: an overview. Clin Chim Acta 286(1–2):145–161

Lane-Donovan C, Philips GT, Herz J (2014) More than cholesterol transporters: lipoprotein receptors in CNS function and neurodegeneration. Neuron 83(4):771–787

Lepedda AJ, Cigliano A, Cherchi GM, Spirito R, Maggioni M, Carta F, Turrini F, Edelstein C, Scanu AM, Formato M (2009) A proteomic approach to differentiate histologically classified stable and unstable plaques from human carotid arteries. Atherosclerosis 203(1):112–118

Lewis TL, Cao D, Lu H, Mans RA, Su YR, Jungbauer L, Linton MF, Fazio S, LaDu MJ, Li L (2010) Overexpression of human apolipoprotein A-I preserves cognitive function and attenuates neuroinflammation and cerebral amyloid angiopathy in a mouse model of Alzheimer disease. J Biol Chem 285(47):36958–36968

Li WH, Tanimura M, Luo CC, Datta S, Chan L (1988) The apolipoprotein multigene family: biosynthesis, structure, structure-function relationships, and evolution. J Lipid Res 29(3):245–271

Liu Y, Atkinson D (2011) Immuno-electron cryomicroscopy imaging reveals a looped topology of apoB at the surface of human LDL. J Lipid Res 52(6):1111–1116

Lu M, Gursky O (2013) Aggregation and fusion of low-density lipoproteins in vivo and in vitro. Biomol Concepts. doi:10.155/bmc-2013-0016

Lu M, Gantz DL, Herscovitz H, Gursky O (2012) Kinetic analysis of thermal stability of human low-density lipoproteins: model for LDL fusion in atherogenesis. J Lipid Res 53(10):2175–2185

Lu JX, Qiang W, Yau WM, Schwieters CD, Meredith SC, Tycko R (2013) Molecular structure of β-amyloid fibrils in Alzheimer's disease brain tissue. Cell 154(6):1257–1268

Lu J, Yu Y, Zhu I, Cheng Y, Sun PD (2014) Structural mechanism of serum amyloid A-mediated inflammatory amyloidosis. Proc Natl Acad Sci U S A 111(14):5189–5194

Lv G, Kumar A, Giller K, Orcellet ML, Riedel D, Fernández CO, Becker S, Lange A (2012) Structural comparison of mouse and human α synuclein amyloid fibrils by solid-state NMR. J Mol Biol 420(1–2):99–111

Lyssenko NN, Hata M, Dhanasekaran P, Nickel M, Nguyen D, Chetty PS, Saito H, Lund-Katz S, Phillips MC (2012) Influence of C-terminal α-helix hydrophobicity and aromatic amino acid content on apolipoprotein A-I functionality. Biochim Biophys Acta 1821(3):456–463

Ma J, Yee A, Brewer HB Jr, Das S, Potter H (1994) Amyloid-associated proteins alpha 1-antichymotrypsin and apolipoprotein E promote assembly of Alzheimer beta-protein into filaments. Nature 372(6501):92–94

MacRaild CA, Hatters DM, Howlett GJ, Gooley PR (2001) NMR structure of human apolipoprotein C-II in the presence of sodium dodecyl sulfate. Biochemistry 40(18):5414–5421

MacRaild CA, Howlett GJ, Gooley PR (2004) The structure and interactions of human apolipoprotein C-II in dodecyl phosphocholine. Biochemistry 43(25):8084–8093

Magy N, Liepnieks JJ, Yazaki M, Kluve-Beckerman B, Benson MD (2003) Renal transplantation for apolipoprotein AII amyloidosis. Amyloid 10(4):224–228

Mahley RW, Innerarity TL, Rall SC Jr, Weisgraber KH (1984) Plasma lipoproteins: apolipoprotein structure and function. J Lipid Res 25(12):1277–1294

Mahley RW, Weisgraber K, Huang Y (2009) Apolipoprotein E: structure determines function, from atherosclerosis to Alzheimer's disease to AIDS. J Lipid Res 50:S183–S18

Maiga SF, Kalopissis AD, Chabert M (2014) Apolipoprotein A-II is a key regulatory factor of HDL metabolism as appears from studies with transgenic animals and clinical outcomes. Biochimie 96:56–66

Marchesi M, Parolini C, Valetti C, Mangione P, Obici L, Giorgetti S, Raimondi S, Donadei S, Gregorini G, Merlini G, Stoppini M, Chiesa G, Bellotti V (2011) The intracellular quality control system down-regulates the secretion of amyloidogenic apolipoprotein A-I variants: a possible impact on the natural history of the disease. Biochim Biophys Acta 1812(1):87–93

Margittai M, Langen R (2008) Fibrils with parallel in-register structure constitute a major class of amyloid fibrils: molecular insights from electron paramagnetic resonance spectroscopy. Rev Biophys 41(3–4):265–297

Mauldin K, Lee BL, Oleszczuk M, Sykes BD, Ryan RO (2010) The carboxyl-terminal segment of apolipoprotein A-V undergoes a lipid-induced conformational change. Biochemistry 49(23):4821–4826

Maurer-Stroh S, Debulpaep M, Kuemmerer N, Lopez de la Paz M, Martins IC, Reumers J, Morris KL, Copland A, Serpell L, Serrano L, Schymkowitz JW, Rousseau F (2010) Exploring the sequence determinants of amyloid structure using position-specific scoring matrices. Nat Methods 7(3):237–242

Mehta R, Gantz DL, Gursky O (2003) Human plasma high-density lipoproteins are stabilized by kinetic factors. J Mol Biol 328(1):183–192

Mei X, Atkinson D (2011) Crystal structure of C-terminal truncated apolipoprotein A-I reveals the assembly of high density lipoprotein (HDL) by dimerization. J Biol Chem 286(44):38570–38582

Mendoza-Espinosa P, Montalvan-Sorrosa D, Garcia-Gonzalez V, Moreno A, Castillo R, Mas-Oliva J (2014) Microenvironmentally controlled secondary structure motifs of apolipoprotein A-I derived peptides. Mol Cell Biochem 393(1–2):99–109

Morrow JA, Segall ML, Lund-Katz S, Phillips MC, Knapp M, Rupp B, Weisgraber KH (2000) Differences in stability among the human apolipoprotein E isoforms determined by the amino-terminal domain. Biochemistry 39(38):11657–11666

Morrow JA, Hatters DM, Lu B, Hochtl P, Oberg KA, Rupp B, Weisgraber KH (2002) Apolipoprotein E4 forms a molten globule. A potential basis for its association with disease. J Biol Chem 277(52):50380–50385

Mucchiano GI, Haggqvist B, Sletten K, Westermark P (2001) Apolipoprotein A-1-derived amyloid in atherosclerotic plaques of the human aorta. J Pathol 193(2):270–275

Mulya A, Lee JY, Gebre AK, Boudyguina EY, Chung SK, Smith TL, Colvin PL, Jiang XC, Parks JS (2008) Initial interaction of apoA-I with ABCA1 impacts in vivo metabolic fate of nascent HDL. J Lipid Res 49(11):2390–2401

Nagao K, Hata M, Tanaka K, Takechi Y, Nguyen D, Dhanasekaran P, Lund-Katz S, Phillips MC, Saito H (2014) The roles of C-terminal helices of human apolipoprotein A-I in formation of high-density lipoprotein particles. Biochim Biophys Acta 1841(1):80–87

Nakajima K, Nagamine T, Fujita MQ, Ai M, Tanaka A, Schaefer E (2014) Apolipoprotein B-48: a unique marker of chylomicron metabolism. Adv Clin Chem 64:117–177

Navab M, Reddy ST, Van Lenten BJ, Fogelman AM (2011) HDL and cardiovascular disease: atherogenic and atheroprotective mechanisms. Nat Rev Cardiol 8(4):222–232

Nguyen D, Dhanasekaran P, Nickel M, Mizuguchi C, Watanabe M, Saito H, Phillips MC, Lund-Katz S (2014) Influence of domain stability on the properties of human apolipoprotein E3 and E4 and mouse Apolipoprotein E. Biochemistry 53(24):4025–4033

Nilsson SK, Christensen S, Raarup MK, Ryan RO, Nielsen MS, Olivecrona G (2008) Endocytosis of apolipoprotein A-V by members of the low density lipoprotein receptor and the VPS10p domain receptor families. J Biol Chem 283(38):25920–25927

Nilsson SK, Heeren J, Olivecrona G, Merkel M (2011) Apolipoprotein A-V; a potent triglyceride reducer. Atherosclerosis 219(1):15–21

Obici L, Franceschini G, Calabresi L, Giorgetti S, Stoppini M, Merlini G, Bellotti V (2006) Structure, function and amyloidogenic propensity of apolipoprotein A-I. Amyloid 13(4):191–205

O'Donnell CW, Waldispühl J, Lis M, Halfmann R, Devadas S, Lindquist S, Berger B (2011) A method for probing the mutational landscape of amyloid structure. Bioinformatics 27(13):i34–i42

Ohta S, Tanaka M, Sakakura K, Kawakami T, Aimoto S, Saito H (2009) Defining lipid-binding regions of human serum amyloid A using its fragment peptides. Chem Phys Lipids 162(1–2):62–68

Parasassi T, De Spirito M, Mei G, Brunelli R, Greco G, Lenzi L, Maulucci G, Nicolai E, Papi M, Arcovito G, Tosatto SC, Ursini F (2008) Low density lipoprotein misfolding and amyloidogenesis. FASEB J 22(7):2350–2356

Patel H, Bramall J, Waters H, De Beer MC, Woo P (1996) Expression of recombinant human serum amyloid A in mammalian cells and demonstration of the region necessary for high-density lipoprotein binding and amyloid fibril formation by site-directed mutagenesis. Biochem J 318(Pt 3):1041–1049

Patel S, Chung SH, White G, Bao S, Celermajer DS (2010) The "atheroprotective" mediators apolipoprotein A-I and Foxp3 are over-abundant in unstable carotid plaques. Int J Cardiol 145(2):183–187

Paz M, Serrano L (2004) Sequence determinants of amyloid fibril formation. Proc Natl Acad Sci U S A 101:87–92

Phillips MC (2013) New insights into the determination of HDL structure by apolipoproteins. J Lipid Res 54(8):2034–2048

Phillips MC (2014) Molecular mechanisms of cellular cholesterol efflux. J Biol Chem 289(35):24020–24029

Rached FH, Chapman MJ, Kontush A (2014) An overview of the new frontiers in the treatment of atherogenic dyslipidemias. Clin Pharmacol Ther 96(1):57–63

Ramella NA, Rimoldi OJ, Prieto ED, Schinella GR, Sanchez SA, Jaureguiberry MS, Vela ME, Ferreira ST, Tricerri MA (2011) Human apolipoprotein A-I-derived amyloid: its association with atherosclerosis. PLoS ONE 6(7):e22532

Ramella NA, Schinella GR, Ferreira ST, Prieto ED, Vela ME, Ríos JL, Tricerri MA, Rimoldi OJ (2012) Human apolipoprotein A-I natural variants: molecular mechanisms underlying amyloidogenic propensity. PLoS ONE 7(8):e43755

Röcken C, Tautenhahn J, Bühling F, Sachwitz D, Vöckler S, Goette A, Bürger T (2006) Prevalence and pathology of amyloid in atherosclerotic arteries. Arterioscler Thromb Vasc Biol 26:676–677

Rothblat GH, Phillips MC (2010) High-density lipoprotein heterogeneity and function in reverse cholesterol transport. Curr Opin Lipidol 21:229–238

Rowczenio D, Dogan A, Theis JD, Vrana JA, Lachmann HJ, Wechalekar AD, Gilbertson JA, Hunt T, Gibbs SD, Sattianayagam PT, Pinney JH, Hawkins PN, Gillmore JD (2011) Amyloidogenicity and clinical phenotype associated with five novel mutations in apolipoprotein A-I. Am J Pathol 179(4):1978–1987

Rozek A, Sparrow JT, Weisgraber KH, Cushley RJ (1999) Conformation of human apolipoprotein C-I in a lipid-mimetic environment determined by CD and NMR spectroscopy. Biochemistry 38(44):14475–14484

Rye KA, Barter PJ (2004) Formation and metabolism of prebeta-migrating, lipid-poor apolipoprotein A-I. Arterioscler Thromb Vasc Biol 24:421–428

Sabate R, Espargaro A, Graña-Montes R, Reverter D, Ventura S (2012) Native structure protects SUMO proteins from aggregation into amyloid fibrils. Biomacromolecules 13(6):1916–1926

Safar JG, Wille H, Geschwind MD, Deering C, Latawiec D, Serban A, King DJ, Legname G, Weisgraber KH, Mahley RW, Miller BL, Dearmond SJ, Prusiner SB (2006) Human prions and plasma lipoproteins. Proc Natl Acad Sci U S A 103(30):11312–11317

Sawashita J, Kametani F, Hasegawa K, Tsutsumi-Yasuhara S, Zhang B, Yan J, Mori M, Naiki H, Higuchi K (2009) Amyloid fibrils formed by selective N-, C-terminal sequences of mouse apolipoprotein A-II. Biochim Biophys Acta 1794(10):1517–1529

Segrest JP, Pownall HJ, Jackson RL, Glenner GG, Pollock PS (1976) Amyloid A: amphipathic helixes and lipid binding. Biochemistry 15(15):3187–3191

Segrest JP, De Loof H, Dohlman JG, Brouillette CG, Anantharamaiah GM (1990) Amphipathic helix motif: classes and properties. Proteins 8:103–117

Segrest JP, Jones MK, De Loof H, Brouillette CG, Venkatachalapathi YV, Anantharamaiah GM (1992) The amphipathic helix in the exchangeable apolipoproteins: a review of secondary structure and function. J Lipid Res 33(2):141–166

Segrest JP, Garber DW, Brouillette CG, Harvey SC, Anantharamaiah GM (1994) The amphipathic alpha helix: a multifunctional structural motif in plasma apolipoproteins. Adv Protein Chem 45:303–369

Segrest JP, Jones MK, Dashti N (1999) N-terminal domain of apolipoprotein B has structural homology to lipovitellin and microsomal triglyceride transfer protein: a "lipid pocket" model for self-assembly of apob-containing lipoprotein particles. J Lipid Res 40(8):1401–1416

Segrest JP, Jones MK, De Loof H, Dashti N (2001) Structure of apolipoprotein B-100 in low density lipoproteins. J Lipid Res 42(9):1346–1367

Sehayek E, Eisenberg S (1991) Mechanisms of inhibition by apolipoprotein C of apolipoprotein E-dependent cellular metabolism of human triglyceride-rich lipoproteins through the low density lipoprotein receptor pathway. J Biol Chem 266(27):18259–18267

Shachter NS (2001) Apolipoproteins C-I and C-III as important modulators of lipoprotein metabolism. Curr Opin Lipidol 12(3):297–304

Sharma V, Forte TM, Ryan RO (2013) Influence of apolipoprotein A-V on the metabolic fate of triacylglycerol. Curr Opin Lipidol 24(2):153–159

Shewmaker F, McGlinchey RP, Wickner RB (2011) Structural insights into functional and pathological amyloid. J Biol Chem 286(19):16533–16540

Shih YH, Tsai KJ, Lee CW, Shiesh SC, Chen WT, Pai MC, Kuo YM (2014) Apolipoprotein C-III is an Amyloid-β-binding protein and an early marker for Alzheimer's disease. J Alzheimers Dis 41(3):855–865

Silva RA, Schneeweis LA, Krishnan SC, Zhang X, Axelsen PH, Davidson WS (2007) The structure of apolipoprotein A-II in discoidal high density lipoproteins. J Biol Chem 282(13):9713–9121

Silva BA, Breydo L, Uversky VN (2013) Targeting the chameleon: a focused look at α-synuclein and its roles in neurodegeneration. Mol Neurobiol 47(2):446–459

Sorci-Thomas MG, Thomas MJ (2002) The effects of altered apolipoprotein A-I structure on plasma HDL concentration. Trends Cardiovasc Med 12(3):121–128

Sorci-Thomas MG, Thomas MJ (2013) Why targeting HDL should work as a therapeutic tool, but has not. J Cardiovasc Pharmacol 62(3):239–246

Srinivasan S, Patke S, Wang Y, Ye Z, Litt J, Srivastava SK, Lopez MM, Kurouski D, Lednev IK, Kane RS, Colón W (2013) Pathogenic serum amyloid A 1.1 shows a long oligomer-rich fibrillation lag phase contrary to the highly amyloidogenic non-pathogenic SAA2.2. J Biol Chem 288(4):2744–2755

Stewart CR, Tseng AA, Mok YF, Staples MK, Schiesser CH, Lawrence LJ, Varghese JN, Moore KJ, Howlett GJ (2005) Oxidation of low-density lipoproteins induces amyloid-like structures that are recognized by macrophages. Biochemistry 44(25):9108–9116

Strittmatter WJ, Saunders AM, Schmechel D, Pericak-Vance M, Enghild J, Salvesen GS, Roses AD (1993) Apolipoprotein E: high-avidity binding to beta-amyloid and increased frequency of type 4 allele in late-onset familial Alzheimer disease. Proc Natl Acad Sci U S A 90(5):1977–1981

Suri S, Heise V, Trachtenberg AJ, Mackay CE (2013) The forgotten APOE allele: a review of the evidence and suggested mechanisms for the protective effect of APOE ε2. Neurosci Biobehav Rev 37(10 Pt 2):2878–2886

Suzuki M, Wada H, Maeda S, Saito K, Minatoguchi S, Saito K, Seishima M (2005) Increased plasma lipid-poor apolipoprotein A-I in patients with coronary artery disease. Clin Chem 51:132–137

Takase H, Tanaka M, Miyagawa S, Yamada T, Mukai T (2014) Effect of amino acid variations in the central region of human serum amyloid A on the amyloidogenic properties. Biochem Biophys Res Commun 444(1):92–97

Teoh CL, Griffin MD, Howlett GJ (2011a) Apolipoproteins and amyloid fibril formation in atherosclerosis. Protein Cell 2(2):116–127

Teoh CL, Pham CL, Todorova N, Hung A, Lincoln CN, Lees E, Lam YH, Binger KJ, Thomson NH, Radford SE, Smith TA, Müller SA, Engel A, Griffin MD, Yarovsky I, Gooley PR, Howlett GJ (2011b) A structural model for apolipoprotein C-II amyloid fibrils: experimental characterization and molecular dynamics simulations. J Mol Biol 405(5):1246–1266

Thompson MJ, Sievers SA, Karanicolas J, Ivanova MI, Baker D, Eisenberg D (2006) The 3D profile method for identifying fibril-forming segments of proteins. Proc Natl Acad Sci U S A 103(11):4074–4078

Tian J, Wu N, Guo J, Fan Y (2009) Prediction of amyloid fibril-forming segments based on a support vector machine. BMC Bioinformatics 10(S1):S45

Toth PP, Barter PJ, Rosenson RS, Boden WE, Chapman MJ, Cuchel M, D'Agostino RB Sr, Davidson MH, Davidson WS, Heinecke JW, Karas RH, Kontush A, Krauss RM, Miller M, Rader DJ (2013) High-density lipoproteins: a consensus statement from the National Lipid Association. J Clin Lipidol 7(5):484–525

Trovato A, Seno F, Tosatto SC (2007) The PASTA server for protein aggregation prediction. Protein Eng Des Sel 20(10):521–523

Tsolis AC, Papandreou NC, Iconomidou VA, Hamodrakas SJ (2013) A consensus method for the prediction of 'aggregation-prone' peptides in globular proteins. PLoS ONE 8(1):e54175

Tubb MR, Silva RA, Pearson KJ, Tso P, Liu M, Davidson WS (2007) Modulation of apolipoprotein A-IV lipid

binding by an interaction between the N and C termini. J Biol Chem 282(39):28385–28394

Tycko R (2006) Molecular structure of amyloid fibrils: insights from solid-state NMR. Q Rev Biophys 39(1):1–55

Tzotzos S, Doig AJ (2010) Amyloidogenic sequences in native protein structures. Protein Sci 19(2):327–348

Ursini F, Davies KJ, Maiorino M, Parasassi T, Sevanian A (2002) Atherosclerosis: another protein misfolding disease? Trends Mol Med 8(8):370–374

van der Hilst JC, Yamada T, Op den Camp HJ, van der Meer JW, Drenth JP, Simon A (2008) Increased susceptibility of serum amyloid A 1.1 to degradation by MMP-1: potential explanation for higher risk of type AA amyloidosis. Rheumatology 47(11):1651–1654

van der Westhuyzen DR, de Beer FC, Webb NR (2007) HDL cholesterol transport during inflammation. Curr Opin Lipidol 18(2):147–151

Ventura S, Zurdo J, Narayanan S, Parreño M, Mangues R, Reif B, Chiti F, Giannoni E, Dobson CM, Aviles FX, Serrano L (2004) Short amino acid stretches can mediate amyloid formation in globular proteins: the Src homology 3 (SH3) case. Proc Natl Acad Sci U S A 101(19):7258–7263

Verghese PB, Castellano JM, Holtzman DM (2011) Apolipoprotein E in Alzheimer's disease and other neurological disorders. Lancet Neurol 10(3):241–252

Walker RG, Deng X, Melchior JT, Morris J, Tso P, Jones MK, Segrest JP, Thompson TB, Davidson WS (2014) The structure of human apolipoprotein A-IV as revealed by stable isotope-assisted cross-linking, molecular dynamics, and small angle x-ray scattering. J Biol Chem 289(9):5596–5608

Walsh I, Seno F, Tosatto SC, Trovato A (2014) PASTA 2.0: an improved server for protein aggregation prediction. Nucleic Acids Res 42:W301–W307 (Web Server issue)

Wang L, Lashuel HA, Colón W (2005) From hexamer to amyloid: marginal stability of apolipoprotein SAA2.2 leads to in vitro fibril formation at physiological temperature. Amyloid 12(3):139–148

Wang L, Jiang ZG, McKnight CJ, Small DM (2010) Interfacial properties of apolipoprotein B292–593 (B6.4–13) and B611–782 (B13–17). Insights into the structure of the lipovitellin homology region in apolipoprotein B. Biochemistry 49(18):3898–3907

Weers PM, Narayanaswami V, Choy N, Luty R, Hicks L, Kay CM, Ryan RO (2003) Lipid binding ability of human apolipoprotein E N-terminal domain isoforms: correlation with protein stability? Biophys Chem 100(1–3):481–492

Weinberg RB, Cook VR, Beckstead JA, Martin DD, Gallagher JW, Shelness GS, Ryan RO (2003) Structure and interfacial properties of human apolipoprotein A-V. J Biol Chem 278(36):34438–3444

Wilson C, Wardell MR, Weisgraber KH, Mahley RW, Agard DA (1991) Three-dimensional structure of the LDL receptor-binding domain of human apolipoprotein E. Science 252(5014):1817–1822

Wilson LM, Mok YF, Binger KJ, Griffin MD, Mertens HD, Lin F, Wade JD, Gooley PR, Howlett GJ (2007) A structural core within apolipoprotein C-II amyloid fibrils identified using hydrogen exchange and proteolysis. J Mol Biol 366(5):1639–1651

Wisniewski T, Lalowski M, Golabek A, Vogel T, Frangione B (1995) Is Alzheimer's disease an apolipoprotein E amyloidosis? Lancet 345:956–958

Wong K, Beckstead JA, Lee D, Weers PM, Guigard E, Kay CM, Ryan RO (2008) The N-terminus of apolipoprotein A-V adopts a helix bundle molecular architecture. Biochemistry 47(33):8768–8774

Wong YQ, Binger KJ, Howlett GJ, Griffin MD (2010) Methionine oxidation induces amyloid fibril formation by full-length apolipoprotein A-I. Proc Natl Acad Sci U S A 107(5):1977–1982

Wong YQ, Binger KJ, Howlett GJ, Griffin MD (2012) Identification of an amyloid fibril forming peptide comprising residues 46–59 of apolipoprotein A-I. FEBS Lett 586:1754–1758

Yamada T, Kluve-Beckerman B, Liepnieks JJ, Benson MD (1995) In vitro degradation of serum amyloid A by cathepsin D and other acid proteases: possible protection against amyloid fibril formation. Scand J Immunol 41(6):570–574

Yao Z, Wang Y (2012) Apolipoprotein C-III and hepatic triglyceride-rich lipoprotein production. Curr Opin Lipidol 23(3):206–212

Zhang Z, Chen H, Lai L (2007) Identification of amyloid fibril-forming segments based on structure and residue-based statistical potential. Bioinformatics 23:2218–2225

Zibaee S, Makin OS, Goedert M, Serpell LC (2007) A simple algorithm locates β-strands in the amyloid fibril core of α-synuclein, Aβ, and tau using the amino acid sequence alone. Protein Sci 16:906–918

Computational Approaches to Identification of Aggregation Sites and the Mechanism of Amyloid Growth

Nikita V. Dovidchenko and Oxana V. Galzitskaya

Abstract

This chapter describes computational approaches to study amyloid formation. The first part addresses identification of potential amyloidogenic regions in the amino acid sequences of proteins and peptides. Next, we discuss nucleation and aggregation sites in protein folding and misfolding. The last part describes up-to-date kinetic models of amyloid fibrils formation. Numerous studies show that protein misfolding is initiated by specific amino acid segments with high amyloid-forming propensity. The ability to identify and, ultimately, block such segments is very important. To this end, many prediction algorithms have been developed which vary greatly in their effectiveness. We compared the predictions for 30 proteins by using different methods and found that, at best, only 50 % of residues in amyloidogenic segments were predicted correctly. The best results were obtained by using the meta-servers that combine several independent approaches, and by the method PASTA2. Thus, correct prediction of amyloidogenic segments remains a difficult task. Additional data and new algorithms that are becoming available are expected to improve the accuracy of the prediction methods, particularly if they use 3D structural information on the target proteins. At the same time, our understanding of the kinetics of fibril formation is more advanced. The current kinetic models outlined in this chapter adequately describe the key features of amyloid nucleation and growth. However, the underlying structural details are less clear, not least because of the apparently different mechanisms of amyloid fibril formation which are discussed. Ultimately, the detailed understanding of the structural basis for amyloidogenesis should help develop rational therapies to block this pathogenic process.

N.V. Dovidchenko • O.V. Galzitskaya (✉)
Institute of Protein Research, Russian Academy
of Sciences, 4 Institutskaya str., Pushchino,
Moscow Region 142290, Russia
e-mail: bones@phys.protres.ru;
ogalzit@vega.protres.ru

© Springer International Publishing Switzerland 2015
O. Gursky (ed.), *Lipids in Protein Misfolding*, Advances in Experimental
Medicine and Biology 855, DOI 10.1007/978-3-319-17344-3_9

Keywords

Amyloidogenic segments • Amyloid prediction methods • Linear and exponential aggregation kinetics • Domain swapping • Nucleation of protein folding and misfolding

9.1   Introduction

More than 40 human diseases are currently known to critically involve protein misfolding and deposition as amyloid fibrils in organs and tissues (Chiti and Dobson 2006). These diseases, collectively called amyloidoses, differ in their etiology and clinical presentation and can be classified as primary *vs.* secondary, acquired *vs.* hereditary, and systemic *vs.* focal diseases. Primary amyloidosis is caused by the deposition of a specific protein. An example is AL amyloidosis caused by deposition of immunoglobulin light chains that are overproduced in plasma cells (Hayman et al. 2001). Secondary amyloidosis occurs as a consequence of another underlying disorder. For example, AA amyloidosis, which is a common complication of chronic inflammation, involves deposition of a proteolytic fragment of serum amyloid A (SAA) that is overproduced in inflammation. In contrast to AL and AA that are acquired diseases, most other amyloidoses have a genetic origin and involve autosomal dominant mutations that make a normally soluble globular protein amyloidogenic. Examples include mutations in proteins such as transthyretin, apolipoproteins A-I and A-II, gelsolin, lysozyme, cystatin, fibrinogen, etc. (Benson 2003). These types of amyloidosis are usually systemic diseases affecting multiple tissues and organs (kidney, liver, heart, etc.). In contrast, focal diseases are localized and affect a single organ where amyloid fibers are deposited, such as brain in neurodegenerative diseases. The best known of such disorders involve depositions of amyloid-beta (Aβ) peptide in Alzheimer's disease and of prion proteins in Creutzfeld-Jakob and Mad Cow diseases. To modulate and, ultimately, block the pathologic transition from the native functional protein conformation into amyloid, it is essential to unravel the properties of the protein sequence and structure underlying this transition.

To form amyloid, a protein molecule must undergo major conformational changes. In fact, the fibril core always consists of β-sheets in which individual β-strands are oriented perpendicular to the main axis of the fibril (Jiménez et al. 1999), whereas the native protein may or may not contain the β-sheet structure. Conversion of amyloidogenic proteins into fibrils is often associated with cytotoxicity (Bucciantini et al. 2004). Notably, immature water-soluble fibrils, pre-fibrillar aggregates and oligomers are typically more toxic to cells than mature insoluble amyloid fibrils (Bucciantini et al. 2004). The structures of fibril precursors, which are rich in β-sheet, and the mechanisms of their toxic action were proposed to be similar for different proteins. This idea is based on the observation that specific antibodies that bind to toxic protofibrils of the Aβ peptide can also bind to fibril precursors formed by other proteins with unrelated amino acid sequences, suggesting structural similarity of these precursors (Kayed et al. 2003).

Importantly, numerous experimental studies show that the ability to form fibrils is not limited to amyloidogenic proteins associated with diseases, but is an inherent property of various structurally unrelated proteins (Chiti et al. 1999; Fändrich et al. 2001). Moreover, an increasing number of proteins are found to form functional amyloid *in vivo* (Fowler et al. 2007). These findings raise a question: what factors trigger protein misfolding into amyloid? Is it a particular amino acid sequence of a protein, its structural properties, interactions with ligands (such as lipids and heparan sulfate proteoglycans, which are ubiquitous components of amyloid deposits *in vivo*), or the lack thereof? What mechanisms can protect normal globular proteins from becoming amyloidogenic?

Evolutionary selection against aggregation resulted in an increased content of amino acids that inhibit protein aggregation (Tartaglia et al. 2005), such as Pro that disrupts the β-sheet structure, Gly that confers mobility to the polypeptide chain and thereby increases the entropic cost for ordering (Rauscher et al. 2006), as well as the increased content of charged residues (Kovacs et al. 2010) that confer protein solubility. However, the demands for protein folding (described in part 4 of this chapter) as well as the functional requirements can make it difficult to fully eliminate protein misfolding. In this chapter, we address the basic properties of the primary, secondary and tertiary protein structure that confer amyloidogenic properties, and describe the computational methods that enable one to predict amyloidogenic regions in proteins and peptides and to understand the kinetic steps and the underlying physical processes involved in amyloid fibril formation.

9.2   Identification of Protein Sites Responsible for Amyloid Formation

9.2.1   Structural Determinants of Amyloidogenic Propensity of Proteins and Peptides

Although it is currently accepted that most if not all proteins can form amyloid fibers under certain conditions *in vitro* (Chiti and Dobson 2006), it remains a major challenge to predict whether or not a given protein or peptide actually forms amyloid at near-physiologic conditions. The difficulty stems, in part, from a wide range of external and internal factors that can influence protein misfolding and amyloid formation *in vivo* and *in vitro*. External factors such as the local pH and the interactions (or lack thereof) with various ligands can critically modulate protein folding and shift the balance towards misfolding and aggregation (Fändrich et al. 2001). Protein mutations and post-translational modifications can also importantly influence amyloid formation. Another crucial determinant is the protein con-

centration determined by the balance between the generation of a potentially amyloidogenic protein or peptide and its proteolysis and clearance. Among the major internal determinants are the overall stability of the native protein conformation (addressed below) and the presence of local protein regions with high propensity to initiate the misfolding. The latter aspect is addressed later in parts 2 and 3 of this chapter.

In vivo and *in vitro* studies consistently show that reduced structural stability of globular proteins tends to promote amyloid fibril formation (Chiti and Dobson 2006). *In vitro* studies demonstrate that mildly denaturing conditions leading to partial protein unfolding promote fibril growth (Fändrich et al. 2003), perhaps due to increased solvent accessibility of the aggregation-prone regions (Dobson 1999). This idea is supported by the observation that many naturally occurring mutations and variations associated with amyloid diseases reduce protein stability (Gertz and Rajkumar 2010). However, exceptions from this general trend have also been observed in various proteins, such as immunoglobulin light chains or apolipoprotein A-I (Sánchez et al. 2006; Klimtchuk et al. 2010; Das et al. 2014), suggesting that fibril formation by a mutant protein does not always correlate with its reduced stability. Thus, although partial protein destabilization is necessary for amyloid fiber formation, it is apparently not sufficient.

Notably, many amyloid diseases involve natively unfolded proteins or peptides. Examples described in this volume include Aβ peptide in Alzheimer's disease, α-synuclein in Parkinson's disease, amylin in type 2 diabetes, serum amyloid A in inflammation-linked amyloidosis, and several small apolipoproteins such as apoA-II or apoC-II that readily form amyloid fibrils *in vivo* and/or *in vitro*. Partial folding of these intrinsically disordered proteins is believed to be prerequisite for their β-aggregation (see Chap. 2 by Uversky in this volume). Thus, the current paradigm in the field is that amyloid formation is initiated by the partially folded structural intermediates. However, such partial folding alone is often insufficient to form amyloid, suggesting that additional factors are involved.

Extensive experimental evidence accumulated since 1990s suggests that protein misfolding is usually initiated by specific amino acid sequence motives that, when exposed to solvent, are more liable to aggregation than the rest of the polypeptide sequence (Tenidis et al. 2000; von Bergen et al. 2000; Ivanova et al. 2004). A protein can become non-amyloidogenic upon deletion of such motifs (Ivanova et al. 2004); in addition, certain mutations in these sensitive motives can either diminish or amplify the amyloidogenic propensity of a protein (Ivanova et al. 2004). Furthermore, synthetic peptide fragments corresponding to these amyloidogenic regions, which are sometimes as short as five residues (López de la Paz and Serrano 2004), can form amyloid fibrils similar to those formed by the full-length proteins (Thompson et al. 2000). Identification of such regions in proteins and peptides is crucial for understanding the mechanism of amyloid formation and, ultimately, for developing targeted therapies to modulate or block amyloidosis.

9.2.2   Development of Prediction Methods for Amyloidogenic Regions

One of the earlier prediction methods of amyloidogenic regions is based on the idea that protein misfolding is initiated by the packing defects in the tertiary structure, which lead to increased solvent accessibility of the backbone hydrogen bonds, termed "insufficient wrapping" (Fernández et al. 2003). The authors reported that 10 % of protein structures deposited to the Protein Data Bank have such packing defects, and proposed an algorithm to search for such potentially labile structural regions. Obviously, this method requires detailed knowledge of the 3D structure of the target protein, which is not always available.

Most other prediction methods do not rely on the 3D structural information of the target protein and use the primary structure as an input. Since fibril formation involves conformational changes leading to an increased β-sheet content (Jiménez et al. 1999; Yoon and Welsh 2004), a computational

algorithm was proposed to search for polypeptide chain segments with high propensity to form β-sheet. This approach can identify short segments with β-sheet propensity, yet it cannot predict whether or not a given protein is likely to form amyloid.

Another method to predict possible amyloidogenic regions in the primary sequence is based on the experimental studies of amyloidogenic properties of six-residue synthetic peptides with various amino acid sequences (López de la Paz and Serrano 2004). The authors determined the hexapeptide sequence (STVIIE) that has the highest propensity to form amyloid fibrils *in vitro*, and used this motif to search for amyloidogenic regions in other proteins. Interestingly, amyloid fiber formation was observed upon insertion of the hexapeptide STVIIE at the N-terminus of the SH3-domain of α-spectrin, a water-soluble protein that normally does not form fibrils (Esteras-Chopo et al. 2005). Thus, the combined computational and experimental studies supported the idea that the amyloidogenic propensity of a protein can be localized in short residue segments.

This idea was further supported in studies of murine β$_2$-microglobulin that normally does not form amyloid. The variable segment in residues 83–89 was substituted for a seven-residue sequence (NHVTLSQ) from human β$_2$-microglobulin that forms amyloid. The substitution induced amyloid formation by the murine protein *in vitro* (Ivanova et al. 2004). Importantly, this synthetic amyloidogenic heptapeptide (NHVTLSQ) forms amyloid in solution, whereas the peptide with a similar amino acid composition but scrambled sequence (QVLHTSN) does not. Studies such as this clearly show that not only the composition of the amino acids but also their order determines the amyloidogenic properties. Furthermore, on the basis of their experimental studies the authors proposed a structural model of β$_2$-microglobulin amyloid in which a β-zipper spine is decorated with the remaining part of the protein molecule that partially retains its native fold (Ivanova et al. 2004).

Many other studies addressed the link between the amino acid sequence of a protein and its ability to form amyloid (Idicula-Thomas and Balaji 2005). The authors reported that proteins with low thermal stability and increased life time have increased propensity to form amyloid fibrils *in vivo*, and determined characteristics of the amino acid sequence which correlate with the fibril formation. Their results showed that a seven-residue peptide with a high β-sheet propensity, which was inserted into an α-helix, increased the propensity of this helix to convert into a β-sheet under mildly denaturing conditions. These studies support the idea that a particular composition and sequence of short amino acid stretches is crucial for amyloid formation. The authors proposed a function for predicting amyloidogenic properties of proteins on the basis of their amino acid sequences, and tested it by using the SwissProtein data base. The results suggested that 32 % of all proteins in the database were amyloidogenic; further, of the 54 proteins that readily formed fibrils, 75 % were correctly identified as amyloidogenic.

9.2.3 FoldAmyloid Algorithm

Several research groups have developed methods for identifying regions within polypeptide chains that are responsible for amyloid formation. One of such methods proposed by our team is FoldAmyloid (Galzitskaya et al. 2006; Garbuzynskiy et al. 2010). FoldAmyloid is based on the well-known concept of enthalpy-entropy compensation stating that a sufficient number of contacts between residues, which provide favorable enthalpy contribution to the free energy of protein stability, is required to compensate for the loss of conformational entropy upon protein arrangement into a more organized compact state (Galzitskaya et al. 2000). Since the enthalpy is determined by the combined strength of the short-range packing interactions, we hypothesized that if the mean expected packing density, which determines the average number of residue contacts within a given distance, is lower than the threshold (i. e. the normal packing density for globular proteins), the protein will remain unfolded. Alternatively, if the mean expected

packing density exceeds the threshold, resulting in an increased number of residue contacts, this will favor amyloid formation. In fact, since amyloid fibrils are thermostable, insensitive to proteases, and rich in β-sheet (Kajava et al. 2004), they are expected to contain such densely packed regions.

We tested this hypothesis computationally and demonstrated that the ability of proteins to fold and form such densely packed regions is often responsible for amyloid formation (Garbuzynskiy et al. 2010). In addition to the packing density for individual amino acids (contact scale), we obtained the probability scales inferred from the statistical analyses of protein structures, such as the scale for the main chain hydrogen bond formation (donor scale). These scales were incorporated into server FoldAmyloid (http://bioinfo. protres.ru/FoldAmyloid) for predicting amyloidogenic regions in a protein sequence.

The server was tested on a database (http:// bioinfo.protres.ru/fold-amyloid/amyloid_base. html) containing 144 peptides that readily form amyloid as well as 263 peptides that do not. The contact scale correctly identified 75 % of amyloid-forming peptides and 74 % of peptides that did not form amyloid. The donor scale correctly determined 69 % of peptides that formed amyloid and 78 % of those that did not. Using these scales with an equal weight, we created a hybrid scale that predicted correctly 80 % of amyloid-forming peptides and 72 % of peptides that did not form amyloid (Garbuzynskiy et al. 2010).

9.2.4 Other Prediction Methods

During the last decade, several algorithms for predicting amyloidogenic segments have been developed and improved. One of such algorithms, Zyggregator (Tartaglia and Vendruscolo 2008), uses side chain hydrophobicity, the tendency to form α-helices and β-sheets, and the protein net charge to determine the aggregation profile of a protein and predict its folding rate (Chiti et al. 2003). This method is available online at http:// www-vendruscolo.ch.cam.ac.uk/ggt23.html.

The method Tango (Fernandez-Escamilla et al. 2004), preceded by Agadir (Muñoz and Serrano 1994), predicts the protein's probability to form a particular secondary structure. Agadir uses a statistical analysis of the empirical properties of amino acids based on 3D protein structures to calculate the relative probability of amino acid stretches to fold into a helical or globular conformation. Tango employs a similar approach but considers four possible structural states: α-helix, β-turn, α-helical and β-sheet aggregates, as well as the unfolded (random coil) state (Fernandez-Escamilla et al. 2004). This method is available online at http://tango.crg.es/protected/academic/calculation.jsp.

Several more recent methods such as Waltz employ machine-learning algorithms and are trained on the data sets of peptides with empirically determined amyloidogenic properties (Maurer-Stroh et al. 2010). Waltz was trained on the database of hexapeptides about one half of which tends to form amyloid at neutral pH. In contrast to black-box methods where the weights assigned to individual amino acids have no physical meaning, the weights is Waltz generally represent the amyloid-forming propensity of amino acids. To obtain the weight values, the authors aligned the training set onto itself and created a position-specific scoring matrix, which reflects the probability of a given type of amino acid to be found in a particular position in the amyloidogenic hexapeptide. Interestingly, the results showed that hydrophobic and aromatic residues are favored in the middle of the hexapeptide; this contrasts with the overall tolerance for placement of charged and polar amino acids. The resultant algorithm is based on a combination of the position-specific scoring matrices and additional empirical information on the amyloid-forming propensity obtained from the physicochemical analyses of the designed set of hexapeptides. In addition, the authors proposed to distinguish the aggregates by their morphology as either fibril-like or amorphous, as the properties of peptides that tend to form such aggregates distinctly differ. In general, Waltz was reported to achieve better prediction results than its predecessor, Tango (Maurer-Stroh et al. 2010).

The SecStr method (Hamodrakas et al. 2007) is based on the hypothesis that regions with a high predisposition to form α-helices as well as β-sheets, as determined by at least three methods for secondary structure prediction, can act as conformation switches that tend to promote amyloid formation. The program is available online at http://biophysics.biol.uoa.gr.

Among the amyloid prediction methods, a special place belongs to AmylPred2, a meta-server for consensus analysis (Tsolis et al. 2013). The server combines 11 methods including FoldAmyloid, Tango, and Waltz. The authors show that, judging from the Matthews correlation coefficient that measures the quality of binary classifications in machine learning, AmylPred2 outperforms its composite methods. However, the results suggest that the overall sensitivity of all methods is relatively low (~50 % for the best cases, ~40 % for AmylPred2) due to significant overprediction of amyloidogenic regions. Other problems emerging from the analysis of certain target proteins where that the predictions by various methods vary greatly. Nevertheless, with inclusion of more experimental data and additional prediction methods, the server can become very useful in finding consensus among various algorithms and predicting amyloidogenic segments with greater reliability. This server is available at http://biophysics.biol.uoa.gr/onlinetools.html.

PASTA2 is a good example of a support vector machine method, which is based on the regression analysis to build a model that divides a given dataset into classes on the basis of the training examples. This method shows an excellent potential for identifying amyloidogenic regions. PASTA2 is an extension of the PASTA method that calculates the propensity of a polypeptide segment to form a cross-β structure on the basis of hydrogen-bonding statistics found in the β-strands. The authors used the training set consisting of ~2,500 protein domains, each under 100 a. a. long, with known high-resolution structures (<1.8 Å resolution). In addition to the energetic parameters, PASTA2 predicts structural features such as the secondary structure (e. g. parallel or antiparallel β-sheet) and intrinsic disorder. According to the authors' data, this approach can outperform AmylPred2; however, overprediction of amyloidogenic segments remains an issue with PASTA (Trovato et al. 2006; Walsh et al. 2014).

The PASTA 2.0 server can be accessed at http://protein.bio.unipd.it/pasta2/.

One of the recently proposed methods, termed GAP, is a result of advances in machine learning (Thangakani et al. 2014). The method is based on the idea of "paired frequencies" postulating that, since the N and N + 2 side chains are similarly oriented in any β-strand, amyloid-forming sequences are expected to have position-specific amino acid propensities similar to those found in the secondary structures of globular proteins. GAP has been tested on 310 amyloid-forming peptides. In spite of its relatively poor performance on amyloidogenic proteins (described in the next section), the method yielded several interesting results. For example, the propensities for β-structure formation in globular proteins differed from those in amyloid-forming peptides (Thangakani et al. 2014). The authors also found that, in spite of the partial overlap, there was a distinct difference in the pairing propensities of the amino acids for the peptides forming amyloid fibrils versus amorphous β-structured aggregates.

The growing volume of experimental data on amyloid formation by proteins and peptides, combined with the development of new computational approaches, leads to continuous improvement in the accuracy of the predictions. Still, these predictions have inherent limitations that are discussed below.

9.3 Experimental Verification of Theoretical Amyloid Predictions

9.3.1 Two Types of Amyloidogenic Segments

The locations of amyloidogenic regions have been determined by experimental methods, such as mass spectrometry, for a number of globular proteins and peptides that readily form fibers *in vivo* or *in vitro*. Many of these proteins and peptides are listed in Table 9.1. In addition, several hundred artificial peptides have been shown to form fibrillar aggregates *in vitro*; most of these peptides were derived from the amyloidogenic segments of natural proteins. Our analysis of the

available experimental data from these proteins and peptides revealed that most amyloidogenic segments have one common property: they have an elevated content of hydrophobic residues (Galzitskaya et al. 2006).

The only notable exception are prion domains of yeast proteins, such as Sup35 and Ure2, which have a high content of polar residues, particularly asparagine (N) and glutamine (Q) (Nelson et al. 2005). The X-ray crystal structure of the GNNQQNY peptide responsible for Sup35 aggregation was determined by (Nelson et al. 2005). The results revealed that hydrogen bonds formed by the protein backbone and the side chains are important for structural stabilization. Aggregates of a similar type can be formed in numerous disease-related proteins containing long polyglutamine tracts (up to several dozen Gln) whose length varies among the patients; particularly long polyQ tracts lead to protein aggregation (Yang et al. 2014) in disorders such as Huntington's disease.

In sum, there are two distinct types of amyloidogenic regions found in naturally occurring proteins. The first type is rich in hydrophobic residues and stabilizes the fibril via the hydrophobic interactions. The second type contributes to fibril formation via the hydrogen bonds formed by polar residues such as Gln and Asn in prion-like domains and in polyQ proteins. Most amyloid prediction methods, including FoldAmyloid (Galzitskaya et al. 2006; Garbuzynskiy et al. 2010), are designed to search for the regions of the first type, which have been subjects of extensive experimental studies. Although FoldAmyloid potentially allows for identification of amyloidogenic regions of the second type, the existing experimental data are insufficient to test this aspect of its performance.

9.3.2 Performance of Prediction Algorithms Using 30 Amyloidogenic Proteins

To test the performance of various sequence-based prediction methods, we used experimentally defined amyloidogenic regions in proteins which were originally compiled and tested by (Walsh et al. 2014). Two methods, Waltz and FoldAmyloid, which were not included in the

Table 9.1 Prediction results for 30 amyloidogenic proteins by using seven methods: Comparison with experimental data

Protein	Experimentally defined amyloidogenic regions	PASTA2	AmylPred2	Tango	MetAmyl	Waltz	FoldAmyloid
Prolactin	7-34	TP:12 FN:16	TP:19 FN:9	TP:9 FN:19	TP:6 FN:22	TP:0 FN:28	TP:15 FN:13
P01236	43-57	TP:0 FN:15	TP:5 FN:10	TP:0 FN:15	TP:6 FN:9	TP:0 FN:15	TP:0 FN:15
[29-227]	#False regions	1	6	1	4	1	5
	#False residue	14	46	9	39	9	47
Calcitonin	15-19	TP:0 FN:6	TP:0 FN:6	TP:0 FN:6	TP:0 FN:6	TP:0 FN:5	TP:0 FN:5
P01258	#FP regions	0	1	0	1	0	0
[85-116]	#FP residues	0	5	0	7	0	0
Apolipoprotein A-I	1-93	TP:14 N:79	TP:14 N:79	TP:10 N:83	TP:39 N:54	TP:0 FN:93	TP:7 FN:86
P02647	#False regions	1	1	1	1	0	1
[25-267]	#False residue	6	9	7	8	0	7
Casein (Bovine)	81-125	TP:4 FN:41	TP:11 N:34	TP:6 FN:39	TP:6 FN:39	TP:6 FN:39	TP:19 FN:26
P02663	#False regions	3	3	0	5	0	0
[16-222]	#False residue	39	15	0	33	0	0
Serum amyloid A1 protein	1-12	TP:5 FN:7	TP:8 FN:4	TP:7 FN:5	TP:0 FN:12	TP:0 FN:12	TP:9 FN:3
P02735	#False regions	0	1	0	0	0	0
[19-122]	#False residue	0	3	0	0	0	0
Transthyretin	10-20	TP:6 FN:5	TP:6 FN:5	TP:6 FN:5	TP:10 FN:1	TP:0 FN:11	TP:7 FN:4
P02766	105-115	TP:0 FN:11	TP:10 FN:1	TP:6 FN:5	TP:0 FN:11	TP:0 FN:11	TP:9 FN:2
[21-147]	#False regions	2	3	3	0	2	1
	#False residue	16	22	16	0	10	7
Lactoferrin	538-545	TP:4 FN:4	TP:4 FN:4	TP:4 FN:4	TP:6 FN:2	TP:6 FN:2	TP:0 FN:8
P02788	#False regions	4	24	7	15	1	7
[20-710]	#False residue	39	121	55	116	7	51
Semenogelin-1	1-142	TP:20 FN:122	TP:15 FN:127	TP:0 FN:142	TP:15 FN:127	TP:0 FN:142	TP:7 FN:135
P04279	#False regions	1	3	0	5	0	1
[24-462]	#False residue	4	9	0	37	0	7
Aβ 42	11-42	TP:26 FN:6	TP:21 FN:11	TP:18 FN:14	TP:20 FN:12	TP:6 FN:26	TP:8 FN:24
P05067	#False regions	0	0	0	0	0	0
[672-713]	#False residue	0	0	0	0	0	0
Gelsolin	173-230	TP:0 FN:58	TP:6 FN:52	TP:0 FN:58	TP:13 FN:45	TP:0 FN:58	TP:0 FN:58
P06396	#False regions	6	20	8	17	2	9

(continued)

Table 9.1 (continued)

[28-782]	#False residue	47	118	60	129	14	73
Tau	589-600	TP:5 FN:7	TP:4 FN:8	TP:0 FN:12	TP:8 FN:4	TP:6 FN:6	TP:0 FN:12
P10636	622-627	TP:5 FN:1	TP:6 FN:0	TP:5 FN:1	TP:6 FN:0	TP:0 FN:6	TP:0 FN:6
[2-758]	#False regions	3	2	0	6	0	1
	#False residue	22	11	0	58	0	7
IAPP (Amylin)	8-37	TP:27 FN:3	TP:15 FN:15	TP:0 FN:30	TP:18 FN:12	TP:0FN:30	TP:7 FN:23
P10997	#False regions	0	0	0	0	0	0
[34-70]	#False residue	0	0	0	0	0	0
Lung	9-34	TP:26 FN:0	TP:22 FN:4	TP:20 FN:6	TP:21 FN:5	TP:10 FN:16	TP:26 FN:0
Surfactant	#False regions	0	0	0	0	0	0
P11686	#False residue	2	0	0	6	0	3
[24-58]							
α-Synuclein	35-44	TP:4 FN:6	TP:6 FN:4	TP:6 FN:4	TP:10 FN:0	TP:10 FN:0	TP:0 FN:10
P37840	49-82	TP:29 FN:5	TP:21 FN:13	TP:16 FN:18	TP:34 FN:0	TP:0 FN:34	TP:0 FN:34
[1-140]	86-95	TP:0 FN:10	TP:8 FN:2	TP:0 FN:10	TP:7 FN:3	TP:0 FN:10	TP:0 FN:10
	#False regions	0	1	1	1	0	0
	#False residue	1	5	6	17	0	0
Lysozyme C	5-14	TP:0 FN:10	TP:0 FN:10	TP:0 FN:10	TP:0 FN:10	TP:0 FN:10	TP:0 FN:10
P61626	25-34	TP:0 FN:10	TP:9 FN:1	TP:0 FN:10	TP:0 FN:10	TP:0 FN:10	TP:9 FN:1
[19-148]	56-61	TP:0 FN:6	TP:6 FN:0	TP:0 FN:6	TP:0 FN:6	TP:0 FN:6	TP:6 FN:0
	#False regions	1	1	0	0	0	2
	#False residue	7	8	0	0	0	22
β2-microglobulin	21-31	TP:8 FN:3	TP:10 FN:1	TP:0 FN:11	TP:10 FN:1	TP:0 FN:11	TP:7 FN:4
P61769	33-41	TP:0 FN:9	TP:0 FN:9	TP:0 FN:9	TP:0 FN:9	TP:0 FN:9	TP:0 FN:9
[21-119]	59-71	TP:11 FN:2	TP:11 FN:2	TP:8 FN:5	TP:4 FN:9	TP:7 FN:6	TP:11 FN:2
	83-89	TP:6 FN:1	TP:5 FN:2	TP:0 FN:7	TP:7 FN:0	TP:0 FN:7	TP:0 FN:7
	91-96	TP:4 FN:2	TP:0 FN:6	TP:0 FN:6	TP:6 FN:0	TP:0 FN:6	TP:0 FN:6
	#False regions	0	0	0	0	0	0
	#False residue	5	3	1	3	0	0
Medin	32-50	TP:19 FN:0	TP:14 FN:5	TP:13 FN:6	TP:16 FN:3	TP:0 FN:19	TP:0 FN:19
Q08431	#False regions	1	1	0	1	0	0
[268-317]	#False residue	7	5	0	7	0	0
Brain natriuretic peptide	66-72	TP:6 FN:1	TP:7 FN:0	TP:5 FN:2	TP:6 FN:1	TP:6 FN:1	TP:7 FN:0
	#False regions	0	1	0	1	3	0
P16860	#False residue	0	1	0	6	21	1

(continued)

Table 9.1 (continued)

[27-134]								
Apolipoprotein C-II P02655 [23-101]	60-70 #False regions #False residue	TP:11 FN:0 0 7	TP:9 FN:2 1 7	TP:0 FN:11 1 10	TP:10 FN:1 1 16	TP:0 FN:11 1 6	TP:0 FN:11 0 0	
Odontogenic ameloblast-associated protein A1E959 [16-279]	112-157 #False regions #False residue	TP:16 FN:30 2 21	TP:16 FN:30 6 24	TP:5 FN:41 0 0	TP:10 FN:36 1 10	TP:6 FN:40 0 0	TP:20 FN:26 1 13	
Cystatin C P01034 [27-146]	98-103 #False regions #False residue	TP:3 FN:3 1 24	TP:6 FN:0 1 15	TP:0 FN:6 1 9	TP:6 FN:0 1 12	TP:6 FN:0 1 7	TP:6 FN:0 1 12	
Insulin chain A P01308 [25-54]	13-18 #False regions #False residue	TP:6 FN:0 0 14	TP:4 FN:2 1 4	TP:0 FN:6 0 0	TP:0 FN:6 0 0	TP:0 FN:6 0 0	TP:6 FN:0 1 9	
Insulin chain B P01308 [90-110]	11-17 #False regions #False residue	TP:7 FN:0 0 20	TP:7 FN:0 1 7	TP:4 FN:3 0 2	TP:3 FN:3 0 11	TP:0 FN:7 0 0	TP:7 FN:0 0 5	
β-lactoglobulin P02754 [17-178]	11-20 101-110 116-126 146-152 #False regions #False residue	TP:0 FN:10 TP:4 FN:6 TP:10 FN:1 TP:0 FN:7 4 36	TP:4 FN:6 TP:8 FN:2 TP:5 FN:6 TP:7 FN:0 7 30	TP:3 FN:7 TP:7 FN:3 TP:0 FN:11 TP:0 FN:7 0 0	TP:0 FN:10 TP:0 FN:10 TP:0 FN:11 TP:0 FN:7 3 20	TP:0 FN:10 TP:0 FN:10 TP:0 FN:11 TP:0 FN:7 0 0	TP:4 FN:6 TP:7 FN:3 TP:0 FN:11 TP:5 FN:2 0 7	
Acylphosphatase-2 P14621 [2-99]	16-31 87-98 #False regions #False residue	TP:1 FN:15 TP:0 FN:12 0 17	TP:8 FN:8 TP:5 FN:7 1 7	TP:0 FN:16 TP:0 FN:12 1 5	TP:7 FN:9 TP:2 FN:10 0 26	TP:0 FN:16 TP:0 FN:12 0 0	TP:7 FN:9 TP:7 FN:5 0 0	
High mobility group protein B1 (Rat) P63159 [2-215]	12-27 #False regions #False residue	TP:5 FN:11 1 21	TP:8 FN:8 2 13	TP:7 FN:9 1 7	TP:9 FN:7 2 12	TP:0 FN:16 1 6	TP:8 FN:8 1 8	
Cold shock protein CspB	1-67 #False regions	TP:20 FN:47 0	TP:13 FN:54 0	TP:0 FN:67 0	TP:18 FN:49 0	TP:9 FN:58 0	TP:0 FN:67 0	

(continued)

Table 9.1 (continued)

P32081 [1-67]	#False residue	0	0	0	0	0	0
Kerato-epithelin Q15582 [24-683]	492-509	TP:0 FN:18	TP:9 FN:9	TP:8 FN:10	TP:9 FN:9	TP:0 FN:18	TP:7 FN:11
	#False regions	2	25	9	17	1	11
	#False residue	58	180	48	163	6	89
Myoglobin (Horse) P68082 [2-154]	1-29	TP:7 FN:22	TP:9 FN:20	TP:3 FN:26	TP:11 FN:18	TP:0 FN:29	TP:10 FN:19
	101-118	TP:17 FN:1	TP:16 FN:2	TP:5 FN:13	TP:15 FN:3	TP:0 FN:18	TP:16 FN:2
	#False regions	1	1	1	1	0	1
	#False residue	11	15	13	6	0	12
Replication protein (P. *Syringae*) Q52546 [23-231]	5-13	TP:9 FN:0	TP:8 FN:1	TP:7 FN:2	TP:0 FN:9	TP:0 FN:9	TP:7 FN:2
	#False regions	4	7	2	3	0	5
	#False residue	51	44	21	19	0	52
Total	Total residues	TP:357 FN:629	TP:405 FN:581	TP:188 FN:798	TP:374 FN:611	TP:74 FN:911	TP:271 FN:714
	#False regions	**38**	121	37	88	8	48
	#False residues	489	727	**269**	761	55	432

The numbers of amyloidogenic residues correctly identified are in blue (True Positives, TP) and the numbers of missed residues are in red (False Negatives, FN). In black are the false positives regions incorrectly identified as amyloidogenic (False Residues). Total sums up the results for all proteins. All methods were used under conditions of optimal specificity; FoldAmyloid was used with a sliding window of seven residues (Note: The original table (Walsh et al. 2014) had a minor error for calcitonin (region 15–19 includes 5 residues, not 6), which did not bias the final results)

original work, were added with some modifications. For MetAmyl, which is a meta-server that predicts amyloidogenic regions by using a combination of four methods, we inherited the results from the original analysis that showed good performance by this method (Emily et al. 2013). We excluded the FISH-Amyloid (Gasior and Kotulska 2014) and FoldAmyloid hybrid methods whose predictions were less accurate. Further, we excluded four prion proteins from the test set since their amyloidogenic regions have a distinct amino acid composition, which necessitates specialized prediction algorithms such as that proposed by (Alberti et al. 2009). The final test set consisted of 30 amyloidogenic proteins and peptides listed in Table 9.1.

Our analysis has two limitations. First, for practical reasons, only a subset of the peptide fragments of the test proteins has been studied experimentally. Therefore, in many cases it is not known whether (i) all predicted non-amyloidogenic residues are truly non-amyloidogenic (except for the experimentally verified amyloidogenic regions predicted as non-amyloidogenic, i. e. false negatives); (ii) all regions predicted to form amyloid actually do so. However, one should keep in mind that proteins are not expected to have many amyloidogenic regions. Second, the average length of the experimentally verified amyloidogenic regions in the test proteins is much larger than the polypeptide segments actually forming the amyloid core (usually 5–11 amino acids) (Gilead and Gazit 2005), which increases the probability of predicting correct amyloidogenic regions.

With these caveats, our analysis clearly showed that, despite high prediction performance stated in many publications, the actual performance is not very good as judged by the number of total true positives. As expected, the best predictors are meta-servers that combine several independent approaches. The top predictor,

MetAmyl, yielded more than half correctly predicted amyloidogenic residues (661 out of 986). However, it also had the highest overprediction rate: almost four times more regions have been predicted to be amyloidogenic than experimentally confirmed. Notably, MetAmyl, which is based on a consensus of four methods, predicts more true positives than another meta-server, AmylPred2, which uses 11 methods. Among the non-meta-servers, PASTA2 and FoldAmyloid yielded the best results; notably, the results for PASTA2 were close to those for the meta-servers. This relatively good performance may be due to the fact that both PASTA2 and FoldAmyloid use averaged structural information on globular proteins to optimize the prediction performance, although the protein datasets for these methods are different.

Several methods such as Waltz, Tango and GAP were trained on the database of peptides that form amyloid fibrils *in vitro*. The description of GAP states that ~90 % true positives were obtained by using the peptides database (Thangakani et al. 2014). However, our test of the proteins listed in Table 9.1 suggests that GAP significantly overpredicts the amyloidogenic regions; for example, GAP predicted ~150 amyloidogenic regions for the first target protein, prolactin, which is unrealistically high. Because of this high overprediction rate, GAP was not included in our final analysis.

Waltz was tested in two regimes, the best performance and the maximal specificity regime. In the regime of maximal specificity, Waltz produced fewer overpredictions than other methods. The obvious drawback of such high specificity is missed amyloidogenic regions, i.e. relatively few true positives and many false negatives. Moreover, comparison of the prediction results by Waltz and its ancestor, Tango, showed that the performance of these two methods is not significantly different (Table 9.1).

The general reason for the limited accuracy of various prediction methods (~50 % correctly predicted residues in amyloidogenic regions by our estimate) is that these methods do not take into account the actual 3D structural information on individual proteins. As a result, the prediction

methods are expected to underestimate the influence of long-range interactions among residues that are distant in the primary sequence but close in 3D space. These and other structural features may be crucial, at least for some proteins (for example, see the lipid surface-binding proteins apoA-I and apoA-IV described in Chap. 8 by Das and Gursky in this volume). The effect is further exacerbated when the training dataset for the machine learning algorithms contains only polypeptides but no globular proteins, as was the case with GAP. Another reason for the limited performance may be attributed to particular amino acid sequences giving rise to various types of aggregates, e.g. β-sheet rich amorphous aggregates versus fibrils, as suggested by the peptide studies (Thangakani et al. 2014).

9.3.3 Peptide Test Case: Huntingtin-Based 17-Residue Sequences

To improve the performance of the prediction methods, their results should be compared with the experimental data obtained from a wide range of proteins and peptides. Such a comparison is reported in many recent studies. One example is the study using the amino acid sequence of a 17-residue N-terminal peptide of huntingtin, a polyQ-containing protein that forms amyloid in Huntington's disease (Roland et al. 2013). The N-terminal 17-mer peptide of huntingtin forms highly α-helical aggregates that do not spontaneously convert into β-sheet-rich fibers under near-physiologic conditions. The authors generated 15 peptides with scrambled sequences and analyzed the aggregation properties of these peptides as well as their ability to form β-sheet-rich fibrils. The experimentally determined amyloidogenic properties were compared with the properties of the scrambled sequences predicted with Zyggregator, Waltz, Zipper, and Tango. Although the peptide that was predicted to be particularly amyloidogenic readily formed amyloid fibrils *in vitro*, the general quality of the predictions was not very high. Individual methods varied in the number of the predicted amyloidogenic sequences and in their ability to correctly identify the fibril-forming peptides. Most methods overpredicted the peptide amyloidogenicity. The authors proposed that this discrepancy between the theory and the experiment may be due, in part, to experimental limitations (e. g. low micromolar peptide concentrations used in these experiments might have been insufficient for fiber formation by some peptides), as well as to the limited fundamental understanding of the link between the primary peptide structure and the process of amyloid formation. At the same time, underprediction of two out of five amyloid-forming peptides by all four methods demonstrated that, evidently, there are certain amino acid sequences whose propensity to form amyloid is not sufficiently accounted for by these prediction methods.

9.3.4 Protein Test Case: Glucan Transferase Bgl2p

Glucan transferase Bgl2p (230 amino acids) is the major thermostable protein in the yeast cell wall. Amyloidogenic regions in Bgl2p were initially predicted by using FoldAmyloid (Galzitskaya et al. 2006; Garbuzynskiy et al. 2010), followed by the experimental demonstration that this protein readily forms amyloid fibrils *in vitro* (Kalebina et al. 2008). The protein was predicted to have a strong amyloid-forming potential (seven amyloidogenic segments); this explains why the mutational analysis of this protein by using several amino acid substitutions could not reduce its amyloidogenic potential *in vitro*.

Next, we used a consensus analysis to predict amyloidogenic peptides in Bgl2p. To do so, we used six programs: PASTA, Tango, Waltz, Aggrescan (de Groot et al. 2012), DHPred (Zimmermann and Hansmann 2006), and FoldAmyloid. Residue segments that were predicted to be amyloidogenic by at least four out of six methods were chosen for the peptide synthesis; another peptide, which was predicted by three methods (Aggrescan, DHPred, and FoldAmyloid), has also been synthesized; finally, an additional synthetic peptide fragment was used as a non-amyloidogenic control. Thus, four 10-residue peptides have been synthesized

(Fig. 9.1, gray highlight): (1) AEGFTIFVGV (residues 80–89, predicted by up to 5 methods), (2) VDSWNVLVAG (residues 166–175, up to 3 methods), (3) VMANAFSYWQ (residues 187–196, up to 4 methods), and (4) NDVRSVVADI (residues 141–150, 0 methods, non-amyloidogenic) (Bezsonov et al. 2013). Aggregation properties of these peptides and of the full-length protein were studied in the pH range 4.4–7.5. Fluorescence spectroscopy (ThT binding) and fluorescence microscopy clearly showed that peptide 4, which was predicted to be non-amyloidogenic, actually did not form fibrils under any experimental conditions explored. Peptides 1 and 3, which were predicted by up to five methods, readily formed amyloid fibrils in all experiments. Peptide 2 predicted by up to three methods formed amyloid at acidic pH but not at pH 7.5. Full-length protein Bgl2p also readily formed amyloid at pH 4.7 but not at pH 7.5, suggesting a positive correlation in the aggregation behavior of Bgl2p and peptide 2 (Bezsonov et al. 2013). Notably, peptide fragments corresponding to all amyloidogenic regions that were predicted by the consensus methods formed amyloid fibrils, demonstrating the utility of the consensus methods for predicting amyloidogenic properties *in vitro*.

A more important and, arguably, more challenging task is to identify amyloidogenic segments that are critical to amyloid formation in a biological context, and to predict the effects of protein mutations and modifications on amyloid-forming propensity *in vivo*. To this end, (Belli et al. 2011) compared the aggregation propensity predicted by different methods with the limited experimental data available on the aggregation propensities of several wild-type and mutant proteins from *E. coli*. The data were based on the quantitative measurements of the levels of mutant proteins found in inclusion bodies relative to the wild type (Carrió et al. 2005; Wang et al. 2008). The authors concluded that, despite limited ability to predict amyloid formation in a biological context, "algorithms that have been developed to predict amyloid formation *in vitro* also offer a considerable degree of accuracy for predicting amyloid propensity *in vivo*."

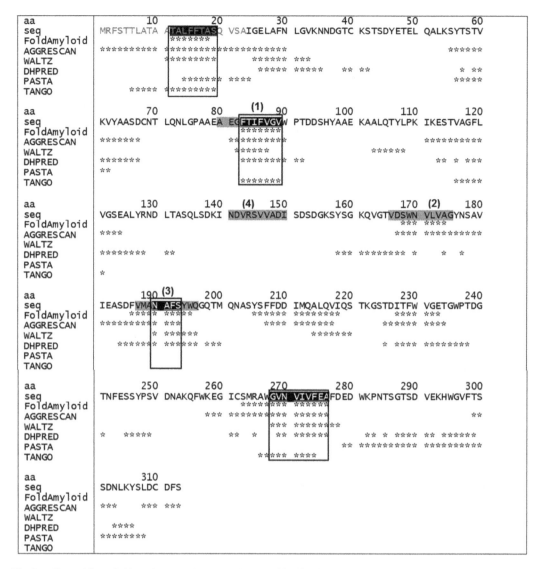

Fig. 9.1 Potential amyloidogenic determinants in glucan transferase Bgl2p from *Saccharomyces cerevisiae* cell wall (UniProtKB/TrEMBL entry number P15703). Amino acids that were predicted by individual algorithms to be located in amyloidogenic segments (*); amino acids that were predicted by four or more methods to be amyloidogenic are *boxed*. The N-terminal signal sequence is in *light gray letters*. Generally, the signal sequences which bind lipid are largely hydrophobic and are predicted to be amyloidogenic. Sequences of the four peptides that have been synthesized and experimentally investigated are highlighted in *gray*. "aa" indicates residue numbers; "seq" shows protein amino acid sequence. The number in brackets relates to the serial number of synthesized peptide mentioned in the text (Figure is modified from (Bezsonov et al. 2013))

9.4   Nucleation and Aggregation Sites in Protein Folding and Misfolding

Since polypeptide chain can either fold into a native structure or misfold and form aggregates or amyloid fibrils, these processes compete, and the outcome can depend on the rate-limiting transition states as well as the folding and misfolding intermediates. For some proteins such as human acylphosphatase, the transition states in folding and aggregation are structurally unrelated (Chiti et al. 2002), suggesting structurally distinct pathways for folding and aggregation. In other proteins such as β_2-microglobulin, partially unfolded species that resemble folding intermediates have been implicated in amyloid formation (Jahn et al. 2006), suggesting that the free-energy landscapes for folding and misfolding of these proteins may be related. Identification of a folding intermediate as the key precursor of β_2-microglobulin fibril elongation under physiological conditions provided direct experimental evidence that the folding and aggregation landscapes for this protein coincide, at least initially, and diverge only at the level of a native-like folding intermediate that resembles the immunoglobulin fold (Jahn et al. 2006). Thus, there is no single rule describing the relationship between the regions important for folding of the native structure and for amyloid formation.

A crucial rate-limiting event in protein folding is the formation of a folding nucleus, which is a structured part of the polypeptide chain in the high-energy transition state. A detailed analysis of the formation and evolution of the folding nucleus in amyloidogenic proteins may help understand what properties make these proteins amyloidogenic. However, experimental data delineating both the folding nucleus and the amyloidogenic regions in the same protein are often lacking. Since the folding nucleus is unstable, it is difficult to investigate it experimentally. An elaborate experimental approach, called Φ-analysis, has been developed to indirectly assess the structure of the folding nuclei (Matouschek et al. 1989). By introducing point mutations into a protein

structure, it is possible to find residues whose mutations have a similar destabilizing effect on the transition state and on the native state. The Φ-value for a mutation in residue r is defined as:

$$\Phi = \Delta_r[F(T) - F(U)] \,/\, \Delta_r[F(N) - F(U)] \quad (9.1)$$

Here $\Delta_r[F(N) - F(U)]$ is the mutation-induced change in the free energy difference between the native (N) and the unfolded (U) state, and $\Delta_r[F(T) - F(U)]$ is the mutation-induced change in the free energy difference between the transition (T) state (which is the high-energy rate-limiting state in protein unfolding) and the unfolded (U) state. Most Φ-values vary from 0 to 1; $\Phi = 1$ indicates that the mutated residue is in the folding nucleus. The values of $\Phi < 0$ or $\Phi > 1$ are rare and indicate non-native contacts in the transition state.

Since the Φ-analysis is very labor-intensive, there is general paucity of experimental data identifying folding nuclei in amyloidogenic proteins. To overcome this problem, we used the available data to compare: (i) the experimentally identified amyloidogenic regions with the predicted folding nuclei (Galzitskaya and Finkelstein 1999; Garbuzynskiy et al. 2004) (for proteins with experimentally identified amyloidogenic regions), and (ii) the experimentally identified folding nuclei with the predicted amyloidogenic regions (for proteins with experimentally identified folding nuclei). The results revealed that most experimentally determined amyloidogenic segments (12 regions, Table 9.2) overlap the predicted folding nuclei (Fig. 9.2), and most predicted amyloidogenic segments overlap the experimentally determined folding nuclei (Galzitskaya and Garbuzynskiy 2008; Galzitskaya 2009, 2011a). On average, Φ-values for residues in amyloidogenic regions were significantly greater than those outside these regions. This implies that the amyloidogenic regions tend to overlap the folding nucleus of a native protein structure. Consequently, amyloidogenic regions can nucleate either the normal protein folding or the misfolding into amyloid fibrils, thus playing a key role in the competition between these processes.

Table 9.2 Proteins with experimentally determined 3D structures and amyloidogenic regions

Protein	PDB ID	No. amino acids		Experimentally determined amyloidogenic regions	Context
		Protein	3D structure used[a]		
Acylphosphatase	1aps[b]	98	98 (1–98)	16–31 (Chiti et al. 2002)	*in vitro*
				87–98 (Chiti et al. 2002)	
β2-microglobulin	1im9	99	99 (1–99)	20–41 (Kozhukh et al. 2002)	*in vivo & in vitro*
				59–71 (Jones et al. 2003)	
				83–89 (Ivanova et al. 2004)	
Gelsolin	1kcq	104	104 (158–261)	52–62 (Maury and Nurmiaho-Lassila 1992)	*in vitro*
Transthyretin	1bm7	127	114 (10–123)	10–19 (Chamberlain et al. 2000)	*in vivo & in vitro*
				105–115 (Jaroniec et al. 2002)	
Lysozyme	193l	130	129 (1–129)	49–64 (Krebs et al. 2000)	*in vivo & in vitro*
Myoglobin	1wla	153	153 (1–153)	7–18 (Picotti et al. 2007)	*in vitro*
				101–118 (Fändrich et al. 2003)	
Human prion	1qm0	253	143 (125–228)	169–213 (Lu et al. 2007)	*in vivo & in vitro*

[a]Numbers in brackets correspond to those in the PDB entry
[b]Amyloidogenic regions were determined experimentally for human acylphosphatase. Although the 3D structure of this protein is unknown, the 3D structure of a highly homologous horse acylphosphatase (95 % sequence identity) has been determined (PDB ID 1aps)

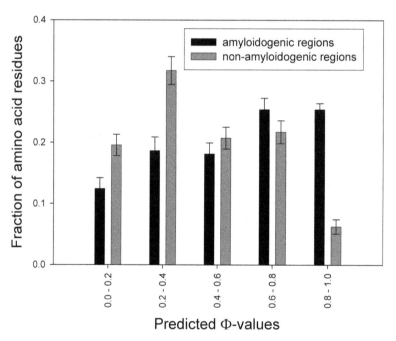

Fig.9.2 Distribution of the predicted Φ-values (Galzitskaya and Finkelstein 1999; Garbuzynskiy et al. 2004) in protein regions that have been demonstrated experimentally to be amyloidogenic (*black bars*) or non-amyloidogenic (*gray bars*). The regions from seven proteins have been used for this analysis: acylphosphatase (1aps), β2-microglobulin (1im9), gelsolin (1kcq), transthyretin (1bm7), lysozyme (193l), myoglobin (1wla), and human prion (1qm0)

Furthermore, our sequence and structural analysis suggests that these regions usually contain clusters of large apolar side chains, leading to restricted motions of the polypeptide backbone and thereby helping nucleate the ordered structure. Since protein folding nuclei are determined based on the Φ-value analysis, while the amyloidogenic regions are predicted from the analysis of the primary structure, the overlap between the two regions observed in our studies enables us to predict the nucleation sites for protein folding on the basis of the primary structure analysis (Galzitskaya and Garbuzynskiy 2008; Galzitskaya 2009, 2011a).

As first proposed by Eisenberg's group, amyloid formation can involve domain swapping whereby two or more polypeptide chains swap identical structural elements to form oligomers (Bennett et al. 1994, 1995). Proteins with a wide range of unrelated amino acid sequences and structures can oligomerize via the domain swapping (Galzitskaya 2011b). The residues from such swapped regions acquire their stable conformation early in the folding process, suggesting that these regions are important for correct protein folding as well as misfolding. We compiled a data base of proteins that contain swapped domains as well as the proteins that have been crystallized in the monomeric form. The folding nuclei were determined based on the monomeric protein structures with the experimental error for Φ-value of ± 0.1, and the amyloidogenic segments were predicted using the amino acid sequence analysis by FoldAmyloid. Together, the results showed that, in 11 out of 17 proteins, the regions with $\Phi > 0.5$ that are probably responsible for folding overlapped with the swapped regions of the polypeptide chain. Furthermore, in 11 out of 17 proteins, the swapped regions overlapped with the predicted amyloidogenic regions (Galzitskaya 2011b). These results support the idea that protein regions undergoing domain swapping are often critical for correct protein folding as well as in misfolding.

9.5 Possible Mechanisms and Kinetic Models of Amyloid Growth

9.5.1 Linear Nucleation-Elongation Model

A commonly used method to study amyloid formation is to monitor the time course of amyloid growth by tracking the binding of diagnostic dyes Thioflavin T (ThT) or Congo Red (CR). Binding to amyloid-like aggregates increases fluorescence intensity of these dyes, which can be used to track the increase in concentration of amyloid-like aggregates in real time (Buxbaum and Linke 2012). Such kinetic experiments using fluorescence can be performed relatively easily (e. g. see the Chap. 4 by Singh et al. on amylin in this volume) and can help dissect the complex multistep pathways of amyloid fiber formation.

Amyloid formation can be considered as a polymerization reaction. Most quantitative models for linear polymerization stem from the work performed more than half a century ago (Oosawa et al. 1959) proposing a kinetic model for actin polymerization. The experimental data showed that actin polymerization is akin to a condensation reaction that takes place only if the concentration of the initial reactant (actin) exceeds the critical threshold. The highly cooperative character of the reaction was supported by two observations: (i) increase in actin concentration resulted in a higher reaction rate at early stages; (ii) addition of a nucleus with a pre-formed aggregate led to rapid polymerization of free actin.

The linear polymerization (i. e. nucleation – consecutive elongation) model was used to explain protofibril formation by hemoglobin in sickle-cell anemia (Hofrichter et al. 1974). The reaction had a high free energy barrier for nucleation. The authors defined a nucleus as the least thermodynamically stable oligomer that can initiate further growth of protofibrils.

Frieden and Goddette (1983) developed additional aspects of the linear model of actin polymerization. They noted that each event of monomer attachment to the growing polymer chain has its own rate constant, and that the reaction begins with the monomer activation, which in the case of actin involved Mg^{2+}-induced conformational changes.

Goldstein and Stryer (1986) further explored the linear model of protein polymerization. They defined the nucleus as a "primer" of a certain size whose formation led to changes in kinetic constants. The goal was to explore numerical methods for optimal fitting of various experimental data to the model; important improvements in the experimental approach have also been proposed.

9.5.2 Exponential Growth Model

Even though the nucleation-consecutive elongation is a common mechanism of protein polymerization, many experimental data on protein polymerization and fibril formation cannot be adequately described by this simple model (Foderà et al. 2008; Xue et al. 2008; Cohen et al. 2013). To describe these data, an "exponential growth" mechanism was proposed. This mechanism reflects an increased number of sites for monomer attachment upon fibril growth in processes such as fibril fragmentation, secondary nucleation, branching, etc. (Fig. 9.3).

The first experiments that were aimed to test the exponential growth model addressed the kinetics of actin polymerization. Reagents such as Ca^{2+} and Mg^{2+} were known to disassemble actin filaments, but the mechanism of their action was unclear. Wegner and Savko demonstrated that actin filaments can undergo spontaneous fragmentation during polymerization reaction. This explained why the nucleation-consecutive elongation model failed to adequately fit the data (Wegner and Savko 1982). Only when fibril fragmentation was accounted for, the model could adequately approximate the kinetic data.

Ferrone et al. (1980) developed a model of heterogeneous nucleation to explain the effect of "extreme autocatalysis" and the strong concentration-dependence observed in aggregation of sickle-cell hemoglobin. This two-step model assumed that, first, regular nucleation leads to the formation of a protofibril; next, additional protofibrils are formed on its surface. This

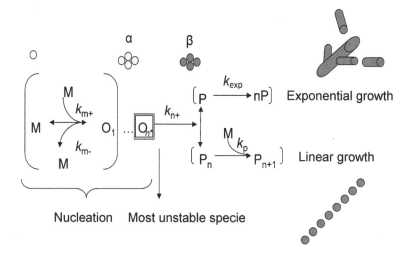

Fig. 9.3 General scheme for amyloid formation depicting linear and exponential growth models. *M* monomer, *O* oligomer, *On** oligomer of critical size *n** that forms the nucleus that is the transient species with high free energy, *P* amyloid aggregate, *km+* rate constant of monomer attachment to the oligomer, *km–* rate constant of monomer dissociation from the oligomer, *kn+* rate constant of amyloid seed formation (the least stable specie on the reaction pathway), *kexp* rate constant of the exponential growth event (bifurcation, fragmentation etc.), *kp* rate constant of monomer attachment to the growing amyloid aggregate, α non-cross-beta aggregate, β cross-beta aggregate

model included two equations describing homogeneous nucleation at the first stage and heterogeneous nucleation at the second stage; the system of equations was solved numerically. The results suggested that the surface available for addition of monomers increases mainly due to the increase in the size of the aggregate, as the protofibril surface can provide new nucleation sites. The concept of heterogeneous nucleation was novel and has importantly contributed to the theory of protein aggregation.

More recently, Miranker and colleagues explored amyloid fibril formation by amylin, a small polypeptide hormone that deposits in type 2 diabetes (also see Chap. 4 by Singh et al. in this volume). Kinetic studies showed that nucleation can proceed via two pathways: protofibril-independent (primary) and protofibril-dependent (secondary). The balance between the two depends on the external interface. In the presence of such an interface, the primary mechanism is dominant; alternatively, the secondary mechanism dominates (Ruschak and Miranker 2007).

Normally, amyloid fibrils are unbranched linear polymers. Interestingly, fibril branching (i. e. growth in a tree-like structure) was observed during amyloid formation by a small hormone glucagon; in these studies, single fibril growth was monitored in real time by using TIRF microscopy (Andersen et al. 2009). Clearly, such branching can lead to exponential growth kinetics.

9.5.3 Mixed Models

Exponential growth model predicts a longer lag phase (nucleation) and/or a faster propagation phase (growth) as compared to the nucleation-consecutive elongation model. Notably, in the latter model, the number of fibers during linear growth is roughly proportional to the number of nuclei formed during the lag phase, because fiber growth is energetically more favorable than the nucleation. Therefore, after a certain time, the number of fibrils becomes constant. In contrast, the number of fibrils continues to increase upon fragmentation (which commonly occurs in amyloid fibrils), branching (which is uncommon in amyloid) and other exponential growth scenarios. This difference is the key distinction between the nucleation-consecutive elongation and the exponential growth models. Since certain experimental data cannot be adequately described by either model alone (Wegner and Savko 1982), several mixed models for protein polymerization have been proposed based on a combination of the nucleation at the first stage and exponential growth at the second stage. In case of amyloid, the most probable mechanism of exponential growth in the second stage is fibril fragmentation (Serio et al. 2000; Xue et al. 2008).

Radford and colleagues approximated the aggregation kinetics of β_2-microglobulin with a modular system of kinetic equations (Xue et al. 2008). A set of modules describing various steps in the aggregation mechanism was selected, and various combinations of these modules were used to fit the experimental data obtained by ThT fluorescence. The best fit was obtained by using a model that included a module for polymerization with a consecutive monomer attachment, and another module for fragmentation.

Morris, Finke and colleagues proposed a simple model to describe the process of amyloid aggregation where exponential growth model incorporated a linear relationship between the "ends" of the growing aggregate and its mass (Morris et al. 2009). Although the non-quadratic mass accumulation during early stages of growth can be described by this model, the analytical solution represents a sigmoid curve. Hence, the model did not apply to proteins displaying non-sigmoid reaction kinetics (Giehm and Otzen 2010).

Knowles and colleagues analytically solved equations describing fiber formation with fragmentation (Knowles et al. 2009). Their model included the nucleation stage and the exponential growth stage with fragmentation; the latter was essential for the accurate approximation of the experimental data. The authors reported that nearly all proteins showed linear scaling in the logarithmic coordinates of relative concentration versus relative lag time, with the constant exponential coefficient (i.e. the dependence $\ln T_{lag} \sim const + \gamma \ln[C]$, with a constant $\gamma = -0.5$). The nature of such "scaling" was addressed by Cohen et al. (2011) who expanded the model by adding secondary nucleation, i.e. nucleus formation on the surface of growing fibrils, and solved the equations analytically. The authors report that γ depends on the size n of the secondary nucleus, $\gamma = -(n+1)/2$. In the exponential growth model in the absence of bifurcation, $n=0$ and hence, $\gamma = -0.5$, which is the "scaling" constant (Knowles et al. 2009). Thus γ reflects the specific mechanism of the amyloid formation and can potentially be used to assess the exponential growth mechanism on the basis of the kinetic data.

Recently, we reported a detailed analysis of various kinetic mechanisms of amyloid growth (Dovidchenko et al. 2014). A useful parameter in this analysis is L_{rel} which describes the ratio between the duration of the lag phase and the time required to include all monomers into the growing polymer (Fig. 9.4). We found that: (i) the linear growth corresponds to a very narrow range of $L_{rel} \leq 0.2$ and occurs only if L_{rel} is independent of the initial monomer concentration; (ii) these limitations do not apply to the exponential growth. Further, we showed that L_{rel} is determined by the size of the primary nucleus (n^*), which is the smallest least stable aggregate on the reaction pathway, and of the secondary nucleus

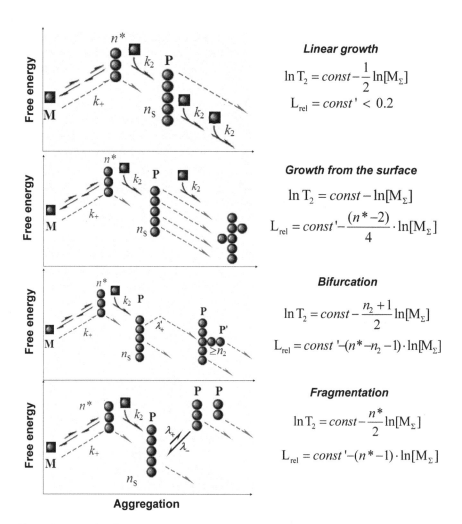

Linear growth

$$\ln T_2 = const - \frac{1}{2}\ln[M_\Sigma]$$

$$L_{rel} = const' < 0.2$$

Growth from the surface

$$\ln T_2 = const - \ln[M_\Sigma]$$

$$L_{rel} = const' - \frac{(n^*-2)}{4}\cdot \ln[M_\Sigma]$$

Bifurcation

$$\ln T_2 = const - \frac{n_2+1}{2}\ln[M_\Sigma]$$

$$L_{rel} = const' - (n^*-n_2-1)\cdot \ln[M_\Sigma]$$

Fragmentation

$$\ln T_2 = const - \frac{n^*}{2}\ln[M_\Sigma]$$

$$L_{rel} = const' - (n^*-1)\cdot \ln[M_\Sigma]$$

Aggregation

Fig. 9.4 Alternative scenarios for amyloid growth and the corresponding kinetic parameters. T_2 is the time of inclusion of all monomers into the aggregate, L_{rel} is the ratio between the duration of the lag phase and the time of inclusion of all monomers into the growing polymer, $[M_\Sigma]$ is the total monomer concentration, n^* is the size of the primary nucleus, and n_2 is the size of the secondary nucleus

Table 9.3 Kinetic parameters of fiber formation

Protein or peptide	L_{rel} (min–max)	$\ln T_2$ (min–max)	$n^* \pm \varepsilon n^*$	$n_2 \pm \varepsilon n^2$
"Exponential" growth with fragmentation/bifurcation				
Insulin[a]	5.17–5.56	−0.49–0.08	**0.81**±0.54	**−0.04**±0.13
β2-microglobulin[b]	1.48–3.86	0.79–2.28	**1.58**±0.58	**−0.06**±0.18
Yeast Prion Sup35[c]	0.29–0.70	4.62–5.52	**1.09**±0.20	**−0.51**±0.45
Yeast Prion Ure2p[d]	0.76–1.02	1.79–2.39	**0.96**±1.52	**−0.23**±1.40
Murine WW domain[e]	1.56–2.10	4.76–6.02	**1.21**±1.27	**0.05**±1.02
"Exponential" growth with bifurcation				
Aβ42[f]	0.53–0.67	−0.51–1.44	**2.64**±0.11	**1.72**±0.05
TI I27[g]	0.14–0.34	5.93–7.85	**2.86**±0.30	**2.04**±0.29
"Linear" growth				
Apolipoprotein C-II[h]	0.06–0.10	2.77–4.82	**4.44**±0.38	–

Tabulated parameters include L_{rel} and $\ln T_2$ described in the text, and the sizes of the nuclei, n^* (primary nucleus) and n^2 (secondary nucleus)

The parameters were determined by applying our kinetic model to approximate the experimental kinetic data recorded by using ThT fluorescence of eight proteins: [a](Selivanova et al. 2014), [b](Xue et al. 2008), [c](Collins et al. 2004), [d](Zhu et al. 2003), [e](Ferguson et al. 2003), [f](Cohen et al. 2013); [g](Wright et al. 2005), [h](Binger et al. 2008)

(n_2), which mediates branching on the surface of the growing fibril. We determined the dependence of L_{rel} on the initial monomer concentration and used it to calculate n^* and n_2. Notably, we found that the scaling effect described by (Knowles et al. 2009) is a general feature of the polymerization reaction which reflects both the nucleus size and the specific scenario for amyloid growth (illustrated in Fig. 9.4).

To determine the sizes of the primary and secondary nuclei and the mechanism of amyloid growth, we used this model to approximate the experimental kinetic data recorded from eight proteins: insulin, β2-microglobulin, yeast prion Sup35, yeast prion Ure2p, murine WW domain, Aβ42, TI I27, and apolipoprotein C-II. The results are summarized in Table 9.3.

Interestingly, in most cases of exponential growth, the size of the primary nucleus, n^*, was close to 1, suggesting that the protein monomer is sufficient to initiate amyloid formation. Alternatively, $n^* \cong 1$ may relate to a group of protein molecules that act as a single entity in solution; another alternative is that a protein monomer initiates fibril growth from an aggregated state. Thus, kinetic data alone are insufficient to unambiguously determine the fibrillation mechanisms. Since molecular mechanisms of amyloid formation can vary from protein to protein, one ought to use additional experimental techniques (e.g.

various types of microscopy, ultracentrifugation, etc.) to carefully rule out alternative scenarios before deciding on the precise mechanism of fibrillation. An example of a combined kinetic and structural approach that utilizes atomic force microscopy to determine the detailed mechanism of amylin fibrillation is described by Jeremic and his team in Chap. 4 of this volume.

In sum, substantial progress has been made in our understanding of the kinetic aspects of amyloid formation. In some cases (such as Aβ42 or apolipoprotein C-II, Table 9.3) it is possible to determine the mechanism of aggregation solely on the basis of the kinetic data, while in many other cases additional structural information is required. By combining computational and experimental approaches, one can determine the size of the primary nucleus, which is the critical state in the fibrillation pathway.

9.6 Concluding Remarks

Despite recent advances in the development and improvement of the sequence-based amyloid prediction algorithms, much remains to be done in this area. The results of our comparative analysis, expanded and averaged in Table 9.4, show that, despite current improvements, individual algorithms have limited accuracy and specificity.

Table 9.4 Averaged results of amyloid predictions by various algorithms. PASTA2, MetAmyl, and Waltz were used in different regimes, including specificity-optimized parameters

Scoring type	PASTA2 90 % specificity	PASTA2 85 % specificity	AmylPred2	Tango	MetAmyl high specificity	MetAmyl global accuracy	Waltz best performance	Waltz high-specificity	FoldAmyloid
Sensitivity	0.36	0.48	0.41	0.19	0.38	0.54	0.19	0.08	0.28
Specificity	0.91	0.85	0.86	0.95	0.86	0.73	0.94	0.99	0.92
False regions predicted as amyloidogenic	38	70	121	37	88	148	37	8	31
No. correctly predicted regions/**total**	33/**46**	37/**46**	42/**46**	17/**46**	33/**46**	41/**46**	22/**46**	11/**46**	29/**46**

The recent improvements are illustrated by comparing two best-performing individual methods, FoldAmyloid (which is a relatively old and simple approach) and PASTA2 (a new more sophisticated approach that performs best among the non-meta-servers). As evident from Table 9.4, the accuracy and sensitivity of PASTA2 predictions are clearly better than those of FoldAmyloid. At the same time, comparison of the overall prediction quality shows only modest improvement, which is due to low sensitivity partly caused by overprediction and partly by the coarseness of the methods. Even the incorporation of several methods into the meta-servers MetAmyl and AmylPred2 does not drastically improve the results. Apparently, the methods that are entirely sequence-based are approaching their limit.

The problem may perhaps be partially overcome by using the training sets containing small proteins and their domains rather than short peptides forming the amyloid core. However, a greater problem is that the existing prediction methods do not incorporate 3D structural information of the target proteins. Because sequence-based prediction of the 3D structure of globular proteins is still unattainable, accurate prediction of the native environment of the peptides forming the amyloid core is also unattainable. The situation is somewhat analogous to the prediction of the antibody-binding protein epitopes on the basis of the primary structure: in both instances, the properties of interest depend on the 3D protein structure and hence, cannot be accurately predicted based on the amino acid sequence alone. We believe that, to qualitatively improve the amyloid prediction methods, it is perhaps necessary to incorporate additional information that critically influences proteins' propensity to form amyloid, such as the native 3D structure. In fact, a prediction test carried out on natively-unfolded amyloidogenic proteins and peptides (Ahmed and Kajava 2013) showed much better performance than the analysis of folded proteins reported in this chapter. Ultimately, in addition to 3D structure, other important factors such as the protein dynamics and the environmental conditions (e. g. the presence of lipid membranes) should also be considered.

Significant progress has been achieved in our understanding of the link between the normal folding and the misfolding of proteins, and in elucidating the kinetic features of protein misfolding and aggregation. Sophisticated kinetic models have been proposed to accurately describe the complex pathways of protein fibrillation, which include nucleation, branching, fragmentation, and growth from the surface. However, detailed structural understanding of these reaction steps is still lacking for most proteins. Such a detailed structural understanding may possibly provide a key element to improve the accuracy of amyloid prediction.

Prediction of protein fibrillation *in vivo* remains a major challenge, since a wide array of environmental factors can influence fiber nucleation and growth in the biological context. For example, most amyloid deposits found *in vivo* contain additional components (such as lipids, cell membrane components such as heparan sulfate proteoglycans, and apolipoproteins), and the complex role of these components in amyloidogenesis is far from clear. Moreover, some proteins form so-called functional amyloids in living cells, wherein the assembly and disassembly of fibrils occurs in response to biological clues (Chiti and Dobson 2006). Understanding the mechanisms of fibril assembly and disassembly *in vivo* and *in vitro* is not only of fundamental scientific importance, but may also help develop new therapeutic targets against amyloidosis.

Acknowledgement We are grateful to Dr. Olga Gursky for her valuable comments and editorial help. This study was supported by the Russian Science Foundation grant №14-14-00536.

References

Ahmed AB, Kajava AV (2013) Breaking the amyloidogenicity code: methods to predict amyloids from amino acid sequence. FEBS Lett 587:1089–1095

Alberti S, Halfmann R, King O, Kapila A, Lindquist S (2009) A systematic survey identifies prions and illuminates sequence features of prionogenic proteins. Cell 137:146–158

Andersen CB, Yagi H, Manno M, Martorana V, Ban T, Christiansen G, Otzen DE, Goto Y, Rischel C (2009)

Branching in amyloid fibril growth. Biophys J 96:1529–1536

Belli M, Ramazzotti M, Chiti F (2011) Prediction of amyloid aggregation in vivo. EMBO Rep 12:657–663

Bennett MJ, Choe S, Eisenberg D (1994) Domain swapping: entangling alliances between proteins. Proc Natl Acad Sci U S A 91:3127–3131

Bennett MJ, Schlunegger MP, Eisenberg D (1995) 3D domain swapping: a mechanism for oligomer assembly. Protein Sci 4:2455–2468

Benson MD (2003) The hereditary amyloidoses. Best Pract Res Clin Rheumatol 17:909–927

Bezsonov EE, Groenning M, Galzitskaya OV, Gorkovskii AA, Semisotnov GV, Selyakh IO, Ziganshin RH, Rekstina VV, Kudryashova IB, Kuznetsov SA, Kulaev IS, Kalebina TS (2013) Amyloidogenic peptides of yeast cell wall glucantransferase Bgl2p as a model for the investigation of its pH-dependent fibril formation. Prion 7:175–184

Binger KJ, Pham CLL, Wilson LM, Bailey MF, Lawrence LJ, Schuck P, Howlett GJ (2008) Apolipoprotein C-II amyloid fibrils assemble via a reversible pathway that includes fibril breaking and rejoining. J Mol Biol 376:1116–1129

Bucciantini M, Calloni G, Chiti F, Formigli L, Nosi D, Dobson CM, Stefani M (2004) Prefibrillar amyloid protein aggregates share common features of cytotoxicity. J Biol Chem 279:31374–31382

Buxbaum JN, Linke RP (2012) A molecular history of the amyloidoses. J Mol Biol 421:142–159

Carrió M, González-Montalbán N, Vera A, Villaverde A, Ventura S (2005) Amyloid-like properties of bacterial inclusion bodies. J Mol Biol 347:1025–1037

Chamberlain AK, MacPhee CE, Zurdo J, Morozova-Roche LA, Hill HA, Dobson CM, Davis JJ (2000) Ultrastructural organization of amyloid fibrils by atomic force microscopy. Biophys J 79:3282–3293

Chiti F, Dobson CM (2006) Protein misfolding, functional amyloid, and human disease. Annu Rev Biochem 75:333–366

Chiti F, Webster P, Taddei N, Clark A, Stefani M, Ramponi G, Dobson CM (1999) Designing conditions for in vitro formation of amyloid protofilaments and fibrils. Proc Natl Acad Sci U S A 96:3590–3594

Chiti F, Taddei N, Baroni F, Capanni C, Stefani M, Ramponi G, Dobson CM (2002) Kinetic partitioning of protein folding and aggregation. Nat Struct Biol 9:137–143

Chiti F, Stefani M, Taddei N, Ramponi G, Dobson CM (2003) Rationalization of the effects of mutations on peptide and protein aggregation rates. Nature 424:805–808

Cohen SIA, Vendruscolo M, Welland ME, Dobson CM, Terentjev EM, Knowles TPJ (2011) Nucleated polymerization with secondary pathways. I. Time evolution of the principal moments. J Chem Phys 135:065105

Cohen SIA, Linse S, Luheshi LM, Hellstrand E, White DA, Rajah L, Otzen DE, Vendruscolo M, Dobson CM, Knowles TPJ (2013) Proliferation of amyloid-β42 aggregates occurs through a secondary nucleation mechanism. Proc Natl Acad Sci U S A 110:9758–9763

Collins SR, Douglass A, Vale RD, Weissman JS (2004) Mechanism of prion propagation: amyloid growth occurs by monomer addition. PLoS Biol 2:e321

Das M, Mei X, Jayaraman S, Atkinson D, Gursky O (2014) Amyloidogenic mutations in human apolipoprotein A-I are not necessarily destabilizing – a common mechanism of apolipoprotein A-I misfolding in familial amyloidosis and atherosclerosis. FEBS J 281:2525–2542

De Groot NS, Castillo V, Graña-Montes R, Ventura S (2012) AGGRESCAN: method, application, and perspectives for drug design. Methods Mol Biol 819:199–220

Dobson CM (1999) Protein misfolding, evolution and disease. Trends Biochem Sci 24:329–332

Dovidchenko NV, Finkelstein AV, Galzitskaya OV (2014) How to determine the size of folding nuclei of protofibrils from the concentration dependence of the rate and lag-time of aggregation. I. Modeling the amyloid protofibril formation. J Phys Chem B 118:1189–1197

Emily M, Talvas A, Delamarche C (2013) MetAmyl: a META-predictor for AMYLoid proteins. PLoS One 8:e79722

Esteras-Chopo A, Serrano L, López de la Paz M (2005) The amyloid stretch hypothesis: recruiting proteins toward the dark side. Proc Natl Acad Sci U S A 102:16672–16677

Fändrich M, Fletcher MA, Dobson CM (2001) Amyloid fibrils from muscle myoglobin. Nature 410:165–166

Fändrich M, Forge V, Buder K, Kittler M, Dobson CM, Diekmann S (2003) Myoglobin forms amyloid fibrils by association of unfolded polypeptide segments. Proc Natl Acad Sci U S A 100:15463–15468

Ferguson N, Berriman J, Petrovich M, Sharpe TD, Finch JT, Fersht AR (2003) Rapid amyloid fiber formation from the fast-folding WW domain FBP28. Proc Natl Acad Sci U S A 100:9814–9819

Fernández A, Kardos J, Scott LR, Goto Y, Berry RS (2003) Structural defects and the diagnosis of amyloidogenic propensity. Proc Natl Acad Sci U S A 100:6446–6451

Fernandez-Escamilla A-M, Rousseau F, Schymkowitz J, Serrano L (2004) Prediction of sequence-dependent and mutational effects on the aggregation of peptides and proteins. Nat Biotechnol 22:1302–1306

Ferrone FA, Hofrichter J, Sunshine HR, Eaton WA (1980) Kinetic studies on photolysis-induced gelation of sickle cell hemoglobin suggest a new mechanism. Biophys J 32:361–380

Foderà V, Librizzi F, Groenning M, van de Weert M, Leone M (2008) Secondary nucleation and accessible surface in insulin amyloid fibril formation. J Phys Chem B 112:3853–3858

Fowler DM, Koulov AV, Balch WE, Kelly JW (2007) Functional amyloid – from bacteria to humans. Trends Biochem Sci 32:217–224

Frieden C, Goddette DW (1983) Polymerization of actin and actin-like systems: evaluation of the time course of polymerization in relation to the mechanism. Biochemistry (Mosc) 22:5836–5843

Galzitskaya OV (2009) Are the same or different amino acid residues responsible for correct and incorrect protein folding? Biochem Biokhimiia 74:186–193

Galzitskaya OV (2011a) Misfolded species involved regions which are involved in an early folding nucleus. In: Haggerty LM (ed) Protein struct. Nova Science Publishers, Hauppauge, pp 1–30

Galzitskaya OV (2011b) Regions which are responsible for swapping are also responsible for folding and misfolding. Open Biochem J 5:27–36

Galzitskaya OV, Finkelstein AV (1999) A theoretical search for folding/unfolding nuclei in three-dimensional protein structures. Proc Natl Acad Sci U S A 96:11299–11304

Galzitskaya OV, Garbuzynskiy SO (2008) Folding and aggregation features of proteins. In: O'Doherty CB, Byrne AC (eds) Protein misfolding. Nova Science Publishers, New York, pp 99–112

Galzitskaya OV, Surin AK, Nakamura H (2000) Optimal region of average side-chain entropy for fast protein folding. Protein Sci 9:580–586

Galzitskaya OV, Garbuzynskiy SO, Lobanov MY (2006) FoldUnfold: web server for the prediction of disordered regions in protein chain. Bioinformatics 22:2948–2949

Garbuzynskiy SO, Finkelstein AV, Galzitskaya OV (2004) Outlining folding nuclei in globular proteins. J Mol Biol 336:509–525

Garbuzynskiy SO, Lobanov MY, Galzitskaya OV (2010) FoldAmyloid: a method of prediction of amyloidogenic regions from protein sequence. Bioinformatics 26:326–332

Gasior P, Kotulska M (2014) FISH Amyloid – a new method for finding amyloidogenic segments in proteins based on site specific co-occurence of aminoacids. BMC Bioinformatics 15:54

Gertz MA, Rajkumar SV (eds) (2010) Amyloidosis: diagnosis and treatment. Humana Press, New York

Giehm L, Otzen DE (2010) Strategies to increase the reproducibility of protein fibrillization in plate reader assays. Anal Biochem 400:270–281

Gilead S, Gazit E (2005) Self-organization of short peptide fragments: from amyloid fibrils to nanoscale supramolecular assemblies. Supramol Chem 17:87–92

Goldstein RF, Stryer L (1986) Cooperative polymerization reactions. Analytical approximations, numerical examples, and experimental strategy. Biophys J 50:583–599

Hamodrakas SJ, Liappa C, Iconomidou VA (2007) Consensus prediction of amyloidogenic determinants in amyloid fibril-forming proteins. Int J Biol Macromol 41:295–300

Hayman SR, Bailey RJ, Jalal SM, Ahmann GJ, Dispenzieri A, Gertz MA, Greipp PR, Kyle RA, Lacy MQ, Rajkumar SV, Witzig TE, Lust JA, Fonseca R (2001) Translocations involving the immunoglobulin heavy-chain locus are possible early genetic events in patients with primary systemic amyloidosis. Blood 98:2266–2268

Hofrichter J, Ross PD, Eaton WA (1974) Kinetics and mechanism of deoxyhemoglobin S gelation: a new approach to understanding sickle cell disease. Proc Natl Acad Sci U S A 71:4864–4868

Idicula-Thomas S, Balaji PV (2005) Understanding the relationship between the primary structure of proteins and their amyloidogenic propensity: clues from inclusion body formation. Protein Eng Des Sel PEDS 18:175–180

Ivanova MI, Sawaya MR, Gingery M, Attinger A, Eisenberg D (2004) An amyloid-forming segment of beta2-microglobulin suggests a molecular model for the fibril. Proc Natl Acad Sci U S A 101:10584–10589

Jahn TR, Parker MJ, Homans SW, Radford SE (2006) Amyloid formation under physiological conditions proceeds via a native-like folding intermediate. Nat Struct Mol Biol 13:195–201

Jaroniec CP, MacPhee CE, Astrof NS, Dobson CM, Griffin RG (2002) Molecular conformation of a peptide fragment of transthyretin in an amyloid fibril. Proc Natl Acad Sci U S A 99:16748–16753

Jiménez JL, Guijarro JI, Orlova E, Zurdo J, Dobson CM, Sunde M, Saibil HR (1999) Cryo-electron microscopy structure of an SH3 amyloid fibril and model of the molecular packing. EMBO J 18:815–821

Jones S, Manning J, Kad NM, Radford SE (2003) Amyloid-forming peptides from beta2-microglobulin-Insights into the mechanism of fibril formation in vitro. J Mol Biol 325:249–257

Kajava AV, Baxa U, Wickner RB, Steven AC (2004) A model for Ure2p prion filaments and other amyloids: the parallel superpleated beta-structure. Proc Natl Acad Sci U S A 101:7885–7890

Kalebina TS, Plotnikova TA, Gorkovskii AA, Selyakh IO, Galzitskaya OV, Bezsonov EE, Gellissen G, Kulaev IS (2008) Amyloid-like properties of Saccharomyces cerevisiae cell wall glucantransferase Bgl2p: prediction and experimental evidences. Prion 2:91–96

Kayed R, Head E, Thompson JL, McIntire TM, Milton SC, Cotman CW, Glabe CG (2003) Common structure of soluble amyloid oligomers implies common mechanism of pathogenesis. Science 300:486–489

Klimtchuk ES, Gursky O, Patel RS, Laporte KL, Connors LH, Skinner M, Seldin DC (2010) The critical role of the constant region in thermal stability and aggregation of amyloidogenic immunoglobulin light chain. Biochemistry (Mosc) 49:9848–9857

Knowles TPJ, Waudby CA, Devlin GL, Cohen SIA, Aguzzi A, Vendruscolo M, Terentjev EM, Welland ME, Dobson CM (2009) An analytical solution to the kinetics of breakable filament assembly. Science 326:1533–1537

Kovacs E, Tompa P, Liliom K, Kalmar L (2010) Dual coding in alternative reading frames correlates with intrinsic protein disorder. Proc Natl Acad Sci U S A 107:5429–5434

Kozhukh GV, Hagihara Y, Kawakami T, Hasegawa K, Naiki H, Goto Y (2002) Investigation of a peptide responsible for amyloid fibril formation of beta 2-microglobulin by achromobacter protease I. J Biol Chem 277:1310–1315

Krebs MR, Wilkins DK, Chung EW, Pitkeathly MC, Chamberlain AK, Zurdo J, Robinson CV, Dobson CM (2000) Formation and seeding of amyloid fibrils from wild-type hen lysozyme and a peptide fragment from the beta-domain. J Mol Biol 300:541–549

López de la Paz M, Serrano L (2004) Sequence determinants of amyloid fibril formation. Proc Natl Acad Sci U S A 101:87–92

Lu X, Wintrode PL, Surewicz WK (2007) Beta-sheet core of human prion protein amyloid fibrils as determined by hydrogen/deuterium exchange. Proc Natl Acad Sci U S A 104:1510–1515

Matouschek A, Kellis JT, Serrano L, Fersht AR (1989) Mapping the transition state and pathway of protein folding by protein engineering. Nature 340:122–126

Maurer-Stroh S, Debulpaep M, Kuemmerer N, Lopez de la Paz M, Martins IC, Reumers J, Morris KL, Copland A, Serpell L, Serrano L, Schymkowitz JWH, Rousseau F (2010) Exploring the sequence determinants of amyloid structure using position-specific scoring matrices. Nat Methods 7:237–242

Maury CP, Nurmiaho-Lassila EL (1992) Creation of amyloid fibrils from mutant Asn187 gelsolin peptides. Biochem Biophys Res Commun 183:227–231

Morris AM, Watzky MA, Finke RG (2009) Protein aggregation kinetics, mechanism, and curve-fitting: a review of the literature. Biochim Biophys Acta 1794:375–397

Muñoz V, Serrano L (1994) Elucidating the folding problem of helical peptides using empirical parameters. Nat Struct Biol 1:399–409

Nelson R, Sawaya MR, Balbirnie M, Madsen A, Riekel C, Grothe R, Eisenberg D (2005) Structure of the cross-beta spine of amyloid-like fibrils. Nature 435:773–778

Oosawa F, Asakura S, Hotta K, Imai N, Ooi T (1959) G-F transformation of actin as a fibrous condensation. J Polym Sci 37:323–336

Picotti P, De Franceschi G, Frare E, Spolaore B, Zambonin M, Chiti F, de Laureto PP, Fontana A (2007) Amyloid fibril formation and disaggregation of fragment 1–29 of apomyoglobin: insights into the effect of pH on protein fibrillogenesis. J Mol Biol 367:1237–1245

Rauscher S, Baud S, Miao M, Keeley FW, Pomès R (2006) Proline and glycine control protein self-organization into elastomeric or amyloid fibrils. Structure 14:1667–1676

Roland BP, Kodali R, Mishra R, Wetzel R (2013) A serendipitous survey of prediction algorithms for amyloidogenicity. Biopolymers 100:780–789

Ruschak AM, Miranker AD (2007) Fiber-dependent amyloid formation as catalysis of an existing reaction pathway. Proc Natl Acad Sci U S A 104:12341–12346

Sánchez IE, Tejero J, Gómez-Moreno C, Medina M, Serrano L (2006) Point mutations in protein globular domains: contributions from function, stability and misfolding. J Mol Biol 363:422–432

Selivanova OM, Suvorina MY, Dovidchenko NV, Eliseeva IA, Surin AK, Finkelstein AV, Schmatchenko VV, Galzitskaya OV (2014) How to determine the size of folding nuclei of protofibrils from the concentration

dependence of the rate and lag-time of aggregation. II. Experimental application for insulin and LysPro insulin: aggregation morphology, kinetics, and sizes of nuclei. J Phys Chem B 118:1198–1206

Serio TR, Cashikar AG, Kowal AS, Sawicki GJ, Moslehi JJ, Serpell L, Arnsdorf MF, Lindquist SL (2000) Nucleated conformational conversion and the replication of conformational information by a prion determinant. Science 289:1317–1321

Tartaglia GG, Vendruscolo M (2008) The Zyggregator method for predicting protein aggregation propensities. Chem Soc Rev 37:1395–1401

Tartaglia GG, Pellarin R, Cavalli A, Caflisch A (2005) Organism complexity anti-correlates with proteomic beta-aggregation propensity. Protein Sci 14:2735–2740

Tenidis K, Waldner M, Bernhagen J, Fischle W, Bergmann M, Weber M, Merkle ML, Voelter W, Brunner H, Kapurniotu A (2000) Identification of a penta- and hexapeptide of islet amyloid polypeptide (IAPP) with amyloidogenic and cytotoxic properties. J Mol Biol 295:1055–1071

Thangakani AM, Kumar S, Nagarajan R, Velmurugan D, Gromiha MM (2014) GAP: towards almost 100 percent prediction for β-strand-mediated aggregating peptides with distinct morphologies. Bioinformatics 30:1983–1990

Thompson A, White AR, McLean C, Masters CL, Cappai R, Barrow CJ (2000) Amyloidogenicity and neurotoxicity of peptides corresponding to the helical regions of PrP(C). J Neurosci Res 62:293–301

Trovato A, Chiti F, Maritan A, Seno F (2006) Insight into the structure of amyloid fibrils from the analysis of globular proteins. PLoS Comput Biol 2:e170

Tsolis AC, Papandreou NC, Iconomidou VA, Hamodrakas SJ (2013) A consensus method for the prediction of "Aggregation-Prone" peptides in globular proteins. PLoS One 8:e54175

Von Bergen M, Friedhoff P, Biernat J, Heberle J, Mandelkow EM, Mandelkow E (2000) Assembly of tau protein into Alzheimer paired helical filaments depends on a local sequence motif ((306)VQIVYK(311)) forming beta structure. Proc Natl Acad Sci U S A 97:5129–5134

Walsh I, Seno F, Tosatto SCE, Trovato A (2014) PASTA 2.0: an improved server for protein aggregation prediction. Nucleic Acids Res 42:W301–W307

Wang Q, Johnson JL, Agar NYR, Agar JN (2008) Protein aggregation and protein instability govern familial amyotrophic lateral sclerosis patient survival. PLoS Biol 6:e170

Wegner A, Savko P (1982) Fragmentation of actin filaments. Biochemistry 21:1909–1913

Wright CF, Teichmann SA, Clarke J, Dobson CM (2005) The importance of sequence diversity in the aggregation and evolution of proteins. Nature 438:878–881

Xue W-F, Homans SW, Radford SE (2008) Systematic analysis of nucleation-dependent polymerization reveals new insights into the mechanism of amyloid self-assembly. Proc Natl Acad Sci U S A 105:8926–8931

Yang H, Li J-J, Liu S, Zhao J, Jiang Y-J, Song A-X, Hu H-Y (2014) Aggregation of polyglutamine-expanded ataxin-3 sequesters its specific interacting partners into inclusions: implication in a loss-of-function pathology. Sci Rep 4:6410

Yoon S, Welsh WJ (2004) Detecting hidden sequence propensity for amyloid fibril formation. Protein Sci 13:2149–2160

Zhu L, Zhang X-J, Wang L-Y, Zhou J-M, Perrett S (2003) Relationship between stability of folding intermediates and amyloid formation for the yeast prion Ure2p: a quantitative analysis of the effects of pH and buffer system. J Mol Biol 328:235–254

Zimmermann O, Hansmann UHE (2006) Support vector machines for prediction of dihedral angle regions. Bioinformatics 22:3009–3015

Role of Syndecans in Lipid Metabolism and Human Diseases

10

Elena I. Leonova and Oxana V. Galzitskaya

Abstract

Syndecans are transmembrane heparan sulfate proteoglycans involved in the regulation of cell growth, differentiation, adhesion, neuronal development, and lipid metabolism. Syndecans are expressed in a tissue-specific manner to facilitate diverse cellular processes. As receptors and co-receptors, syndecans provide promising therapeutic targets that bind to a variety of physiologically important ligands. Negatively charged glycosaminoglycan chains of syndecans, located in the extracellular compartment, are critical for such binding. Functions of syndecans are as diverse as their ligands. For example, hepatic syndecan-1 mediates clearance of triglyceride-rich lipoproteins. Syndecan-2 promotes localization of Alzheimer's amyloid Aβ peptide to the cell surface, which is proposed to contribute to amyloid plaque formation. Syndecan-3 helps co-localize the appetite-regulating melanocortin-4 receptor with its agonist, leading to an increased appetite. Finally, syndecan-4 initiates the capture of modified low-density lipoproteins by macrophages and thereby promotes the atheroma formation. We hypothesize that syndecan modifications such as desulfation of glycosaminoglycan chains may contribute to a wide range of diseases, from atherosclerosis to type 2 diabetes. At the same time, desulfated syndecans may have beneficial effects, as they can inhibit amyloid plaque formation or decrease the appetite. Despite considerable progress in understanding diverse functions of syndecans, the complex physiological roles of this intriguing family of proteoglycans are far from

E.I. Leonova
Institute of Protein Research, Russian Academy of Sciences, Pushchino, Moscow Region 142290, Russia

Institute of Biochemistry and Physiology of Microorganisms, Russian Academy of Sciences, Pushchino, Moscow Region 142290, Russia
e-mail: 1102.elena@gmail.com

O.V. Galzitskaya (✉)
Institute of Protein Research, Russian Academy of Sciences, Pushchino, Moscow Region 142290, Russia
e-mail: ogalzit@vega.protres.ru

© Springer International Publishing Switzerland 2015
O. Gursky (ed.), *Lipids in Protein Misfolding*, Advances in Experimental Medicine and Biology 855, DOI 10.1007/978-3-319-17344-3_10

clear. Additional studies of syndecans may potentially help develop novel therapeutic approaches and diagnostic tools to alleviate complex human diseases such as cardiovascular and Alzheimer's diseases.

Keywords

Apolipoproteins • Heparan sulfate proteoglycans • Atherosclerosis • Alzheimer's disease • Obesity • Shedding of syndecans

Abbreviations

AD	Alzheimer's disease
AgRP	Agouti-related protein
apo	Apolipoprotein
APP	Amyloid-β precursor protein
Aβ	Amyloid-β peptide
DPAB	Dense peripheral actin bands
EG	Endothelial glycocalyx
HDL	High-density lipoproteins
HSPG	Heparan sulfate proteoglycan
IDL	Intermediate-density lipoproteins
LDL	Low-density lipoproteins
MAP	mitogen-activated protein kinase
MC4R	Melanocortin-4 receptor
MMP	Matrix metalloproteinase
PTK	Protein tyrosine kinase
SULF2	Heparan sulfate glucosamine-6-O-endosulfatase-2
TG	Triacylglycerols
VLDL	Very low-density lipoproteins
α-MSH	α-melanocyte-stimulating anorexigenic hormone

10.1 Introduction

Heparan sulfate proteoglycans (HSPGs) are glycoproteins containing covalently attached long chains of negatively charged heparan sulfates. Three groups of HSPGs differ in their location in the cell plasma membrane (syndecans and glypicans), in the extracellular matrix (agrin, perlecan, type XVIII collagen), or in the secretory vesicles (serglycin) (Sarrazin et al. 2011). This chapter is devoted to membrane HSPGs, particularly syndecans. In contrast to glypicans that are attached to the membrane via a lipid anchor, syndecans contain a membrane-spanning domain, as well as the extracellular and cytoplasmic domains in their core protein. This allows syndecans to bind various extracellular ligands and modulate their downstream signaling.

Vertebrates have four types of syndecans that are located on the surfaces of most cell types, including fibroblasts and epithelial cells. Syndecans function as membrane receptors and co-receptors that can undergo ligand-induced endocytosis. The endocytic activity of syndecan-1 plays a crucial physiological role in lipid metabolism (Stanford et al. 2009), while syndecan-4 is important in capturing of modified low-density lipoproteins (LDL) by macrophages (Boyanovsky et al. 2009), which contributes to formation of atherosclerotic plaques. As a co-receptor, syndecan-2 contributes to the accumulation of amyloid-beta (Aβ) peptide on the cell surface and hence, to the development of Alzheimer's disease (AD) (Narindrasorasak et al. 1991; Li et al. 2005; Zhang et al. 2014), while syndecan-3 helps regulate the appetite (Karlsson-Lindahl et al. 2012). An excellent review on the mechanisms of action of HSPGs such as syndecan-2 in AD was recently published by Zhang et al. (2014). Here, we provide a broad overview of the roles of syndecans in various physiological processes including, but not limited to, AD and other disorders of lipid metabolism such as type 2 diabetes, obesity and cardiovascular disease.

Lipid metabolism involves complex biochemical processes that occur in living organisms and are controlled by the endocrine glands. A variety of biochemical mechanisms underlie the action of hormones produced by these glands. One example is regulation of the body weight by the hypothalamus. Arguably, the best-studied pathway to influence hypothalamic appetite involves a G-protein coupled receptor melanocortin-4 (MC4R), its agonist α-melanocyte-stimulating hormone (α-MSH), its antagonist agouti-related protein (AgRP), and its co-receptor syndecan-3. Mutations in MC4R lead to obesity in humans, and modifications in syndecan-3 can also influence the body weight (Lubrano-Berthelier et al. 2003; Reizes et al. 2003).

Perhaps the most common cause of impaired lipid metabolism involves aberrant utilization of lipoproteins. These water-soluble nanoparticles, which include high-, low- intermediate- and very-low-density lipoproteins (HDL, LDL, IDL, VLDL) and chylomicrons, are comprised of lipids (phospholipids, cholesterol, cholesterol esters, triacylglycerides) and specific proteins termed apolipoproteins (apos) that solubilize lipids and mediate their transport and metabolism in circulation (see Chap. 8 by Das and Gursky in this volume). Aberrant lipoprotein metabolism is involved in major human diseases such as diabetes, hypercholesterolemia, atherosclerosis and obesity, and also contributes to various kinds of dementia including AD. The link between the elevated plasma levels of cholesterol and the severity of AD is well established ((Puglielli et al. 2003; Barrett et al. 2012) and references therein), leading to a hypothesis that impaired lipid metabolism is a major hallmark of AD (see Sect. 7). An important link between lipoprotein metabolism and AD is provided by apolipoprotein E (apoE). ApoE, which circulates mainly on HDL and VLDL, is synthesized by the liver, brain, kidney, and spleen and mediates cholesterol transport and metabolism in plasma and cerebrospinal fluid; apoE is an important risk factor for AD where it acts in an isoform-specific manner (Liu et al. 2013). Various cell receptors bind apoE and internalize apoE-containing lipoproteins, and one of such receptors is syndecan-1.

In this chapter, we summarize the current knowledge of the roles of syndecans in lipid metabolism and other vital processes, as well as their potential involvement in major human diseases including atherosclerosis and AD.

10.2 Biosynthesis, Structure and Function of Syndecans

10.2.1 Syndecans: Primary Structure, Biosynthesis and Signalling

Syndecans were discovered in 1989 by Merton Bernfield's group at Stanford during studies of the cell surface proteins containing heparan sulfate and chondroitin sulfate chains. Such proteins were called "syndeins", which means "to bind" in Greek (Saunders et al. 1989). Syndecans contain a core protein consisting of a long external domain (ectodomain) with covalently attached glycosaminoglycan chains, a single-pass transmembrane domain, and a short cytoplasmic domain. During evolution, the ectodomains have undergone numerous mutations, whereas the cytoplasmic domains have not changed much and contain highly conserved functional regions.

In vertebrates, there are four types of syndecans that originate from a single invertebrate precursor protein. Following gene duplication, two groups of syndecans were formed, one including syndecans-1 and -3 and the other syndecans-2 and -4 (Chakravarti and Adams 2006). Another round of gene duplication occurred in each group, explaining high homology within the groups (Chen et al. 2002). Syndecans-2 and -4 are shorter and contain only heparan sulfate chains, while syndecans-1 and -3 are longer and contain both heparan sulfate and chondroitin sulfate chains (Fig. 10.1).

The core protein of syndecans is synthesized on the membrane-bound ribosomes followed by translocation to the endoplasmic reticulum lumen. Covalent attachment of glycosaminoglycan chains to the core protein takes place in Golgi. Finally, syndecans are transported to the cell surface via exocytosis. Attachment of the heparan sulfate and chondroitin sulfate to the

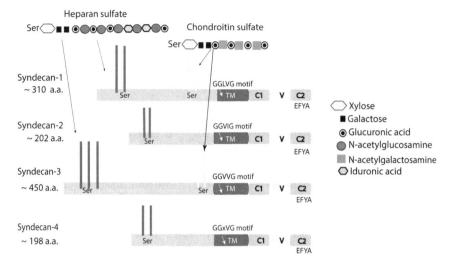

Fig. 10.1 Schematic representation of four types of syndecans found in vertebrates. Syndecans-1 and -3 are longer than syndecans-2 and -4. Glycosaminoglycan chains that are covalently attached to the core protein consist of serine, xylose, two galactoses, and glucuronic acid. This is followed by repeating disaccharides of heparan sulfate, one of which is N-acetylglucosamine and glucuronic acid, and the other is N-acetylglucosamine and iduronic acid. Hondroitin sulfate is composed of N-acetylgalactosamine and glucuronic acid. The cytoplasmic domain contains two conserved regions (*C1* and *C2*) flanking the variable region (*V*) that is different for different types of syndecans. Transmembrane domain (*TM*) contains GxxxG motif that is unique in each type of syndecan and is essential for dimerization and, potentially, for cholesterol binding. C2 domain contains EFYA motif that binds PDZ domains (Modified from Manon-Jensen et al. 2010)

syndecan core protein occurs via a common mechanism (Manon-Jensen et al. 2010).

Glycosaminoglycan chains can contain 50–200 disaccharides joined by the negatively charged sulfate groups, which allows these chains to bind a large number of positively charged ligands and numerous plasma proteins, as well as the proteins secreted by the extracellular matrix. These proteins include various fibroblast growth factors, transforming growth factors, receptor tyrosine kinases, chemokines and interleukins, lipases and apolipoproteins (Bishop et al. 2007). Electrostatic repulsion among the negatively charged chains causes glycosaminoglycans to stretch in space, increasing cell surface coverage. Therefore, the formation and function of syndecan complexes with their partners may depend on the spatial distribution of the negatively charged heparan sulfates (Kreuger et al. 2006). Notably, sulfation of glycosaminoglycans is non-uniform: areas with a small content of sulfate groups are scattered among those with a high content, with less sulfated sites usually located closer to the core protein.

Direct interactions of the core protein with various proteins also contribute to syndecan function. For example, the cytoplasmic domains of the core protein can bind cytoskeletal proteins and kinases, while the ectodomains can act in concert with other cell-surface receptors such as integrins to activate cell attachment, cell spreading and motility (Couchman et al. 2001). Cytoplasmic domains of syndecans contain two conserved regions, C1 (consisting of 11 amino acids) and C2 (4 amino acids), which flank the variable V region (20 amino acids) (Fig. 10.1). Four amino acids (EFYA) of the C2 region bind to PDZ-binding proteins, including synbindin, synectin, calcium/calmodulin-dependent kinase, and syntenin. This binding is important in vesicle transport and in cancer cells proliferation (Beauvais and Rapraeger 2004). The proximal C1 region interacts with actin-binding proteins ezrin, radixin, and moesin, which control the organization of the actin cytoskeleton. The variable region is unique for each type of syndecan and determines its ligand binding specificity (Brown 2011) (Fig. 10.1).

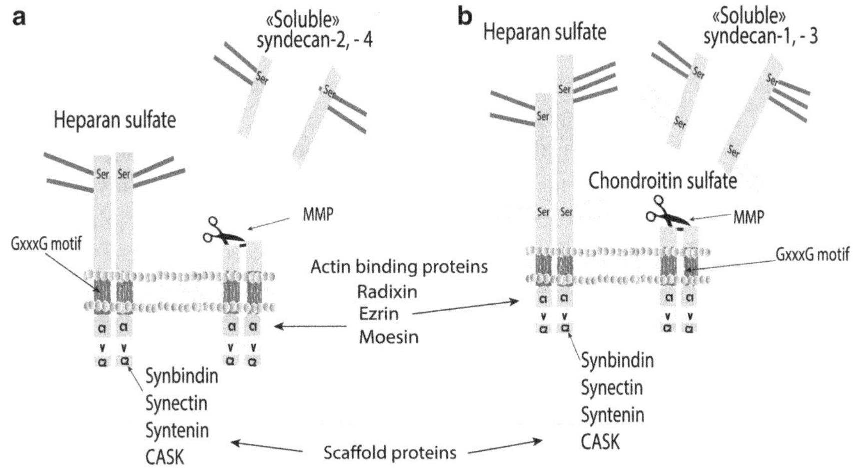

Fig. 10.2 Schematic representation of the structures of syndecan-1 and -3 (**a**) and syndecan-2 and -4 (**b**) and their shedding. During shedding, the ectodomains are cleaved from the membrane surface by matrix metallo-proteinases (*MMPs*), generating a "soluble" form of syndecans. The C2 region of the cytoplasmic domain contains a PDZ binding motif that binds the scaffold proteins of the cell (*synbindin*, *synectin*, etc.). The C1 region interacts with the actin-binding proteins in the cell (*radixin, exrin, moesin*). The GxxxG motif in the transmembrane domain is necessary for syndecan dimerization as well as for its partitioning into choles-terol-rich membrane domains (Modified from Manon-Jensen et al. 2010)

10.2.2 Shedding of Syndecans

There are two different forms of syndecans: the full-length form that spans the membrane and the "soluble" form containing just the cleaved ectodomain dissociated from the membrane. Shedding of the full-length form of syndecans to produce the soluble form is a highly regulated process that can occur upon various extracellular stimuli (including growth factors, chemokines, bacterial virulence factors, trypsin, heparanase, insulin) and under cellular stress (Manon-Jensen et al. 2010). Soluble ectodomains can function as paracrine or autocrine effectors that compete with transmembrane syndecans for extracellular ligands. Ectodomains of syndecans are constitu-tively shed by the extracellular zinc-dependent endopeptidase and by matrix metalloproteinases (MMPs) (Fig. 10.2). The shedding enhances in response to inflammation and other disruptive processes. For example, large amounts of soluble syndecan-1 accumulate in the tissue fluid sur-rounding the wounds. Upon action of thrombin and epidermal growth factor, which are espe-cially active in wound healing, the shedding of syndecanes intensifies (Subramanian et al. 1997).

Phosphorylation of conserved tyrosines in the cytoplasmic domain of syndecans also stimulates the shedding (Manon-Jensen et al. 2010).

Interestingly, enhanced shedding of syndecans induces their expression. Therefore, the amount of heparan sulfate chains present on the syndecan core proteins was proposed to control both the rate of shedding and the biosynthesis of syn-decans (Ramani et al. 2012). Despite its impor-tance in pathological states such as inflammation and cell stress, the mechanism of syndecan shed-ding is not known in detail and needs further investigation.

10.2.3 Functions of Individual Syndecans in Vertebrates

Syndecan-1 plays important roles in lipid metab-olism (van Barlingen et al. 1996), wound healing (Vanhoutte et al. 2007), regulation of the leuko-cyte migration (Savery et al. 2013), and in reor-ganization of the endothelial cytoskeleton in response to fluid shear stress (Thi et al. 2004). The latter effect helps to protect the vessel wall from damage, thereby preventing the development of atherosclerosis. However, increased gene expression of syndecan-1 in macrophages counteracts this protective effect and promotes formation of atherosclerotic plaques via an unknown mechanism (Asplund et al. 2009). Furthermore, the increased content of soluble syndecan-1 in serum is a biomarker of multiple myeloma (Kim et al. 2010).

Syndecan-2 plays a major role in the formation of the nervous system during embryogenesis, as well as in angiogenesis (Lin et al. 2007; Noguer et al. 2009; De Rossi et al. 2014). However, syndecan-2 is also implicated in amyloid plaque formation in neurodegenerative diseases, as it helps retain Aβ peptide on the cell surface, which is proposed to trigger the development of AD (Snow and Wight 1989; Manich et al. 2011; Zhang et al. 2014 and references therein). In addition, syndecan-2 contributes to the migration of cancer cells during the development of intestinal tumors (Lin et al. 2007; Choi et al. 2012).

Syndecan-3 is required for normal brain function, regulation of appetite, memory improvement, as well as for the development of skeletal muscles (Kaksonen et al. 2002; Cornelison et al. 2004; Karlsson-Lindahl et al. 2012). The latter is supported by the animal model studies showing that syndecan-3-null mice are thin and weak (Kaksonen et al. 2002; Cornelison et al. 2004; Karlsson-Lindahl et al. 2012).

Syndecan-4 plays important roles in angiogenesis, focal adhesion, cell migration, in reducing blood pressure, as well as in LDL uptake by the arterial macrophages and formation of atherosclerotic plaques (Denhez et al. 2002; Boyanovsky et al. 2009). Because LDL uptake by the cells is a major determinant of cholesterol concentration in the plasma membrane, which importantly influences the development of cardiovascular as well as Alzheimer's diseases, we speculate that syndecan-4 may indirectly influence the development of these major disorders.

10.3 Endothelial Glycocalyx and Its Major Glycosaminoglycans

Glycosaminoglycans are associated with the arterial lumen and contain mainly heparan sulfates, chondroitin sulfates, and hyaluronic acids. Heparan sulfate comprises 50–90 % of the total glycosaminoglycans. Two major core proteins in EG have attached heparan sulfate chains: syndecan-1 and glypican-1. These two proteins have distinct functions. For example, syndecan-1 can transmit a signal into the cell to reconstruct cytoskeleton, while glypican-1 is involved in signaling to activate nitric oxide synthesis (Kokenyesi and Bernfield 1994; Weinbaum et al. 2007).

Syndecan-1 (33 kDa) contains three sites for covalent attachment of heparan sulfate which are located close to the N-terminus of the core protein, and two sites for attachment of chondroitin sulfate located close to the transmembrane domain (Kokenyesi and Bernfield 1994). The cytoplasmic domain of syndecan-1 interacts with structural proteins in the cells.

Glypican-1 (64 kDa) contains three to four sites for covalent attachment of heparan sulfate which are located close to the membrane (Fransson et al. 2004). In contrast to syndecan-1 that is anchored to the membrane via the transmembrane protein, glypican is anchored through glycosyl phosphatidyl inositol (Fransson et al. 2004). This lipid anchor contributes to the partitioning of glypican into caveolae, which are much larger than the lipid rafts and form 50–100 nm pits reinforced by the cytoskeleton (van Deurs et al. 2003). Caveolae (Latin for "little pits") are essential in cell signalling and endocytosis.

Compared to heparan sulfate and chondroitin sulfate, hyaluronic acid is much longer and may reach 1,000 kDa (Fig. 10.3, green line). It is located on the cell surface and is not bound covalently to the core protein (Laurent and Fraser 1992). Hyaluronic acid contains no sulfate groups, and its negative charge results solely

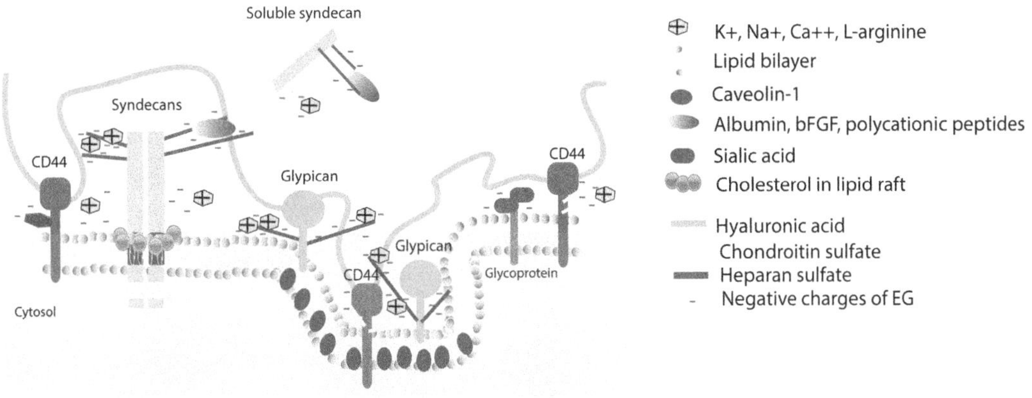

Fig. 10.3 Schematic representation of the endothelial glycocalyx components. Caveolin, which is associates with membrane areas rich in cholesterol and sphingolipids (*lipid rafts*), facilitates formation of caveolae. Glypican is associated with heparan sulfate and is usually located in the caveolae. Syndecan-1 is shown as a dimer located in the lipid rafts. Syndecan-1 contains heparan sulfate and chondroitin sulfate chains. The short cytoplasmic domain of syndecan-1 interacts with the structural proteins of the cell. Sialic acid is bound to the glycoprotein, and hyaluronic acid is bound to the transmembrane receptor CD44, which is also bound to chondroitin sulfate and is often located in the caveolae. Numerous plasma proteins, growth factors and cations are retained in EG via the ionic interactions with the glycosaminoglycans (Modified from Weinbaum et al. 2007)

from the carboxyl groups that also confer its hydrophilic properties. Hyaluronan interacts with transmembrane receptor CD44 and with chondroitin sulfate that are preferentially located in caveolae (van Deurs et al. 2003; Singleton and Bourguignon 2004).

The endothelial glycocalyx (EG) forms an extracellular layer on the luminal surface of endothelial cells. EG is the main site of triacylglycerols (TG) hydrolysis in TG-rich lipoproteins such as VLDL by lipoprotein lipase. EG has a multicomponent structure that contains proteoglycans anchored to the plasma membrane and glycoproteins with glycosaminoglycan chains attached to their ectodomains (Fig. 10.3). The complex structure of EG and its location at the interface between the blood flow and the endothelium determine its various functions. For example, EG protects the endothelium from pathogens such as bacterial virulence factors, and is important in selective infiltration of blood components and in protection against the mechanical impact of the blood flow (Weinbaum et al. 2007). The acidic character of

EG facilitates interactions with the basic moieties on plasma proteins, growth factors, cytokines, as well as with various cations and water. In addition, EG contains various cell receptors (e.g., integrins), cell adhesion molecules (selectins), and immunoglobulins. These molecules can bind oligosaccharides such as sialic acid, and thereby contribute an additional negative charge to EG (Pries et al. 2000).

10.4 Role of Syndecan-1 in Reconstruction of Endothelial Cell Cytoskeleton Under Fluid Shear Stress

Many studies have shown that EG plays an important role in signal transduction resulting from the shear stress on the actin cytoskeletons. This transduction can involve various signaling cascades. The "bumper car" model postulates that the mechanical pressure of the blood flow resulting from the high fluid shear stress is

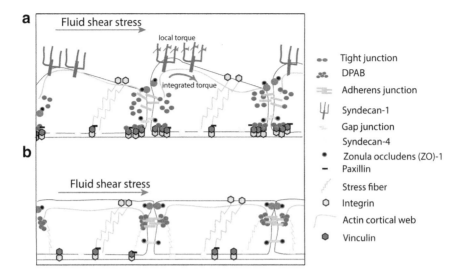

Fig. 10.4 Scheme showing reconstruction of cytoskeleton in response to high fluid shear stress. (**a**) The core protein of syndecan-1 under the influence of the local torque transmits clockwise rotation into the cell, resulting in reorganization of the cytoskeleton, altered cell-cell contacts, and changes in the cell shape. (**b**) In the absence of syndecan-1, there is no cytoskeleton reorganization and no rearrangement of the cell-cell contacts. The fluid sheer stress acts directly on the apical surface of the endothelial cells (Modified from Thi et al. 2004)

dissipated by the core protein of syndecan-1, which protects the apical membrane of the vascular endothelial cells (Thi et al. 2004; Weinbaum et al. 2007). The core protein of syndecan-1 causes a torque on the actin cortical web that produces a clockwise rotation of the cell (Fig. 10.4). Though the bending moment on each core protein is small, all together they promote clockwise rotation of the dense peripheral actin bands (DPAB), which disrupts cell junctions, retaining only tight junctions. Such disruption serves as a mechanism of cell protection, just as the disruption of a bumper protects the car.

The "bumper car" model also explains why the cells are elongated and oriented along the direction of the flow. Reorganization of DPAB by fluid shear stress lowers the impact from the force on the tight junctions, and thereby provides greater stability in response to the blood flow. Once DPAB are reorganized, they can form new contacts and thus adapt to the new conditions of shear stress (Thi et al. 2004).

Under the impact of the laminar blood flow and high shear stress, as well as in the presence of HSPGs, the proliferation of elongated endothelial cells is suppressed. However, upon damage of the EG, endothelial cells acquire the ability to proliferate, which is important in wound healing (Yao et al. 2007).

10.5 Syndecan Dimerization and Clustering in Lipid Rafts

Biological activity of syndecans and other specific receptors and co-receptors involves their dimerization and clustering in lipid rafts. Lipid rafts, enriched in cholesterol and sphingolipids, form densely packed areas in the cell membrane. Typically, lipid rafts are not fixed in the membrane. Specific membrane proteins such as syndecans contribute to raft stabilization. The transmembrane domain of all syndecans is a single membrane spanning α-helix that contains a conserved dimerization motif GXXXG (Figs. 10.1 and 10.2). This motif places two Gly residues next to each other on the same side of the α-helix, forming a flat surface that is well suited for dimerization as well as for cholesterol binding. Sanders and colleagues (Barrett et al. 2012) demonstrated that this motif plays a key role in cholesterol binding to the transmembrane portion of the amyloid-β precursor protein (APP) (also see Chap. 1 by Hong in this volume). Such binding was proposed to lead to APP partitioning into the lipid rafts and thereby substantially influence its proteolytic processing by γ-secretase and generation of Aβ (Barrett et al. 2012). Similarly, the GXXXG motif in the transmembrane region of syndecans is implicated in the protein dimerization as well as in its partitioning into lipid rafts via the binding of membrane cholesterol (Dews and Mackenzie 2007; Barrett et al. 2012). Both dimerization and location in lipid rafts are essential for the syndecan action in transmembrane signaling, endocytosis, and other functions (Chen and Williams 2013). The biological significance of these interactions is supported by the strong sequence conservation of syndecan GxxxG motifs across different species (Leonova and Galzitskaya 2014).

Syndecan homodimers significantly differ in their strength of self-association: syndecan-1 homodimer is very weak, syndecan-3 and syndecan-4 homodimers are much stronger, and syndecan-2 homodimer is very strong (Dews and Mackenzie 2007). Different types of syndecans can also form heterodimers and even heterotrimers, except for syndecan-1 and -4 that do not form a heterodimer with each other (Dews and Mackenzie 2007). This may lead to formation of various syndecan complexes in cells that express more than one syndecan type. Dimerization and clustering of syndecans in lipid rafts plays a crucial role in various aspects of syndecan-dependent signal transduction and raft-mediated endocytosis.

10.6 ApoE and Syndecan-1 Mediate Hepatic Clearance of Triglyceride-Rich Lipoproteins

Liver is the key organ that regulates lipid metabolism, from secretion of various lipoproteins to their hepatic clearance, cholesterol uptake and synthesis of bile acids. In 1985, Brown and Goldstein won the Nobel Prize for their discovery of the LDL receptor and the receptor-mediated LDL endocytosis (Goldstein et al. 1985). Two types of hepatic lipoprotein receptors were identified, one binding to apolipoproteins B and E on the lipoprotein surface and the other only to apoE (Mahley et al. 1981). ApoB/apoE receptors are responsible for binding and endocytosis of LDL, while apoE receptors selectively bind triglyceride-rich lipoproteins (such as IDL, VLDL and their remnants) and mediate their endocytosis.

ApoE in association with HDL is the major apolipoprotein that mediates lipid transport and metabolism in the brain; moreover, apoE on plasma lipoproteins such as VLDL directs the delivery of the lipoprotein lipids to the liver (Getz and Reardon 2008). Several apoE receptors on the surface of hepatocytes are involved in the latter process, one of which is syndecan-1 (MacArthur et al. 2007). IDL, which are transient products of VLDL metabolism, are converted into LDL upon TG hydrolysis by lipoprotein lipase, which is accompanied by dissociation of apoE and other related proteins from the lipoprotein surface. Lipoprotein lipase that is anchored to the arterial wall was proposed to contribute to the retention of LDL in the arterial endothelium, thereby promoting the development of atherosclerosis (Pentikäinen et al. 2002). Furthermore, the thickness of the EG in the affected areas of the coronary arteries in mice was reported to be much smaller than that in the neighboring areas that were not affected by atheromas (van den Berg et al. 2006). Moreover, the thickness of EG was reported to decrease in the presence of high plasma levels of cholesterol, especially in the presence of oxidized LDL, and significantly increase by the action of high shear stress (Weinbaum et al. 2007). These changes in EG thickness critically involve syndecan-1.

Binding of apoE- and apoB-containing lipoproteins to heparan sulfates has long been known as a major mechanism of retention of these lipoproteins in the arterial endothelium, which is an important early trigger of atherosclerosis. Several stretches of basic residues in apoB and one such stretch in apoE (see Chap. 8 by Das and Gursky in this volume) are implicated in binding of LDL, VLDL and related

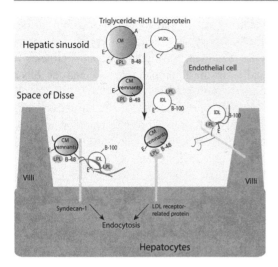

Fig. 10.5 Cartoon illustrating the role of syndecan-1 in clearance of triglyceride-rich lipoproteins. Chylomicrons (*CM*) are found in the intestine, and very-low, intermediate- and low-density lipoproteins (*VLDL, IDL* and *LDL*, respectively) circulate in plasma. Lipoprotein lipase (*LPL*) hydrolyses TG in these lipoproteins. Apolipoproteins A, E, Cs and two forms of apolipoprotein B (apoB-48 found on CM and apoB-100 found on VLDL and their metabolic products) are indicated (also see Chap. 8 by Das and Gursky in this volume) (Modified from MacArthur et al. 2007)

lipoproteins to their receptors, as well as in lipoprotein binding to HSPGs (Mahley et al. 1981; Chen and Williams 2013). More recently it has been proposed that syndecan-1 utilizes binding to heparan sulfate chains to provide a new receptor for the apoE-containing lipoproteins (Wilsie et al. 2006) (Fig. 10.5). This idea was supported by the observation that syndecan-1-deficient mice have high plasma levels of TG-rich lipoproteins, suggesting reduced clearance of these lipoproteins in the absence of syndecan-1 (Stanford et al. 2009).

Syndecan-1 is normally synthesized by the sub-endothelial hepatocytes. Numerous villi on the hepatocytes extend into the pre-capillary space of Disse (MacArthur et al. 2007). Plasma lipoproteins are small enough to enter the space of Disse and bind to the hepatocyte receptors. Similar to other apoE receptors, syndecan-1 mediates the uptake of TG-rich lipoproteins via the whole-particle endocytosis (Chen and Williams 2013). These lipoproteins contain multiple copies

of apoE, thus forming multivalent ligands for syndecan-1 chains. Interestingly, hepatocytes express genes of syndecan-1, -2, -4, yet only syndecan-1 binds lipoproteins. To explain this binding selectivity, it was proposed that heparan sulfate chains in syndecan-1, which contain more sulfated sugars than any other types of syndecans, mediate lipoprotein binding by interacting with the basic stretches from multiple copies of apoE on the lipoprotein surface (Stanford et al. 2009).

Aberrant clearance of triglyceride-rich lipoproteins occurs in type 2 diabetes characterized by elevated levels of plasma TG. These high levels can result from impaired clearance of TG-rich lipoproteins such as VLDL via the apoE receptors. Decreased sulfation of HSPG can decrease such clearance. Heparan sulfate glucosamine-6-O-endosulfatase-2 (SULF2) removes the 6-O-sulfate group from HSPG (Chen et al. 2010). In a mouse model of type 2 diabetes, the inhibition of SULF2 normalizes the clearance of TG-rich lipoproteins in spite of extensive deregulation of lipid metabolism in these animals (Hassing et al. 2012). Hormones adiponectin and insulin also inhibit SULF2 in cultured cells, resulting in enhanced lipoprotein catabolism. Taken together, these studies indicate that desulfation of heparan sulfate chains of syndecan-1 impairs the clearance of TG-rich lipoproteins and contributes to insulin resistance in the mouse model of diabetes 2, and that these pathogenic effects can be corrected by increased sulfation of HSPGs (Chen et al. 2010; Williams and Chen 2010).

10.7 Role of ApoE4 and Syndecan-2 in Amyloid Plaque Formation

Alzheimer's disease (AD) is the most common form of dementia characterized by the extracellular accumulation of Aβ peptide followed by the neuronal death and brain atrophy. This debilitating disease presents a growing public health challenge in the developed countries, with millions of patients affected worldwide. There is no cure for AD and the treatment options are limited, which makes it imperative to elucidate the factors that can promote or prevent neuronal damage in AD and the underlying molecular mechanisms. The "amyloid cascade" hypothesis postulates that accumulation of the neurotoxic Aβ peptide leads to the formation of amyloid plaques and causes aberrant cellular functions, including disruption of the plasma membrane or activation of apoptotic genes (Hardy and Higgins 1992). High concentrations of Aβ peptide, which are toxic to cells, can result from the peptide overproduction (in the early-onset hereditary AD) or its impaired clearance (in the late-onset sporadic AD that constitutes the majority of cases) (Prasansuklab and Tencomnao 2013). Aβ is a normally soluble 40–42 residue peptide generated from the large insoluble amyloid-β precursor protein (APP) upon cleavage of its transmembrane domain by specific secretases. Cholesterol binding via the transmembrane GxxxG motif of APP leads to its preferential partitioning into lipid rafts, which co-localizes APP with its processing γ-sectretase leading to increased production of Aβ (Barrett et al. 2012). This process may contribute to the direct correlation between the plasma levels of cholesterol and AD (see Chap. 3 by Morgado and Garvey and Chap. 1 by Hong in this volume).

Aβ circulates in the cerebrospinal fluid and in plasma (Pirttilä et al. 1994) mainly in association with HDL. This association led to a proposal that a major hallmark of AD is the disorder of lipid metabolism (Biere et al. 1996; Koudinov et al. 1994, 2001). Partial support for this idea comes from later studies elucidating the link between AD and elevated levels of cholesterol (Puglielli et al. 2003; Barrett et al. 2012). Another related hypothesis postulates the compensatory mechanism of neurotoxicity, and is focused on comparing normal functions of Aβ in healthy subjects and in AD patients (Koudinov and Berezov 2004). Aβ is produced in low amounts in normal brains and is important in synaptic functions ((Puzzo et al. 2008) and references therein). One of these functions includes the effect of Aβ on the synaptic plasticity through various receptor signaling pathways, which is critical for learning and memory. In fact, mouse models including APP knock-out reportedly show impaired long-

term potentiation and memory loss (Dawson et al. 1999; Laird et al. 2005). These findings strongly support the idea that, in contrast to high concentrations of Aβ in the brain, which contribute to the development of dementia in AD, low (picomolar) concentrations of Aβ play an important modulatory role in neurotransmission and memory.

ApoE, which is the major apolipoprotein in the central nervous system, provides an important link between lipoprotein transport and AD (Poirier 2000). Humans have three apoE gene alleles: e2, e3, and e4. The corresponding protein isoforms differ by amino acid substitutions at two sites: apoE2 (Cys112, Cys158), apoE3 (Cys112, Arg 158), and apoE4 (Arg112, Arg158) (Davignon et al. 1988; LaDu et al. 1997). ApoE3 and, to a less extent, apoE2 are more common in humans as compared to apoE4, yet nearly 25 % of the population have at least one copy of the e4 allele (Puglielli et al. 2003). The presence of one and, particularly, two e4 alleles is the strongest established genetic risk factor for AD, whereas e2 is protective (Liu et al. 2013). The origin of these differences is not entirely clear and is generally attributed to the isoform-specific interactions between apoE and Aβ, which influence the clearance of Aβ (Deane et al. 2008). Contrary to this prevailing idea, no significant isoform-dependent association of apoE with Aβ has been detected in several studies; in fact, minimal direct interactions between apoE and soluble Aβ in cerebrospinal fluid were reported, suggesting that such interactions may not significantly influence the metabolism of soluble Aβ (Verghese et al. 2013) and references therein). The authors proposed an alternative mechanism in which HDL-bound apoE and soluble Aβ directly compete for the binding to lipoprotein receptors in the brain. Regardless of the exact mechanism, soluble Aβ is cleared more effectively from the brain interstitial fluid in mice carrying apoE2 and apoE3 isoforms as compared to apoE4 (Castellano et al. 2011), supporting the idea that clearance of Aβ is influenced by apoE in an isoform-specific manner.

Another possible link between lipid metabolism, apoE, and accumulation or clearance of Aβ is via the interactions with HSPGs (Fig. 10.6).

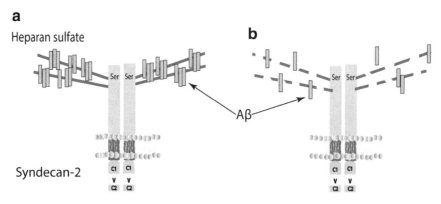

Fig. 10.6 Cartoon illustrating interactions of Aβ peptide with heparan sulfate chains of syndecan-2 in normal and in heparanase-overexpressing tissues. (**a**) Binding of Aβ to heparan sulfate of syndecan-2 promotes Aβ accumula-tion on the cell surface and amyloid deposition. (**b**) This pathogenic deposition is abolished if heparan sulfates are degraded by heparanase, which diminishes syndecan binding to Aβ (Figure modified from Li et al. 2005)

Highly sulfated HSPGs have been found in most amyloid deposits, including AD plaques (Snow et al. 1987, 1988; Aguilera et al. 2014; Zhang et al. 2014). Although the complex role of HSPGs in amyloid formation is far from clear, specific interaction between APP and HSPG were proposed to be involved in the earliest stages of AD as well as in Down syndrome (Narindrasorasak et al. 1991; Li et al. 2005). Moreover, the accumulation of Aβ on the cell surface is promoted by direct binding of the peptide N-terminal HHQK motif to heparan sulfate chains (Giulian et al. 1998), which was reported to inhibit Aβ clearance and promote amyloid fibril formation (Snow et al. 1987; Li et al. 2005). The role of HSPGs in AD is further supported by the observation of reduced amyloid deposition in transgenic mice overexpressing human heparanase that cleaves the heparan sulfate chains (Li et al. 2005). Furthermore, in vitro studies demonstrated that complete removal of all sulfates from the heparin chains completely blocks amyloid fibrils formation by Aβ, whereas partial removal blocks it only partially (Castillo et al. 1999).

In line with this idea, full-length form of syndecan-2 binds Aβ peptide and retains it on the cell surface, thereby promoting amyloid plaque formation in AD (Watanabe et al. 2004). Several earlier studies showed co-deposition of Aβ and HSPGs, such as syndecans 1–3, in AD plaques ((Zhang et al. 2014) and references

therein). Recent analysis of Aβ plaques in SAMP8 mice, which are often used as a model of AD, has revealed the presence of syndecan-2 and perlecan associated with Aβ (Manich et al. 2011) (Fig. 10.6). Furthermore, transgenic mice overexpressing heparanase that decreases syndecan shedding are resistant to induction of amyloid plaque formation in vivo (Li et al. 2005). Together, these studies support the idea that the binding of APP or Aβ to HSPGs, including syndecan-2, can importantly influence Aβ generation and/or clearance and thereby affect the development of AD. Therefore, blocking the binding of Aβ to heparan sulfates may provide a potential therapeutic strategy against amyloid plaque formation.

10.8 Regulation of Appetite by Syndecan-3

Obesity, which is a major public health problem in many developed countries, greatly increases the risk of cardiovascular disease, type 2 diabetes, and certain types of cancer (Daousi et al. 2006; Schuster 2010; Khandekar et al. 2011). The balance between consumed food and the energy expenditure is regulated by the central nervous system. The cerebral cortex has a trophic effect on the adipose tissue, either through sympathetic and parasympathetic systems or through

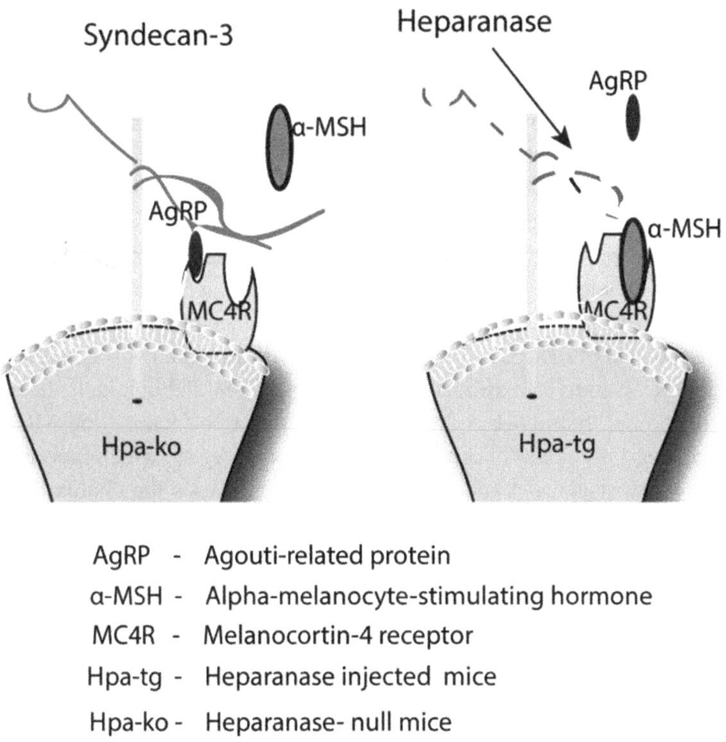

AgRP   - Agouti-related protein

α-MSH - Alpha-melanocyte-stimulating hormone

MC4R   - Melanocortin-4 receptor

Hpa-tg - Heparanase injected mice

Hpa-ko - Heparanase- null mice

Fig. 10.7 Schematic representation of the MC4R signaling. Syndecan-3 serves as a co-receptor for AgRP while also blocking the binding of its competitive antagonist, α-MSH. (**a**) In heparanase-null mice (*Hpa-ko*), syndecan-3 helps co-localize AgRP with MC4R and thereby promote their binding, which leads to increased food consumption. (**b**) Heparanase injected in these mice (*Hpa-tg*) cleaves heparan sulfate chains, thus allowing the MC4R receptor to interact with α-MSH, which leads to reduced appetite (Modified from Karlsson-Lindahl et al. 2012)

the endocrine glands. Body weight is regulated by the hypothalamus. The best-understood pathway via which hypothalamus regulates the appetite is through the melanocortin system, which includes G-protein coupled receptor MC4R, the appetite-reducing melanocyte hormone α-MSH, and its competitive antagonist that is the obesity-inducing agouti-related protein AgRP (Gantz and Fong 2003). The α-MSH hormones are produced by neurons in the arcuate nucleus of the hypothalamus and are transferred to the adjacent paraventricular nucleus where they concurrently activate MC4R. Mutations in this receptor are among the most frequent causes of obesity in humans (Mergen et al. 2001).

Syndecan-3 acts as a co-receptor that regulates the appetite and body weight by promoting the co-localization and binding of AgRP to MC4R, which contributes to increased appetite (Reizes et al. 2003; Madonna et al. 2012). Polysaccharide chains of syndecan-3 play a key role in these interactions by binding to the positively charged AgRP. The role of syndecan-3 in the body weight regulation is supported by the observation that syndecan-3-null mice are resistant to obesity (Reizes et al. 2001). Furthermore, shedding of syndecan-3 provides an important mechanism in appetite regulation. In fact, the tissue inhibitor of mealloprotease-3 (TIMP-3) that inhibits the shedding is increased in food deprivation in mice (Reizes et al. 2003). Moreover, cleavage of heparane sulfates on syndecan-3 by heparanase abolishes the interaction of AgRP with MC4R (Fig. 10.7). As a result, heparanase-deficient mice develop increased appetite and consume much more food that their wild-type counterparts (Karlsson-Lindahl et al. 2012). Taken together, these studies show that syndecan-3 is an important co-receptor in regulating body weight and appetite in mice and, potentially, in humans.

10.9 Conclusions

Syndecans are responsible for diverse physiological processes in healthy state and are implicated in several major human diseases including disorders of lipid metabolism (atherosclerosis, type 2 diabetes), AD and certain types of cancer. This highly conserved family of proteoglycans has evolved from the single syndecan precursor found in invertebrates to four different types found in vertebrates. Syndecans function as co-receptors and receptors for a variety of extracellular ligands that are bound mainly by the glycosaminoglycan chains. Functions of syndecans are involved in every stage of organism development, from the embryonic to adult state. Due to their strategic membrane-spanning location and their ability to bind numerous intra- and extracellular ligands, syndecans can transmit signals across the plasma membrane and can also help maintain the cytoskeleton under stress conditions. To activate different signaling cascades, ectodomains of syndecans bind multiple positively charged ligands through their glycosaminoglycan chains outside the cell, while the conserved cytoplasmic domains bind various adaptor proteins and enzymes inside the cell. Clustering of syndecans in lipid rafts, together with their dimerization, is critical for inducing signaling cascades. In this important process, the GxxxG motif in the transmembrane helical domain of the core protein is responsible for syndecan dimerization as well as for cholesterol binding and partitioning in lipid rafts (Barrett et al. 2012; Leonova and Galzitskaya 2013).

Notably, both APP and syndecans contain a cholesterol-binding GxxxG motif and, as a result, preferentially partition into lipid rafts (Barrett et al. 2012; Leonova and Galzitskaya 2013). This co-localization, together with the direct involvement of this motif in formation of homo- and heterodimers of syndecans and other transmembrane α-helical domains (described by Hong in Chap. 1 of this volume), prompts us to speculate that this transmembrane motif may potentially mediate direct interactions between APP and syndecans and thereby influence APP processing and generation of Aβ.

A wide range of syndecan functions involves ligand binding by heparan sulfate and chondroitin sulfate chains. Syndecan-1 utilizes this binding for clearance of apoE-containing TG-rich lipoproteins.

Heparan sulfate chains are utilized by all syndecan. In syndecan-2, they facilitate Aβ retention on the cell surface, which is proposed to contribute to the development of amyloid plaques in AD. In syndecan-3, binding to heparan sulfate chains helps to co-localize the MC4R receptor with its agonist AgRP, which increases the appetite. Syndecan-4 uses the chains to initiate capture of modified LDL by macrophages in the process that promotes atheroma formation, increases cholesterol levels in the plasma membrane, and thereby adversely affects the development of cardiovascular disease and AD.

As a result, desulfation of glycosaminoglycan chains of syndecans can contribute to the development of various diseases such as hypercholesterolemia, type 2 diabetes and obesity. On the other hand, desulfation of syndecans can have beneficial effects, such as inhibition of amyloid plaque formation in AD, inhibition of atheroma formation in cardiovascular disease, and decreased appetite.

Despite significant progress in elucidating multiple functions of syndecans, their mechanisms of action and biomedical significance are not fully understood. Further studies may ultimately help develop new syndecan-based therapeutic targets. For example, inhibition of syndecan-3 may enhance the binding of AgRP to MC4R and thereby reduce the appetite and help alleviate obesity, while desulfation of syndecan-2 may inhibit amyloid plaque formation in AD. Moreover, early detection of genetic defects and predispositions to certain diseases may also potentially include syndecans. We speculate that populational studies addressing the roles of syndecans in human disease may provide an alternative approach to understanding the etiology of major complex diseases, including cardiovascular disease, Alzheimer's disease and other disorders of lipid metabolism.

Acknowledgement We are grateful Olga Gursky for her expert assistance and valuable comments. This study was supported by the Russian Science Foundation grant №14-14-00536 to O. V. G. and by the Russian Academy of Sciences (Molecular and Cell Biology program (grant 01201353567) to E. I. L.

References

Aguilera JJ, Zhang F, Beaudet JM, Linhardt RJ, Colón W (2014) Divergent effect of glycosaminoglycans on the in vitro aggregation of serum amyloid A. Biochimie 104:70–80

Asplund A, Ostergren-Lundén G, Camejo G, Stillemark-Billton P, Bondjers G (2009) Hypoxia increases macrophage motility, possibly by decreasing the heparan sulfate proteoglycan biosynthesis. J Leukoc Biol 86(2):381–388

Barrett PJ, Song Y, Van Horn WD, Hustedt EJ, Schafer JM, Hadziselimovic A, Beel AJ, Sanders CR (2012) The amyloid precursor protein has a flexible transmembrane domain and binds cholesterol. Science 336(6085):1168–1171

Beauvais DM, Rapraeger AC (2004) Syndecans in tumor cell adhesion and signaling. Reprod Biol Endocrinol RBE 2:3

Biere AL, Ostaszewski B, Stimson ER, Hyman BT, Maggio JE, Selkoe DJ (1996) Amyloid beta-peptide is transported on lipoproteins and albumin in human plasma. J Biol Chem 271(51):32916–32922

Bishop JR, Schuksz M, Esko JD (2007) Heparan sulphate proteoglycans fine-tune mammalian physiology. Nature 446(7139):1030–1037

Boyanovsky BB, Shridas P, Simons M, van der Westhuyzen DR, Webb NR (2009) Syndecan-4 mediates macrophage uptake of group V secretory phospholipase A2-modified LDL. J Lipid Res 50(4):641–650

Brown NH (2011) Extracellular matrix in development: insights from mechanisms conserved between invertebrates and vertebrates. Cold Spring Harb Perspect Biol 3(12):a005082

Castellano JM, Kim J, Stewart FR, Jiang H, DeMattos RB, Patterson BW, Fagan AM, Morris JC, Mawuenyega KG, Cruchaga C, Goate AM, Bales KR, Paul SM, Bateman RJ, Holtzman DM (2011) Human apoE isoforms differentially regulate brain amyloid- peptide clearance. Sci Transl Med 3(89):89ra57–ra89ra57

Castillo GM, Lukito W, Wight TN, Snow AD (1999) The sulfate moieties of glycosaminoglycans are critical for the enhancement of beta-amyloid protein fibril formation. J Neurochem 72(4):1681–1687

Chakravarti R, Adams JC (2006) Comparative genomics of the syndecans defines an ancestral genomic context associated with matrilins in vertebrates. BMC Genomics 7:83

Chen K, Williams KJ (2013) Molecular mediators for raft-dependent endocytosis of syndecan-1, a highly conserved, multifunctional receptor. J Biol Chem 288(20):13988–13999

Chen L, Couchman JR, Smith J, Woods A (2002) Molecular characterization of chicken syndecan-2 proteoglycan. Biochem J 366(Pt 2):481–490

Chen K, Liu M-L, Schaffer L, Li M, Boden G, Wu X, Williams KJ (2010) Type 2 diabetes in mice induces hepatic overexpression of sulfatase 2, a novel factor that suppresses uptake of remnant lipoproteins. Hepatology 52(6):1957–1967

Choi S, Kim J-Y, Park JH, Lee S-T, Han I-O, Oh E-S (2012) The matrix metalloproteinase-7 regulates the extracellular shedding of syndecan-2 from colon cancer cells. Biochem Biophys Res Commun 417(4):1260–1264

Cornelison DDW, Wilcox-Adelman SA, Goetinck PF, Rauvala H, Rapraeger AC, Olwin BB (2004) Essential and separable roles for syndecan-3 and syndecan-4 in skeletal muscle development and regeneration. Genes Dev 18(18):2231–2236

Couchman JR, Chen L, Woods A (2001) Syndecans and cell adhesion. Int Rev Cytol 207:113–150

Daousi C, Casson IF, Gill GV, MacFarlane IA, Wilding JPH, Pinkney JH (2006) Prevalence of obesity in type 2 diabetes in secondary care: association with cardiovascular risk factors. Postgrad Med J 82(966):280–284

Davignon J, Gregg RE, Sing CF (1988) Apolipoprotein E polymorphism and atherosclerosis. Arterioscler Dallas Tex 8(1):1–21

Dawson GR, Seabrook GR, Zheng H, Smith DW, Graham S, O'Dowd G, Bowery BJ, Boyce S, Trumbauer ME, Chen HY, Van der Ploeg LH, Sirinathsinghji DJ (1999) Age-related cognitive deficits, impaired long-term potentiation and reduction in synaptic marker density in mice lacking the beta-amyloid precursor protein. Neuroscience 90(1):1–13

De Rossi G, Evans AR, Kay E, Woodfin A, McKay TR, Nourshargh S, Whiteford JR (2014) Shed syndecan-2 inhibits angiogenesis. J Cell Sci 127(21):4788–4799

Deane R, Sagare A, Hamm K, Parisi M, Lane S, Finn MB, Holtzman DM, Zlokovic BV (2008) ApoE isoform–specific disruption of amyloid β peptide clearance from mouse brain. J Clin Invest 118(12):4002–4013

Denhez F, Wilcox-Adelman SA, Baciu PC, Saoncella S, Lee S, French B, Neveu W, Goetinck PF (2002) Syndesmos, a syndecan-4 cytoplasmic domain interactor, binds to the focal adhesion adaptor proteins paxillin and Hic-5. J Biol Chem 277(14):12270–12274

Dews IC, Mackenzie KR (2007) Transmembrane domains of the syndecan family of growth factor coreceptors display a hierarchy of homotypic and heterotypic interactions. Proc Natl Acad Sci U S A 104(52): 20782–20787

Fransson L-A, Belting M, Cheng F, Jönsson M, Mani K, Sandgren S (2004) Novel aspects of glypican glycobiology. Cell Mol Life Sci CMLS 61(9):1016–1024

Gantz I, Fong TM (2003) The melanocortin system. Am J Physiol Endocrinol Metab 284(3):E468–E474

Getz GS, Reardon CA (2008) Apoprotein E as a lipid transport and signaling protein in the blood, liver, and artery wall. J Lipid Res 50(Supplement):S156–S161

Giulian D, Haverkamp LJ, Yu J, Karshin W, Tom D, Li J, Kazanskaia A, Kirkpatrick J, Roher AE (1998) The HHQK domain of beta-amyloid provides a structural basis for the immunopathology of Alzheimer's disease. J Biol Chem 273(45):29719–29726

Goldstein JL, Brown MS, Anderson RG, Russell DW, Schneider WJ (1985) Receptor-mediated endocytosis: concepts emerging from the LDL receptor system. Annu Rev Cell Biol 1:1–39

Hardy JA, Higgins GA (1992) Alzheimer's disease: the amyloid cascade hypothesis. Science 256(5054): 184–185

Hassing HC, Mooij H, Guo S, Monia BP, Chen K, Kulik W, Dallinga-Thie GM, Nieuwdorp M, Stroes ES, Williams KJ (2012) Inhibition of hepatic sulfatase-2 in vivo: a novel strategy to correct diabetic dyslipidemia. Hepatology 55(6):1746–1753

Kaksonen M, Pavlov I, Võikar V, Lauri SE, Hienola A, Riekki R, Lakso M, Taira T, Rauvala H (2002) Syndecan-3-deficient mice exhibit enhanced LTP and impaired hippocampus-dependent memory. Mol Cell Neurosci 21(1):158–172

Karlsson-Lindahl L, Schmidt L, Haage D, Hansson C, Taube M, Egeciouglu E, Tan Y, Admyre T, Jansson J-O, Vlodavsky I, Li J-P, Lindahl U, Dickson SL (2012) Heparanase affects food intake and regulates energy balance in mice. PLoS One 7(3):e34313

Khandekar MJ, Cohen P, Spiegelman BM (2011) Molecular mechanisms of cancer development in obesity. Nat Rev Cancer 11(12):886–895

Kim JM, Lee JA, Cho IS, Ihm CH (2010) Soluble syndecan-1 at diagnosis and during follow up of multiple myeloma: a single institution study. Korean J Hematol 45(2):115–119

Kokenyesi R, Bernfield M (1994) Core protein structure and sequence determine the site and presence of heparan sulfate and chondroitin sulfate on syndecan-1. J Biol Chem 269(16):12304–12309

Koudinov AR, Berezov TT (2004) Alzheimer's amyloid-beta (A beta) is an essential synaptic protein, not neurotoxic junk. Acta Neurobiol Exp (Wars) 64(1):71–79

Koudinov A, Matsubara E, Frangione B, Ghiso J (1994) The soluble form of Alzheimer's amyloid beta protein is complexed to high density lipoprotein 3 and very high density lipoprotein in normal human plasma. Biochem Biophys Res Commun 205(2):1164–1171

Koudinov AR, Berezov TT, Koudinova NV (2001) The levels of soluble amyloid beta in different high density lipoprotein subfractions distinguish Alzheimer's and normal aging cerebrospinal fluid: implication for brain cholesterol pathology? Neurosci Lett 314(3):115–118

Kreuger J, Spillmann D, Li J, Lindahl U (2006) Interactions between heparan sulfate and proteins: the concept of specificity. J Cell Biol 174(3):323–327

LaDu MJ, Lukens JR, Reardon CA, Getz GS (1997) Association of human, rat, and rabbit apolipoprotein E with beta-amyloid. J Neurosci Res 49(1):9–18

Laird FM, Cai H, Savonenko AV, Farah MH, He K, Melnikova T, Wen H, Chiang H-C, Xu G, Koliatsos VE, Borchelt DR, Price DL, Lee H-K, Wong PC (2005) BACE1, a major determinant of selective vulnerability of the brain to amyloid-beta amyloidogenesis, is essential for cognitive, emotional, and synaptic

functions. J Neurosci Off J Soc Neurosci 25(50):11693–11709

Laurent TC, Fraser JR (1992) Hyaluronan. FASEB J 6(7):2397–2404

Leonova EI, Galzitskaya OV (2013) Structure and functions of syndecans in vertebrates. Biochemistry (Mosc) 78(10):1071–1085

Leonova EI, Galzitskaya OV (2015) Cell communication using intrinsically disordered proteins: what can syndecans say? J Biomol Struct Dyn 33(5):1037–1050

Li J-P, Galvis MLE, Gong F, Zhang X, Zcharia E, Metzger S, Vlodavsky I, Kisilevsky R, Lindahl U (2005) In vivo fragmentation of heparan sulfate by heparanase overexpression renders mice resistant to amyloid protein A amyloidosis. Proc Natl Acad Sci U S A 102(18): 6473–6477

Lin Y-L, Lei Y-T, Hong C-J, Hsueh Y-P (2007) Syndecan-2 induces filopodia and dendritic spine formation via the neurofibromin-PKA-Ena/VASP pathway. J Cell Biol 177(5):829–841

Liu C-C, Kanekiyo T, Xu H, Bu G (2013) Apolipoprotein E and Alzheimer disease: risk, mechanisms and therapy. Nat Rev Neurol 9(2):106–118

Lubrano-Berthelier C, Cavazos M, Dubern B, Shapiro A, Stunff CLE, Zhang S, Picart F, Govaerts C, Froguel P, Bougneres P, Clement K, Vaisse C (2003) Molecular genetics of human obesity-associated MC4R mutations. Ann N Y Acad Sci 994:49–57

MacArthur JM, Bishop JR, Stanford KI, Wang L, Bensadoun A, Witztum JL, Esko JD (2007) Liver heparan sulfate proteoglycans mediate clearance of triglyceride-rich lipoproteins independently of LDL receptor family members. J Clin Invest 117(1):153–164

Madonna ME, Schurdak J, Yang Y, Benoit S, Millhauser GL (2012) Agouti-related protein segments outside of the receptor binding core are required for enhanced short- and long-term feeding stimulation. ACS Chem Biol 7(2):395–402

Mahley RW, Hui DY, Innerarity TL, Weisgraber KH (1981) Two independent lipoprotein receptors on hepatic membranes of dog, swine, and man. Apo-B, E and apo-E receptors. J Clin Invest 68(5):1197–1206

Manich G, Mercader C, del Valle J, Duran-Vilaregut J, Camins A, Pallàs M, Vilaplana J, Pelegrí C (2011) Characterization of amyloid-β granules in the hippocampus of SAMP8 mice. J Alzheimers Dis JAD 25(3):535–546

Manon-Jensen T, Itoh Y, Couchman JR (2010) Proteoglycans in health and disease: the multiple roles of syndecan shedding. FEBS J 277(19):3876–3889

Mergen M, Mergen H, Ozata M, Oner R, Oner C (2001) A novel melanocortin 4 receptor (MC4R) gene mutation associated with morbid obesity. J Clin Endocrinol Metab 86(7):3448

Narindrasorasak S, Lowery D, Gonzalez-DeWhitt P, Poorman RA, Greenberg B, Kisilevsky R (1991) High affinity interactions between the Alzheimer's beta-amyloid precursor proteins and the basement membrane form of heparan sulfate proteoglycan. J Biol Chem 266(20):12878–12883

Noguer O, Villena J, Lorita J, Vilaró S, Reina M (2009) Syndecan-2 downregulation impairs angiogenesis in human microvascular endothelial cells. Exp Cell Res 315(5):795–808

Pentikäinen MO, Oksjoki R, Öörni K, Kovanen PT (2002) Lipoprotein lipase in the arterial wall: linking LDL to the arterial extracellular matrix and much more. Arterioscler Thromb Vasc Biol 22(2):211–217

Pirttilä T, Kim KS, Mehta PD, Frey H, Wisniewski HM (1994) Soluble amyloid beta-protein in the cerebrospinal fluid from patients with Alzheimer's disease, vascular dementia and controls. J Neurol Sci 127(1): 90–95

Poirier J (2000) Apolipoprotein E and Alzheimer's disease. A role in amyloid catabolism. Ann N Y Acad Sci 924:81–90

Prasansuklab A, Tencomnao T (2013) Amyloidosis in Alzheimer's disease: the toxicity of amyloid beta (Aβ), mechanisms of its accumulation and implications of medicinal plants for therapy. Evid Based Complement Alternat Med 2013:1–10

Pries AR, Secomb TW, Gaehtgens P (2000) The endothelial surface layer. Pflügers Arch Eur J Physiol 440(5):653–666

Puglielli L, Tanzi RE, Kovacs DM (2003) Alzheimer's disease: the cholesterol connection. Nat Neurosci 6(4):345–351

Puzzo D, Privitera L, Leznik E, Fà M, Staniszewski A, Palmeri A, Arancio O (2008) Picomolar amyloid-beta positively modulates synaptic plasticity and memory in hippocampus. J Neurosci 28(53):14537–14545

Ramani VC, Pruett PS, Thompson CA, DeLucas LD, Sanderson RD (2012) Heparan sulfate chains of syndecan-1 regulate ectodomain shedding. J Biol Chem 287(13):9952–9961

Reizes O, Lincecum J, Wang Z, Goldberger O, Huang L, Kaksonen M, Ahima R, Hinkes MT, Barsh GS, Rauvala H, Bernfield M (2001) Transgenic expression of syndecan-1 uncovers a physiological control of feeding behavior by syndecan-3. Cell 106(1): 105–116

Reizes O, Benoit SC, Strader AD, Clegg DJ, Akunuru S, Seeley RJ (2003) Syndecan-3 modulates food intake by interacting with the melanocortin/AgRP pathway. Ann N Y Acad Sci 994:66–73

Sarrazin S, Lamanna WC, Esko JD (2011) Heparan sulfate proteoglycans. Cold Spring Harb Perspect Biol 3(7):a004952–a004952

Saunders S, Jalkanen M, O'Farrell S, Bernfield M (1989) Molecular cloning of syndecan, an integral membrane proteoglycan. J Cell Biol 108(4):1547–1556

Savery MD, Jiang JX, Park PW, Damiano ER (2013) The endothelial glycocalyx in syndecan-1 deficient mice. Microvasc Res 87:83–91

Schuster DP (2010) Obesity and the development of type 2 diabetes: the effects of fatty tissue inflammation. Diabetes Metab Syndr Obes 3:253–262

Singleton PA, Bourguignon LYW (2004) CD44 interaction with ankyrin and IP3 receptor in lipid rafts promotes hyaluronan-mediated Ca2+ signaling

leading to nitric oxide production and endothelial cell adhesion and proliferation. Exp Cell Res 295(1):102–118

Snow AD, Wight TN (1989) Proteoglycans in the pathogenesis of Alzheimer's disease and other amyloidoses. Neurobiol Aging 10(5):481–497

Snow AD, Willmer J, Kisilevsky R (1987) A close ultrastructural relationship between sulfated proteoglycans and AA amyloid fibrils. Lab Investig 57(6):687–698

Snow AD, Mar H, Nochlin D, Kimata K, Kato M, Suzuki S, Hassell J, Wight TN (1988) The presence of heparan sulfate proteoglycans in the neuritic plaques and congophilic angiopathy in Alzheimer's disease. Am J Pathol 133(3):456–463

Stanford KI, Bishop JR, Foley EM, Gonzales JC, Niesman IR, Witztum JL, Esko JD (2009) Syndecan-1 is the primary heparan sulfate proteoglycan mediating hepatic clearance of triglyceride-rich lipoproteins in mice. J Clin Invest 119(11):3236–3245

Subramanian SV, Fitzgerald ML, Bernfield M (1997) Regulated shedding of syndecan-1 and -4 ectodomains by thrombin and growth factor receptor activation. J Biol Chem 272(23):14713–14720

Thi MM, Tarbell JM, Weinbaum S, Spray DC (2004) The role of the glycocalyx in reorganization of the actin cytoskeleton under fluid shear stress: a "bumper-car" model. Proc Natl Acad Sci 101(47):16483–16488

Van Barlingen HH, de Jong H, Erkelens DW, de Bruin TW (1996) Lipoprotein lipase-enhanced binding of human triglyceride-rich lipoproteins to heparan sulfate: modulation by apolipoprotein E and apolipoprotein C. J Lipid Res 37(4):754–763

Van den Berg BM, Spaan JAE, Rolf TM, Vink H (2006) Atherogenic region and diet diminish glycocalyx dimension and increase intima-to-media ratios at murine carotid artery bifurcation. Am J Physiol Heart Circ Physiol 290(2):H915–H920

Van Deurs B, Roepstorff K, Hommelgaard AM, Sandvig K (2003) Caveolae: anchored, multifunctional platforms in the lipid ocean. Trends Cell Biol 13(2):92–100

Vanhoutte D, Schellings MWM, Götte M, Swinnen M, Herias V, Wild MK, Vestweber D, Chorianopoulos E, Cortés V, Rigotti A, Stepp M-A, Van de Werf F, Carmeliet P, Pinto YM, Heymans S (2007) Increased expression of syndecan-1 protects against cardiac dilatation and dysfunction after myocardial infarction. Circulation 115(4):475–482

Verghese PB, Castellano JM, Garai K, Wang Y, Jiang H, Shah A, Bu G, Frieden C, Holtzman DM (2013) ApoE influences amyloid- (A) clearance despite minimal apoE/A association in physiological conditions. Proc Natl Acad Sci 110(19):E1807–E1816

Watanabe N, Araki W, Chui D-H, Makifuchi T, Ihara Y, Tabira T (2004) Glypican-1 as an Abeta binding HSPG in the human brain: its localization in DIG domains and possible roles in the pathogenesis of Alzheimer's disease. FASEB J 18(9):1013–1015

Weinbaum S, Tarbell JM, Damiano ER (2007) The structure and function of the endothelial glycocalyx layer. Annu Rev Biomed Eng 9:121–167

Williams KJ, Chen K (2010) Recent insights into factors affecting remnant lipoprotein uptake. Curr Opin Lipidol 21(3):218–228

Wilsie LC, Gonzales AM, Orlando RA (2006) Syndecan-1 mediates internalization of apoE-VLDL through a low density lipoprotein receptor-related protein (LRP)-independent, non-clathrin-mediated pathway. Lipids Health Dis 5:23

Yao Y, Rabodzey A, Dewey CF Jr (2007) Glycocalyx modulates the motility and proliferative response of vascular endothelium to fluid shear stress. Am J Physiol Heart Circ Physiol 293(2):H1023–H1030

Zhang G-L, Zhang X, Wang X-M, Li J-P (2014) Towards understanding the roles of heparan sulfate proteoglycans in Alzheimer's disease. BioMed Res Int 2014:516028

Index

© Springer International Publishing Switzerland 2015
O. Gursky (ed.), *Lipids in Protein Misfolding*, Advances in Experimental
Medicine and Biology 855, DOI 10.1007/978-3-319-17344-3

CPSIA information can be obtained
at www.ICGtesting.com
Printed in the USA
LVHW061140280620
659213LV00007B/537